T0269302

This book is a comprehensive text on the theory of the magnetic recording process. It gives the reader a fundamental, in-depth understanding of all the essential features of the writing and retrieval of information for both high density disk recording and tape recording. The material is timely because magnetic recording technology is currently undergoing rapid advancements in systems capacity and data rate.

A major contribution to the growth in magnetic recording technology is the recent development of advanced thin film materials for both recording media and transducers. In addition, sophisticated signal processing schemes are being implemented in order to achieve ultra large recording densities as well as high data transfer rates. A comprehension of the basic physics and engineering aspects of magnetic recording is essential if these developments are to be understood and to be utilized in the design of future systems. This text gives a thorough grounding in four basic areas of magnetic recording: structure and fields of heads and media, the replay process, the recording process, and medium noise analysis. In addition to these fundamental issues, key systems questions of non-linearities and overwrite are also discussed. A complete chapter is devoted to the emerging technology of magneto-resistive heads. A parallel treatment of time and frequency response is given to facilitate the understanding and evaluation of signal processing schemes. Using the information presented in this text, the reader should be able to design and analyze key experiments for head and medium evaluation and for overall system testing.

This text, which is unique in its scope, will be valuable for both senior undergraduates and graduate students taking courses in magnetic recording. It will also be of value to research-and-development scientists in the magnetic recording industry. An important element of the book is the inclusion of a large number of homework problems. The author assumes that the reader has had basic introductory courses in physics, in electricity and magnetism, and in applied mathematics.

THEORY OF MAGNETIC RECORDING

THEORY OF
MAGNETIC RECORDING

H. NEAL BERTRAM
University of California at San Diego

CAMBRIDGE
UNIVERSITY PRESS

Published by the Press Syndicate of the University of Cambridge
The Pitt Building, Trumpington Street, Cambridge CB2 1RP
40 West 20th Street, New York, NY 10011–4211, USA
10 Stamford Road, Oakleigh, Melbourne 3166, Australia

First published 1994

A catalogue record for this book is available from the British Library

Library of Congress cataloguing in publication data
Bertram, H. Neal.
Theory of magnetic recording / H. Neal Bertram.
p.   cm.
Includes bibliographical references and index.
ISBN 0–521–44512–4. — ISBN 0–521–44973–1 (pbk.)
1. Magnetic recorders and recording.   I. Title
TK7881.6.B47   1994
621.382′34–dc20   93–29978 CIP

ISBN 0 521 44512 4 hardback
ISBN 0 521 44973 1 paperback

Transferred to digital printing 2003

KW

To my beloved Ann,
to our dear son Seth,
and with love and gratitude to my parents,
Manya and Barry Bertram

Contents

Preface

Magnetic recording is a technology that has continually undergone steady and substantial advancement throughout its history. Typically, in the last two decades areal densities in computer disk recording have increased by over two orders of magnitude. This development has occurred via the simultaneous growth in new materials for heads and media, advanced signal processing schemes, and mechanical engineering of the head–medium interface. In addition, there has been substantial growth in the theoretical understanding of the magnetic behavior of heads, media and, in general, the magnetic recording process. Fundamental understanding of the physics of magnetic recording is necessary not only for system design, but so that the specific behaviour of magnetic components can be analyzed, either analytically or numerically, saving time-consuming and expensive experimentation. For example, it is difficult to produce all the media variations required to perform a thorough comparison of different modes of recording, such as longitudinal and perpendicular recording.

In recent years there have been many publications that cover the fundamentals and applications of the magnetic recording process. These books or papers have been either technically oriented discussions of specific topics or have provided an introduction at an elementary level to the basics of magnetic recording. The philosophy of this book is to provide a pedagogical introduction to the physics of magnetic recording. The level is advanced and all basic aspects of magnetic recording are included: magnetic fields of heads and media, the linear replay process, the non-linear recording process including interferences, and medium noise. The basic mathematical tools for magnetic recording analysis will be developed, however, this text will not simply be a presentation of results and convenient formulae, it will attempt to give a thorough understanding, offering the reader the tools necessary to understand why results occur, to develop models, to design experiments, or to understand in-depth theoretical analyses in the current literature.

Because this text will emphasize the physical sense of processes and phenomena, the mathematical development will focus on analytic expressions rather than numerical results. General formulations will be given at the beginning of each chapter or section, however, simplified examples will be presented so that physical insight into the magnetic recording process can be stressed. When complete analytical expressions are not possible, approximate approaches will be emphasized, such as in the slope model for digital recording, or fields and transforms of thin film heads. Over the last decade there has been considerable development of media and heads for perpendicular recording. Even though conventional longitudinal recording has remained the dominant mode of high density recording (for reasons which are not entirely related to magnetic phenomena), in this text results for both modes are given in parallel. Not only does such a presentation give physical insight into magnetic phenomena, it provides the background in case perpendicular recording is utilized in the future or for the development of other modes such as canted grains utilized in metallic tape for 8mm video.

In the development of the recording process in this book, a parallel presentation of temporal and spectral analysis will be given. Magnetic recording systems are generally analyzed in the time domain. In digital magnetic recording a magnetization transition from one saturation direction to the opposite corresponds to an encoded '1' of information in a bit cell. Errors occur, for example with simple peak detection, if noise and interferences shift the apparent zero-crossing or replay pulse peak to an adjacent cell. Thus the temporal analysis of recording phenomena, such as bit shift is important. However, the frequency response of the recording channel is also useful. The spectral envelope of the fundamental component of square wave recording is the system transfer function, which is necessary for any applications where equalization is employed. Spectral analysis is also useful for analysis of recording parameters. Playback expressions in magnetic recording involve convolutions of head fields and recorded magnetization profiles, so that Fourier transforms or frequency analysis allow for the separation of head parameters, such as gap length and spacing, and medium parameters, such as transition length.

This text is divided into four primary sections plus an overview. In Chapter 1 a review of the technology, including materials utilized for heads and media is given. In addition a review of cgs and MKS unit systems is given, although the text is entirely in the MKS system. Chapters 2–4 comprise an in-depth discussion of magnetic fields for heads and media. This first section is designed to give the reader a solid foundation in magnetostatic fields, in particular the magnetostatic fields associated with magnetization patterns in media. In the second section comprising Chapters 5–6, the linear theory for calculating playback voltages from general magnetization patterns is presented. In addition to an in-depth development of the reciprocity principle, general and specific expressions are derived for single and multiple recorded transitions. In particular peak response ('roll-off curve') and spectral analysis of square wave recording are compared. In Chapters 7–9 the record process is presented for both thin films and tapes, including a discussion of non-linear phenomena such as bit shift and overwrite. In Chapters 10–12 modeling of varieties of medium noise: additive, modulation, transition, and correlation effects, is covered for both thin films and thick particulate tape. In the final chapter varieties of signal-to-noise expressions are derived along with error rates for simple zero-crossing detection. At the end of each chapter 'problems' are given that test the reader's understanding of the material presented.

This text grew out of the first year graduate course that the author has given for many years at the University of California at San Diego, in the Department of Electrical and Computer Engineering and as member of the Center for Magnetic Recording Research. This material is reasonably self-contained, however, an undergraduate-level background in static electromagnetic fields, mathematical methods such as complex variables and Fourier transforms, and random processes would facilitate a reading of this text.

The writing of this text occurred primarily during two sabbatical periods, in Paris, France and Cambridge, Massachusetts. I am grateful to my hosts during these periods who provided an atmosphere conducive to productive work. I would also like to thank my many colleagues over the years of my career in magnetic recording, who stimulated and, indeed, educated me on many aspects of the physics of magnetic recording. Without these interactions this book could never have been written. In addition, there were many people who contributed substantial time in the preparation of the figures and in general assistance in the preparation of the final copy. Particular gratitude goes to Dr Samuel Yuan and Betty Manoulian. Without their dedicated assistance during the final months of preparation of the manuscript, the book could not have been completed.

Dr Neil Smith carefully read Chapter 7 and made many helpful suggestions. He is to be especially thanked, not only for illuminating discussions, but also for providing many figures for that chapter. In addition, thanks go to Dr Yashwant Gupta, for preparation of many of the figures in the early stages of writing. I am grateful to Dr Lineu Barbosa and to Peng Qingzhi for carefully reading Chapters 10–12 and to Herbert Lin for examining Chapters 8–9. I would like to express special thanks to Dr Giora Tarnopolsky for his invaluable help during the final stages of preparation of this book. Finally, I would like to thank my wife, Ann, not only for the moral support and encouragement to see the project through, but also for a careful editing of the manuscript.

The theory of magnetic recording processes is in large part an application of the general theory of electromagnetic fields. It is a very beautiful mathematical, as well as physical subject, and it is hoped that the reader will find the same aesthetic pleasure in this material that the author has found.

H. Neal Bertram

1
Overview

Magnetic recording is the central technology of information storage. Utilization of hard disk drives as well as flexible tape and disk systems provides, inexpensively and reliably, all features essential to this technology. A data record can be easily written and read with exceedingly fast transfer rates and access times. Information can be permanent or readily overwritten to store new data. Digital recording is the predominant form of magnetic storage, although frequency modulation for video recording and ac bias for analog recording may persist in consumer applications. Data storage is universally digital. Superb areal densities for disk drives and volumetric densities for tape systems are achievable with extremely low error rates. In the last decade there have been extraordinary advances in magnetic recording technology. Current densities and transfer rates for disk systems are typically 60Mbits/in^2 and 10Mhz, respectively, but systems with densities of 1–2Gbits/in^2 are realizable (Wood, 1990; Howell et $al.$, 1990; Takano, et $al.$, 1991). The ability to coat tape with extremely smooth surfaces has permitted the development of very high density helical scan products (S-VHS, 8mm video) (Mallinson, 1990). The digital audio helical scan recorder (DAT) is representative of very high density tape recording with linear densities of greater than 60kbits/in, track densities of 250 tpi, and volumetric densities on the order of 50Gbits/in^3 (Ohtake, et $al.$, 1986). High data rate tape recording systems near 150MHz have been developed (Ash, et $al.$, 1990; Coleman, et $al.$, 1984). In general, densities and data rates of magnetic recording systems have been increasing at a rate exceeding a doubling every three years.

For rigid disk computer applications, thin film media with their high magnetizations and extremely thin coatings are now utilized in all new computer drives. Understanding micromagnetic processes in these materials enabled development labs to produce films with a substantial reduction in noise or magnetization fluctuations (Yogi, *et al.*, 1990a,b). These improved thin film materials have made possible the evolution in storage. Magnetic particles that make up tape have evolved into ever smaller and more uniform grains with high magnetizations and coercivities. In addition metal evaporated thin film tape is utilized in advanced video products (Chiba, *et al.*, 1989; Luitjens, 1990). Thin films also provide advanced technology for soft materials that comprise magnetic heads for all recording systems. Advanced heads are either entirely thin film or utilize thin films at the gap region of a ferrite core structure. For extremely advanced recording in the $1-2Gbits/in^2$ range, thin film magnetoresistive heads are required in order to achieve signal to noise ratios limited only by medium noise. Magnetoresistive heads, due to their high, speed-independent, playback sensitivity, will eventually become commonplace in recording systems. The first commercial application in a high density disk drive is a 1 Gigabyte drive (IBM Corsair). In general, the science of multilayer films plays a dominant role in the formation of optimum thin film recording media and advanced inductive and magnetoresistive heads. For a review of materials for magnetic recording see Berkowitz, 1990.

Materials and magnetization processes

All advanced recording media are made of distinct fine grains. Granular structures yield the large coercivities and permanent magnetizations required of 'hard' magnetic storage materials. These properties coupled with extremely small particle size yield excellent signal to medium noise ratios, as well as minimal interference in recorded data patterns from non-linearities, residual overwrite or erasure, and extraneous fields. Magnetic tape is composed of elongated distinct particles of dimensions typically $\sim 250\text{Å} \times 250\text{Å} \times 1000\text{Å}$ with net anisotropy axis along the particle long axis. Tape materials include γFe_2O_3, Co-γFe_2O_3, (surface passivated) Fe, CrO_2 or BaFe with particle magnetizations on the order of 370–500 kA/m (370–500 emu/cc) and tape coercivities in the range 24–120 kA/m (300–1500 Oe) (Köester & Arnoldussen, 1989). Tape coatings are extremely smooth with surface roughness on the order of 50nm (Wierenga, *et al.*, 1985; Robinson, *et al.*, 1985), and particles are usually well oriented in the direction of relative head medium motion (Fig. 1.1). Tape M-H major loops exhibit typical squarenesses S, S* ~ 0.85. Particle volume loading is $\sim 30\%$ by volume, and saturation remanent

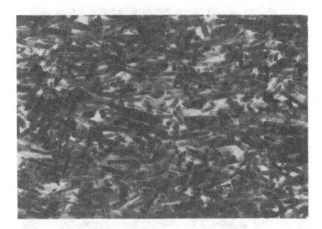

Fig. 1.1. TEM of cross-section of tape composed of oriented CrO_2 particles. Courtesy of Eberhard Köester, BASF, Ludwigshaven, Germany.

magnetizations (B_r) of 0.15T (1500 G) for oxide materials to 0.25T (2500 G) for metallic particles are achieved. Tape media is coated on a 'polyurethane base-film' varying in thickness from 10–12μm. Although the magnetic layers are typically 3–5μm thick, optimized high density digital recording utilizes only about 0.5μm of the surface layer for recording magnetic signals.

High quality magnetic thin films utilized as recording media are generally composed of a Co dominant alloy sputtered on a suitable growth enhancing underlayer. The films are polycrystalline with the grains or 'particles' on the order of 200–500Å in diameter in the film plane (Fig. 1.2). For longitudinal films the grain easy axes (Co hcp axis) are either random in the plane or oriented along the head–medium motion direction. For perpendicular films the Co easy axes are perpendicular to the film plane within a cone of $\sim 5°$. In evaporated metal tape the grains grow at an angle to the film plane of $\sim 30°$. In all film recording media the grains are believed to extend uniformly through the film thickness, which is ~ 200–500Å for longitudinal films, ~ 2000–5000Å for perpendicular films and ~ 1500Å for metal evaporated tape (Luitjens, 1990; Köester & Arnoldussen, 1989). In all cases high quality films are produced only if the grains are magnetically well segregated with negligible magnetic or intergranular exchange coupling (Bertram & Zhu, 1992).

The magnetization of recording media averaged over regions that comprise many grains is in general a vector relation of the applied field

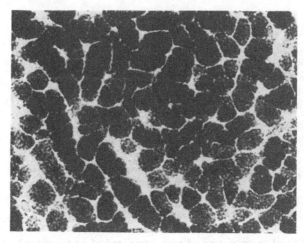

Fig. 1.2.   TEM image of typical polycrystalline thin film recording media. This top view shows a system of well-defined single crystallite grains, which are continuous through the film thickness. Taken from Johnson *et al.* (1990).

history, which is spatially uniform for a bulk magnetic measurement or is a spatially non-uniform, rotating vector during the recording process as each element of media passes the recording head. In recording materials each grain is assumed (designed) to be a single domain so that the magnetic state is a magnetization of fixed magnitude equal to the grain saturation magnetization and with orientation dependent on the field history, grain anisotropy, and magnetic interactions. In all magnetic materials, these interactions include the long range magnetostatic interaction between every pair of grains and possible intergranular exchange energy between every adjacent pair of grains. Induced stress anisotropy may occur in films via the magnetostatic interaction. In general, the magnetization state of a collection of interacting grains is a complicated micromagnetic process involving dynamic gyromagnetic magnetization processes driven by all the above 'effective fields' as well as thermal agitation. The resulting coupled equations are damped by magnetization relaxation processes to the grain lattice (Brown, 1962; Haas & Callen, 1963). The time scale of magnetization reversal is on the order of 10^{-9}s and temporal cell intervals are, currently, on the order of 10^{-7}s (30 Mbits/s for computer and 100 Mbits/s for tape storage).

The difficulty of numerical simulation of the magnetization process in recording materials is due to the long-range magnetostatic interaction energy. Since every pair of grains is coupled, large scale computations utilizing supercomputers are required. In general, magnetization processes in these materials are quite complicated because the intergranular interactions yield collective reversal. These processes depend on the relative strength of the interactions and the intrinsic grain anisotropy magnitudes and dispersions. A general rule is that in order for collective processes to occur the interaction strength must exceed the dispersion in the individual grain energies. In magnetic tape the distribution in individual particle anisotropy magnitudes as well as location is believed to exceed the average magnetostatic interaction strength so that the magnetic history may not be dominated by interparticle interactions (Parker & Berkowitz, 1990). However, in tape systems perfect particle dispersion does not occur, so that the coercivity is reduced from that of individual grains not only by an average mean interaction field but predominantly by the collective reversal of interacting particle pairs (Salling, *et al.*, 1991). In thin films the lack of non-magnetic space between particles, the (probable) relative uniformity in crystalline anisotropy magnitude and the large grain magnetizations lead to strong collective reversal processes (Zhu & Bertram, 1991b).

Magnetic materials may be evaluated by a VSM (vibrating sample magnetometer) measurement to determine the basic quantities of magnetization, coercivity, and squarenesses (Monson, 1989). A uniform field is applied to a sample, and the net magnetization component along the magnetic field is measured. The field direction is generally taken along the net easy axis (e.g., the particle orientation direction in tape), and, if necessary, sample shape demagnetization effects are removed. The result is an 'M–H' loop as illustrated in Fig. 1.3. This loop is generated by

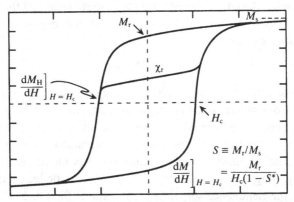

Fig. 1.3.   Typical hysteresis loop of high density recording media. M_s, M_r, H_c, S, S^* are the medium saturation magnetization, saturation remanence, coercivity, remanent squareness, loop squareness, respectively. χ_r is the reversible susceptibility along a minor loop. (Courtesy of Hewlett Packard.)

first applying a saturating field in one direction that is large enough to reverse all grains, or grain clusters, so that they are parallel to the direction of the applied field. The field is reduced to zero, reversed and increased to saturation in the opposite direction. Sweeping the field in the opposite sense gives an identical curve. The magnetization component along the applied field direction M_H is given generally for 'hard' granular recording media by:

$$M_H = \frac{1}{V} \sum_i M_i v_i \cos \theta_i \tag{1.1}$$

where M_i and v_i are the saturation magnetization and volume of the ith grain, respectively, and θ_i is the angle between the magnetization of the ith grain and the applied field direction. The sum is over all the grains in the sample of volume V. The magnetization M_i in each grain is given by the saturation magnetization at the temperature of measurement and generally does not vary from grain to grain.

In Fig. 1.3 the primary quantities are the saturation magnetization M_s, the saturation remanent magnetization M_r, the coercivity H_c, the remanent squareness $S = M_r/M_s$, and the loop squareness S^* defined by the loop slope at the coercive state:

$$\frac{dM_H}{dH} = \frac{M_r}{H_c(1 - S^*)} \tag{1.2}$$

The M–H loop in Fig. 1.3 is of thin film recording media. The loop is fairly square with values of $S, S^* \sim 0.85$ and is typical of thin film media as well as oriented particulate tape. In the saturated state all the direction cosines are equal to 1.0 ($\cos \theta_i = 1$) so that the medium saturation magnetization is given by:

$$M_s = p M_{grain} \tag{1.3}$$

where p is the volume packing fraction of the grains. As the applied field is reduced from saturation, the magnetizations rotate back, reversibly, toward their nearest individual easy axis direction, but are affected by the interparticle interactions. In the saturation remanent state in oriented tape, the particle magnetizations lie approximately along their respective easy axes yielding $S \sim 0.85$. Interparticle magnetostatic interactions tend to increase the remanence somewhat (Bertram & Bhatia, 1973). In thin film media with random easy axes, the strong magnetostatic coupling, as well as possible neighboring exchange interactions, increases the remanent squareness substantially from the non-interacting value of

S = 0.5 to S ~0.85 (Zhu & Bertram, 1988a). The magnetostatic fields coupled with the random anisotropy yield a 'ripple' pattern of remanent magnetization whose scale depends on the degree of exchange coupling.

As the field is increased in the reverse direction, the magnetizations initially rotate reversibly toward the field direction. Eventually irreversible processes occur in which the grain magnetizations switch direction by large angles into the reverse direction. For a 'square' loop, nucleation of reversal occurs over a very small field range for the grains, yielding a steep portion of the loop in the vicinity of the coercivity. For oriented magnetic tape the switching of particle magnetization directions is essentially from one direction along the easy axis nearest the field direction ($\cos\theta$ ~ 1) to the opposite easy axis direction ($\cos\theta$ ~ -1). The switching order is dominated by the distribution in magnitudes of particle anisotropy fields, modified by magnetostatic interactions and nearest neighbor clustering. For unoriented thin films nucleation is dominated by the formation of vortex structures from the ripple state (Zhu & Bertram, 1991b). The vortex shape minimizes the magnetostatic energy, although their size is increased by the presence of intergranular exchange. With vanishing intergranular exchange the reversal process is by the formation and annihilation of vortices in relatively fixed locations. With intergranular exchange the vortices form and then translate to line up along 'domain walls'. The reversal process is completed by the motion and subsequent interaction annihilation of these vortex walls.

The test of irreversibility or nucleation processes is that if along any portion of the loop the field is changed in the opposite direction, the magnetization does not retrace the previous history. A variety of minor loops can be generated. An important variant is the 'remanence loop' where at each step along the 'major' loop the field is removed (to zero) and the remanence magnetization versus applied field is plotted (Cullity, 1972). As the field is decreased from a point along the major loop, the magnetization decreases due to reversible rotation of each particle magnetization toward the respective easy axes (Fig. 1.3). This reversible phenomenon is characterized by a reversible susceptibility χ_r, which to first order is independent of the specific position along the major loop where the field is reversed. Thus, the remanence loop does not contain reversible rotation and simply counts the net number of grain magnetizations oriented in the field direction. Clearly the remanence curve is complicated for highly interacting grains in polycrystalline thin films, but, in general for 'square' loops as illustrated in Fig. 1.3, the remanence curve is not too different from the 'major' loop. Torque curves that result from changing the direction of the applied field are useful in the analysis of the switching mechanisms as well as for a measurement of the grain anisotropy fields (Cullity, 1972; Chikazumi, 1964).

Magnetic transducers or heads for recording and playback are ideally 'soft' materials with vanishing coercivity, as large a saturation magnetization as possible, and large initial permeability. Since the rate of data transfer is important, these heads must maintain their properties to very high frequencies (~ 100MHz). The geometry of the transducers should be designed so that they are 'efficient' in terms of the translation of input current to recording head field (or head gap magnetomotive force) at all operating frequencies.

Historically the most widely utilized head material has been hot pressed MnZn with saturation flux density about 4500 G, coercivity ~ 0.5 Oe and initial susceptibility ($\sim B_s/H_c$) ~ 2000 (Smit, & Wijn, 1959; Jones & Mee, 1990). Ferrite materials are polycrystalline with grain sizes on the order of 0.05µm and are advantageous, in particular for contact tape recording, due to their hardness and durability. Two disadvantages to ferrite heads have led to the development of new materials. A low saturation flux density restricts their use to recording on media with coercivities no greater than approximately 80kA/m (1000 Oe). In addition magnetization processes in ferrite are dominated by domain wall motion that results in a permeability spectrum that decreases in amplitude (and develops significant loss components) near 1–2MHz. Even though high-efficiency head structures can be designed that yield a head efficiency spectrum that does not decrease until the range 20–40MHz, the rapid increase of data rate in recording systems requires transducers that are efficient to at least 100–200MHz.

To accommodate the increase in recording medium coercivity beyond 80kA/m, metallic films have been developed. Common materials are Sendust (AlFeSi) and Permalloy (NiFe) with $B_s \sim 1$T (10,000G); however, higher saturation materials are continually under investigation such as CoZr ($B_s \sim 1.4$T) and Co-Nitride Alloy films ($B_s \sim 2$–3T) (Jagielinski, 1990). For contact tape recording, metallic films are utilized to form the gap region of durable ferrite heads, 'metal-in-gap' heads (Jeffers, 1986; Ash, et al., 1990). For example all DAT audio and 8mm video recorders utilize metal-in-gap heads with Fe or metal evaporated tape. For disk systems with 'flying' heads, metal-in-gap heads are also utilized, however, thin film heads of permalloy predominate. In Fig. 1.4 the structure of a typical thin film inductive head is shown. The general structure of a 'gapped torroid' of magnetic material applies to all

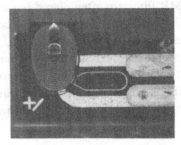

Fig. 1.4. Structure of inductive thin film head. Current applied to the windings yields recording fields in the gap or tip region. Flux into the gap region during replay yields time changing flux in the poles threading the windings to produce a voltage at the head terminals. An MR playhead may use the poles of a thin film head as a shield. The MR film would then be deposited centered in the gap extending across the head a width slightly less than the pole width with a depth from the gap surface of the order of 1μm. (Courtesy of Applied Magnetics Corporation.)

inductive heads. Current is applied to the turns that results in flux circulating the structure. In the head core the flux density is predominately due to the magnetization since the internal fields are small. In the tip region the flux density is largest because of the decrease in cross-sectional area, and saturation generally begins in this region. In the gap the flux density is due solely to the magnetic field. Flux that fringes into the region above the gap where the medium passes provides the recording field.

Thin film heads can be designed with suitable geometry in the gap region to exhibit high efficiency and saturation gap fields approaching the B_s of permalloy. In addition, if the domain structure in the tip region is such that the magnetic fields in the tip are perpendicular to the domain walls, the playback magnetization response is by domain magnetization rotation rather than by domain wall motion, and response to at least 100MHz is obtained (Kasiraj, *et al.*, 1986, Kasiraj & Holmes, 1990; Hoyt, *et al.*, 1984).

The magnetoresistive phenomenon provides an extremely efficient playback transducer whose enhanced sensitivity over inductive heads provides system signal-to-noise ratios that are limited only by medium noise (Jeffers, 1986). In the magnetoresistance effect a current is applied to a film (usually permalloy), and the magnetization is biased by a variety of techniques to lie at approximately 45° to the field from the recorded medium in the film plane. Rotation of the magnetization in the film changes the resistance and thus induces a voltage across the ends of the film. MR films are often placed in the gap (parallel to the gap face) of a thin film head to form a shielded structure and extend slightly less across the head than the recording width to assist in the reduction of crosstalk from neighboring tracks. Use of a magnetoresistive playback transducer permits the utilization of larger record gaps in order to reduce head saturation effects. MR heads, in contrast to inductive heads, yield a playback voltage independent of head medium relative motion.

The magnetic recording channel

Information storage involves a data input and retrieval scheme of which the magnetic recording process is a central part. Digital signal processing is the dominant form of information transfer at high data rates and is utilized for high density magnetic recording (Wolf, 1990). A schematic of the digital magnetic recording channel is given in Fig. 1.5. The input data may be either digital or analog that is first converted to digital by an A/D converter. The subsequent digital data stream is encoded by a variety of modulation processes to an encoded data sequence (Patel, 1989; Cannon & Seger, 1989; Yaskawa & Heath, 1989; Mallinson & Miller, 1976). The encoded data modulates voltage reversals that are applied to the recording head amplifier. This last stage of modulation is generally 'NRZI' in which a voltage reversal occurs corresponding to the center of each data cell where a '1' occurs. If the data cell contains an '0' no voltage reversal occurs. Commonly, group {d,k} codes are utilized, such as 2/3{1,7} or 1/2{2,7} where the timing cell length with respect to the original data before encoding is 2/3 or 1/2, respectively. In these codes the minimum separation between '1's is 2 or 3, respectively; thus, the

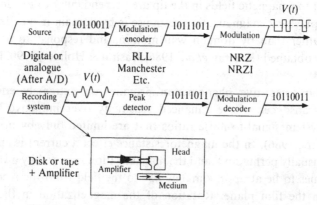

Fig. 1.5. Schematic of the digital recording channel.

minimum temporal separation between cells containing '1's compared to the incoming data cell length is 1.33 or 1.5, respectively.

The alternating voltage pattern corresponding to the encoded input data stream is applied via a write amplifier to the recording head. The write amplifier produces an alternating current sequence corresponding to the timing of the input voltage changes. The write current is applied to the recording head in order to record encoded data patterns onto the media as a spatial stream of magnetization transitions. Since each '1' of encoded data corresponds to a current reversal applied to the head, the subsequent reversal of head magnetic field writes a magnetization transition in the medium. The write current magnitude is set to optimize the recording process and is maintained at a fixed level. The temporal changes of write current can be modified by a 'precompensation' scheme that assists in the linearization of the recording channel and which limits the maximum magnetic field in a magnetoresistive play head (Schneider, 1985).

In the magnetic recording process a medium (tape or disk) moves at a fixed speed past a transducer that writes information over a track width W by applying temporally varying fields to the magnetic medium (Fig. 1.6). This relative motion of the head and medium translates temporal variations to spatial variations ($x = vt$ or $\lambda = v/f$) so that a changing pattern of magnetization is written along the track (Fig. 1.7). In the digital magnetic recording process current applied to the head yields a corresponding changing pattern of head field polarities applied to the medium. The instant of field reversal writes a magnetic transition into the medium by saturating the magnetization above the gap with respect to an oppositely saturated magnetization beyond the gap edge. A magnetization transition of finite length occurs whose length is set primarily by a balance of the head field gradient and the demagnetizing field gradient at the transition center. The field magnitude is fixed at a value large enough to saturate the magnetization or to 'close the loop' ($\sim 2H_c$, Fig. 1.3) and in general is set to optimize the recording process. The temporal sequence of applied current polarity changes that occur in the center of 'bit' cells of fixed time size are translated to a spatial sequence of bit cells of length B along the medium where a magnetic transition occurs corresponding to each encoded '1' of information. The physical 'bit' cell area as recorded in the medium is $W \times B$ where B denotes the cell length in the medium (or the minimum transition spacing for {0,k} codes). A numerical example is the National Storage Industrial Consortium project (NSIC) to explore disk recording at 10Gbits/in^2. A reasonable linear density is 400kbits/in.

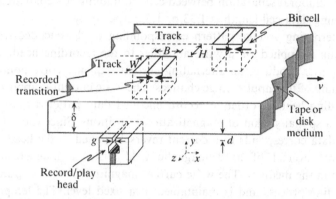

Fig. 1.6. Diagram of recording geometry. B is the size of the flux reversal cell or, alternatively, the distance between recorded transitions without minimum transition separation code constraints. W is the recorded track width and δ is the magnetic thickness of the recorded layer. For magnetic tape δ refers to the depth of signal recording, which is much less than the coated depth (the total magnetic layer is utilized in noise analysis). d is the head–media spacing and g is the gap length of the recording head. The medium moves with velocity v with respect to the head. The coordinate system (x,y,z) utilized here is x along the recording direction, y normal to the film or tape plane and z along the cross-track direction. Parallel tracks are recorded at lateral separation 'H' forming an information 'crosstalk' guard band. For shielded MR playback the MR element lies in the gap as indicated and the 'effective' gap length is $g/2$.

At a data rate of 100Mbits/sec the data temporal cell interval would be 10 nanosec at a relative head-medium speed of $v = 250$ in/sec. The track density would be 25ktpi so that allowing 33% for guard band (H in Fig. 1.6) gives a track width $W = 0.7\mu m$. Utilizing a 2/3 [1,7] code yields a transition cell length of $B = 1.33 \times 2.5 \ \mu in = 3.32 \ \mu in$ (83nm) with a minimum transition spacing of 166nm.

In the following chapters the emphasis will be on magnetic phenomena of the recording process so that details about transition spacing with group codes will not be emphasized. The focus will be on writing a transition or multiple transitions of given separation. B will be both the recorded flux cell size as well as the flux cell separation (tacitly assuming $\{0,k\}$ codes where k $\rightarrow \infty$ for the analysis of square wave recording). Thus, the terms 'flux cell' and 'bit' will be used interchangeably.

Playback of recorded information is accomplished by a transducer that is either a conventional inductive head or a magnetoresistive sensor. The playback process translates the spatial magnetization data into a

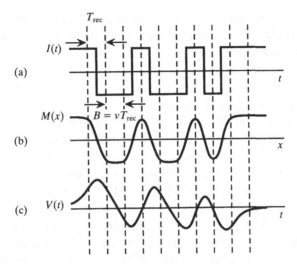

Fig. 1.7. Schematic of translation of temporal data stream into a sequence of magnetization transitions. The current or head field maintains a fixed magnitude but alternates in polarity at the center of cells of encoded time interval T_{rec} where a '1' of encoded information occurs (a). The temporal change of head field coupled with relative head-to-medium motion yields a magnetization transition at the center of spatial cell of length $B = vT_{rec}$ where in (b) $M(x)$ represents the track-width averaged magnetization profile. During playback the head yields a voltage pulse corresponding to each transition change (c). In this example an 'isolated' transition and a 'dibit' are illustrated. The dibit corresponds to two adjacent transitions that are reasonably well isolated from the other transitions in the data pattern. The dibit voltage amplitude is reduced from that of the isolated transition due primarily to linear superposition of the individual pulses.

temporal sequence of alternating voltage pulses (Fig. 1.7) that occur at temporal separations corresponding to the input encoded data sequence. The pulse shape can vary depending on the medium microstructure as well as the head design; however, most commonly a somewhat symmetric voltage pulse occurs. This stream of voltage pulses can be peak detected to ascertain the occurrence of a transition in a given cell (Fig. 1.5). The resulting sequence is decoded to obtain the original input data. In high density systems detection schemes such as partial response rather than peak detection can be utilized (Wood & Peterson, 1986; Howell, *et al.*, 1990).

Errors can occur in data recovery due to a variety of mechanisms. Fundamental is noise due to the granularity of the medium, as well as playback amplifier and transducer noise. In addition, errors result from non-linearities (transition shift and magnetization reduction) in the recording process, insufficient overwrite of previous recording data, and playback interference of adjacent track information. Finite amplifier rise times can broaden the written transition as well as the playback voltage pulse; however, neglecting that effect as well as velocity modulation that can occur with elastic tape, the process of recording transitions on the medium is purely a spatial phenomenon.

The only part of the magnetic recording process where 'time' is important is the response of the recording head either during recording or playback (Wood, *et al.*, 1986). The temporal dependence or frequency response of the head efficiency causes abrupt changes in current not to yield abrupt changes in field. This can cause the amplitude of the head field between bit cells to depend on the data pattern. Generally, for high data rate systems where the bit cell length is on the order of the head field rise time the write current amplifier is 'overdriven' so that the head field reaches sufficient magnitude to saturate the media in-between transitions for every pattern (Coleman, *et al.*, 1984). Temporal head effects also alter the playback pulse shape in addition to band limiting of the playback amplifier. Temporal head phenomena are not considered in this work. It is assumed that the head possesses a fixed head efficiency, independent of frequency.

Thin films currently in use for digital disk recording are 'longitudinal' films characterized by a net in-plane anisotropy. Thus, individual grain magnetizations lie predominantly in the film plane with the average film magnetization along the head–medium motion direction. Perpendicular recording, where the net grain anisotropy and magnetization lie perpendicular to the film plane, has been proposed for extremely high density recording. However, to date, perpendicular recording media (with associated heads and signal processing schemes) have not been introduced into a product.

A precise study of the magnetic recording process involves the application of a spatially varying head field to granular media. Computationally intensive numerical micromagnetics is then required to solve for the magnetizations of the interacting grains in the medium in order to determine the average magnetization transition, the fluctuations leading to noise, as well as a variety of interference phenomena (Bertram, *et al.*, 1992; Victora & Peng, 1989; Zhu, 1993). In this book the recording process will be discussed in terms of simplified analytic models or general formulations that permit physical insight into the magnetic recording process. The results of micromagnetic modelling will be referred to in appropriate sections for emphasis.

Units

All the analysis in the subsequent chapters of this book will be in the MKS or the SI system of units. However, since the cgs system is still utilized, in particular for material characterization, a discussion of these two major systems will be given here.

The three basic quantities utilized in the description of macroscopic magnetic phenomena are the magnetization M, the magnetic field H and the flux density or magnetic induction B. All three quantities are 'vector fields': they possess magnitude and direction that, in general, change from point to point in space. Magnetism is a relativistic phenomenon that results from the motion of charged particles. In magnetic recording two types of motion are of interest: the motion of electrons along wires and the orbital and spin atomic motion of electrons in solids. Current is characterized by a current density J or current I for thin wires. The atomic motion may be characterized by a magnetic moment for each electron, or in less detail (for oxides with localized spins) by a net vector moment per atom. For atomic moments the magnetic moment characterizes the circulating or rotating charge motion and is, in essence, a charge angular momentum. The magnetic field may be written in terms of all these current sources, using either a differential equation or an integral formulation. The dimensions of magnetic field are current per length, and of magnetic moment are current times (length)^2.

Magnetic recording analysis may generally be treated as a macroscopic phenomenon: detailed knowledge of the atomic spin structure is required only for the understanding of advanced materials. In this case it is convenient to define the magnetization:

$$M \equiv \lim_{V \to 0} \frac{1}{V} \sum_i \mu_i \qquad (1.4)$$

The magnetization is the net vector moment per unit volume. It exists in space only where there is magnetic material and is a specific quantity independent of the sample volume. In (1.4) the rule is to take a volume V, sum the net vector moment, and divide by the volume. The limit notes that the volume is to be made small compared to the macroscopic scale under study, but large compared to the detailed atomic structure so that a meaningful average exists. Magnetization is a vector quantity that can vary both in direction and magnitude from 'macroscopic' point to point in a material; the dimensions are current per length.

The magnetic flux density B is defined by a constitutive relation as the sum of the magnetic field H and the magnetization M. The two most commonly utilized units are (rationalized) MKS or SI where $B = \mu_0(H + M)$ or the (unrationalized) cgs system where $B = H + 4\pi M$. The units for both these systems as well as equivalents are given in Table 1.1. Note that in general the constitutive relation is a vector condition that applies at every point r:

$$B(r) = \mu_0(H(r) + M(r)) \tag{1.5}$$

For an example consider the remanent magnetization of a magnetic tape where the particles are packed at a volume fraction of 33% and the orientation yields a squareness of $S = 0.8$. If the particle magnetization is 370emu/cc in cgs units, then the saturation magnetization of the tape (1.3) is $M_s = 370/3 \sim 124$emu/cc or 124kA/m in the MKS system. The saturation remanence is $M_r = SM_s \sim 100$emu/cc $= 100$kA/m. The remanent flux density in the cgs system is $B_r = 4\pi M_r = 1200$G $= 0.12$T or, directly in the MKS $B_r = \mu_0 M_r = 4\pi \times 10^{-7} \times 100 \times 10^3 \sim 0.12$T.

Table 1.1. *Units for MKS(SI) and cgs system*

Quantity	SI(MKS rationalized)	CGS (unrationalized)	Equivalents
B	tesla (T)	gauss (G)	1T $= 10^4$ G
H	amps/meter (A/m)	oersted (Oe)	$\dfrac{10^3}{4\pi}(\sim 80)$ A/m $= 1$Oe
M	amps/meter (A/m)	emu/cc	1kA/m $= 1$emu/cc

Constitutive relation:
$$B = \mu_0(H + M) \qquad B = H + 4\pi M$$
$$\mu_0 = 4\pi \times 10^{-7} \text{(Henrys/m)}$$

2

Review of magnetostatic fields

Introduction

Chapter 2 provides a review of basic electromagnetic theory as applied to static magnetic fields. Neither wave motion nor eddy current effects caused by conductive media are considered: it is presumed that all time scales are long compared to these phenomena. Time enters only through the constant head-to-medium relative speed, v, so that all temporal information is transformed immediately into the fundamental spatial recording process by $x = vt$ or dx/dt. The purpose of this chapter is to provide useful relations for the determination of fields, both from integrals over field sources and solutions of differential equations for field potentials. Thus, the framework will be provided for the determination of magnetic fields from magnetized heads (Chapter 3) and media (Chapter 4), and in addition, expressions for Fourier and Hilbert transforms will be presented for utilization in spectral analysis of recording signals and noise. These transforms involve operations on the spatial variable, x, which represent the head-to-medium motion direction. In particular, the Fourier transform will be expressed in terms of the spatial transform variable, k: the wavenumber or inverse wavelength $(2\pi/\lambda)$. For direct correspondence with measured frequency the simple transformation $f = vk/2\pi$ can be utilized.

In magnetic recording the track width is generally large with respect to dimensions in the nominal recording plane, which includes the head-to-tape motion direction and the direction perpendicular to the medium surface (thickness direction). Therefore, two-dimensional field expressions are useful and will be given explicitly in this chapter. In two dimensions the Fourier transforms acquire a particularly simple form yielding the familiar exponential spacing loss. The Hilbert transform

17

becomes a simple relation between the longitudinal, or x, directed fields and the vertical, y, field components. Three-dimensional fields, including Fourier transforms, will be discussed in subsequent chapters in an approximate form as applied to head edge effects. Green's function techniques will be referred to, but not discussed: a good introduction may be found in Jackson, 1975.

Basic field expressions

For continuous media and static fields, Maxwell's differential equations for the flux density B and magnetic field H apply:

$$\nabla \cdot B = 0 \tag{2.1}$$

$$\nabla \times H = J \tag{2.2}$$

where J is the current density. Since the constitutive relation is:

$$B = \mu_0(H + M) \tag{1.5}$$

(2.1) can be written as:

$$\nabla \cdot H = -\nabla \cdot M = \rho_m \tag{2.3}$$

where $\rho_m \equiv -\nabla \cdot M$ is an effective magnetic charge (pole) density. Equations (2.2) and (2.3) give the magnetic field due to electric charges in motion represented by the continuous variables: J (real current density) and magnetization M (atomic current or charge angular momentum density).

Note that for a material with a permanent magnetization M_0 and linear susceptibility χ, the net magnetization is given by:

$$M = \chi H + M_0 \tag{2.4}$$

and the net induction B is given by

$$B = \mu H + M_0 \tag{2.5}$$

where μ is the permeability:

$$\mu = \mu_0(1 + \chi) \tag{2.6}$$

The magnetic field may be written directly as an integral over sources $J(r)$ and $\rho_m(r) = -\nabla \cdot M(r)$. In terms of the current density, an integral form of (2.2) is:

$$H(r) = \frac{1}{4\pi} \int d^3 r' \, J(r') \times (r - r') / |r - r'|^3 \qquad (2.7)$$

where the integral is taken of all space where J does not vanish (Fig. 2.1). Note that (2.7) includes all currents due to wires as well as induced currents occurring in conductive media. For fields due to magnetized media, the magnetostatic fields, a useful integral form of (2.3) is:

$$H(r) = -\frac{1}{4\pi} \int_V d^3 r' \nabla \cdot M(r')(r - r') / |r - r'|^3$$

$$+ \frac{1}{4\pi} \int_S d^2 r' \hat{n}' \cdot M(r')(r - r') / |r - r'|^3 \qquad (2.8)$$

The total vector magnetic field is the sum of contributions from both current (2.7) and magnetization (2.8) sources. In (2.7) and (2.8) the explicit dependence of the vector fields on the position variables r, r' are shown, where r' refers to the (distributed) field sources and r refers to the location of determination of the magnetic field.

In general the vector form of (2.8) is such that the field from a small element of equivalent magnetic charge is directed from the source to the observation point with a magnitude proportional to the incremental charge $(-\nabla \cdot M(r) d^3 r'$ or $\hat{n}' \cdot M(r) d^2 r')$ divided by the square of the distance. The first integral in (2.8) is taken over the volume of magnetic material while the second integral is over the surface of the magnetic body (Fig. 2.2). $\hat{n}'$ represents the outward surface normal at r' and the form of (2.8) presumes that $M(r')$ vanishes discontinuously at the surface. Thus, $\hat{n}' \cdot M(r')$ represents an equivalent surface charge density

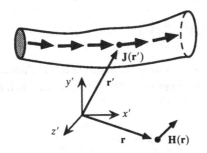

Fig. 2.1. Illustration of current source for magnetic field. r' is the distributed vector position of the current source, and r is the location of the desired magnetic field.

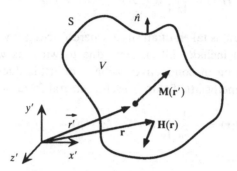

Fig. 2.2.   Illustration of integration geometry for magnetic field from magnetized media (2.8).

and accounts for an infinity in $-\nabla \cdot M(r')$ if (2.8) had been expressed solely by the first term with the volume integral taken over all space (Brown, 1962).

Both (2.2) and (2.3) can be recovered from (2.7) and (2.8) by use of the relation:

$$\nabla \cdot \left(\frac{r}{|r|^3} \right) = 4\pi \delta^3(r) \tag{2.9}$$

where $\delta^3(r)$ is the three-dimensional delta function.

In magnetic recording, fields due to currents are generally generated by simple wire sources. The field external to the wire is desired and the wires can be approximated as carrying a current I and having negligible cross-section. The current density is then represented by a (two-dimensional) delta function, and (2.7) can be simplified to:

$$H(r) = \frac{I}{4\pi} \int_P dr' \hat{\imath} \times (r - r')/|r - r'|^3 \tag{2.10}$$

where the single integral is along the wire path. I is the current magnitude and $\hat{\imath}$ is a unit vector tangent to the current path at each r'. An extremely useful integral form of (2.10) is

$$\oint_c H \cdot dl = I_{total} \tag{2.11}$$

where the integral is taken over a closed path and I_{total} is the total current threading the path. (2.11) can be derived from (2.2) or (2.7) so that I_{total}

represents an area integration normal to and over the current threading the closed path $(I_{total} = \oint \boldsymbol{J} \cdot \hat{n} dA)$. The line integral always vanishes for magnetic fields that arise from magnetizations, so that (2.11) holds even in magnetized media, and the field utilized in the line integral can be the total magnetic field due to all sources.

For magnetic recording applications it is useful to note that instead of (2.8), where the fields are expressed in terms of effective charge densities, an alternate form exists utilizing (2.7) and magnetic equivalent current densities. For a magnetized body with magnetization $\boldsymbol{M}(\boldsymbol{r}')$, the equivalent volume and surface currents, respectively, are:

$$\boldsymbol{J}_{vol}(\boldsymbol{r}_s) = \nabla \times \boldsymbol{M}(\boldsymbol{r}_s)$$
$$\boldsymbol{J}_s(\boldsymbol{r}_s) = \boldsymbol{M}(\boldsymbol{r}_s) \times \hat{n}$$

(2.12)

where $\boldsymbol{r}_s$ denotes a point on the object surface. These equivalent currents when utilized in (2.7) yield $\boldsymbol{B}(\boldsymbol{r})$ everywhere:

$$\boldsymbol{B}(\boldsymbol{r}) = \frac{\mu_0}{4\pi} \int\int\int d^3 r' \boldsymbol{J}_{vol}(\boldsymbol{r}') \times (\boldsymbol{r} - \boldsymbol{r}')/|\boldsymbol{r} - \boldsymbol{r}'|^3$$

(2.13)

$$+ \frac{\mu_0}{4\pi} \int\int d^2 r' \boldsymbol{J}_s(\boldsymbol{r}) \times (\boldsymbol{r} - \boldsymbol{r}')/|\boldsymbol{r} - \boldsymbol{r}'|^3$$

where the second integral is over the surface. Exterior to the body, $\boldsymbol{B}$ and $\boldsymbol{H}$ are identical, apart from the constant μ_0. Again fields from real currents must be added to (2.13) to obtain the total flux density.

A comparison of (2.13) with (2.7) shows that from field (H) measurements external to an object, it is impossible to tell whether the source is magnetized media or a real current distribution. Therefore, recording media magnetized with a recording pattern can always be represented by an equivalent current distribution (generally a distribution of wires wrapped around the medium circling above and below and coiling along the head–medium motion direction). In addition, an energized head can be considered to be driven by a permanently magnetized thin disk placed at each wire turn (magnetized normal to the disk plane), as long as interest is only in fields exterior to the minimum volume containing the wires. Note that (2.13) may be derived directly from (2.2) for zero real current and substitution of $\boldsymbol{H} = \boldsymbol{B}/\mu_0 - \boldsymbol{M}$.

At boundaries where there is a distinct change in magnetic properties (e.g. permeability or simply permanent magnetization magnitude or direction) conventional boundary conditions apply. Across a boundary the tangential component H is discontinuous by the amount of any *real* surface current and the normal component of B is always continuous. These relations can be easily derived either from Maxwell's differential equations or the integral expressions given above.

These fields and equivalent sources are illustrated in Fig. 2.3 for the case of a uniformly magnetized (very wide) rectangular bar. In Fig. 2.3(a) the equivalent source surface pole densities and surface currents are shown (there are no volume magnetic equivalent charges or currents). Magnetic surface charge density $(\hat{n} \cdot M(r))$ are distributed uniformly across the end surfaces; their sign is positive and negative, respectively, at the end surfaces where the vector magnetization ends and begins. In Fig. 2.3(b) flux lines (B fields) are plotted. By symmetry only one quarter section is shown, as well as for Fig. 2.3(c). Flux plots are defined such that the direction of the lines is parallel to the B field at each point, and the separation between lines is inversely proportional to the flux density magnitude. In the center of the bar the flux density is parallel to the source magnetization and virtually equal in magnitude to the magnetization M. At the ends of the rod the B field concentrates at the sharp corners and, in fact, becomes infinite at the corners (see section on 2D fields). Exterior to the rod the flux 'flows' out the right side, divides, and circles back around to enter the left face in a symmetrical fashion. Note that the normal component of B is continuous across all boundaries. At the top surface the tangential component of B is discontinuous since there is a discontinuity in the (tangential) magnetization. At the end surfaces B is continuous in both components. Use of (2.13) in an equivalent current formulation yields no 'current surfaces' at the ends to give a discontinuity. Alternatively, the tangential magnetic field is continuous, and at the ends there is no tangential magnetization discontinuity.

In Fig. 2.3(c) the lines of magnetic field H are plotted. If magnetic field is considered in terms of equivalent magnetic charge, then the field flows continuously from sources at either end of the rod. In contrast to the B field, the vector H field is continuous across the top and bottom boundaries: it is discontinuous at the ends. Exterior to the rod the field, apart from a scaling constant, the 'fringing' field exactly correspond to the flux plots of B in Fig. 2.3(b). Inside the rod the field is given by the difference between the flux density and the magnetization. Generally, the field opposes the magnetization and thus is a 'demagnetization' field, in particular near the center of the rod. Near the corners, however, the field changes sign, is directed towards the corners, and becomes infinite in

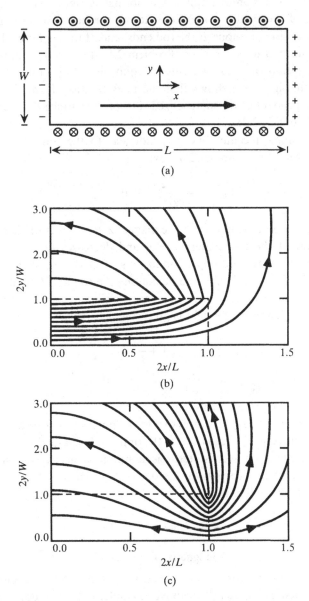

Fig. 2.3. Illustration of **B** and **H** fields from a uniformly magnetized wide bar of length L and depth W. (a) indicates the direction of magnetization and sketches the equivalent field sources. The magnetic charges are only on the end surfaces, and the equivalent surface current wraps around (top and bottom) the bar. (b) is a plot of flux lines (**B**) and (c) gives a plot of magnetic field (**H**) lines. In (b) and (c), due to symmetry, only one quarter of the rod is shown.

magnitude at the corners. In this example it is the tangential component at the ends that becomes infinite; the normal is finite. The tangential component of the field is continuous across the boundary; the normal component is discontinuous at the rod ends, and at the center of the ends the field is normal to the surface (Problem 2.5). It should be noted that the interpretation of field in terms of equivalent magnetic charges as illustrated in Fig. 2.3(c) shows that the field H arises from the source charges independent of the 'magnetized side' of the surface that generates the poles (Problem 2.3).

The field along a center line perpendicular to the rod axis is given simply (utilizing, for example (2.8)) by:

$$H_x(0 \cdot y) = -\frac{M}{\pi}\left(\tan^{-1}\left(\frac{W-2y}{L}\right) + \tan^{-1}\left(\frac{W+2y}{L}\right)\right) \qquad (2.14)$$

The field has a component only in the magnetization direction, is greatest in magnitude at the rod center ($y = 0$) and decreases uniformly away from the center. At the rod upper and lower surface ($y = \pm W/2$) the field is tangential to the boundary. The field along the rod axis ($y = 0$) is also 'longitudinal' and can be written in a form similar to (2.14). At the rod ends ($x = \pm L/2$, $y = 0$) the field is given by:

$$H_x^{inside} = -\frac{M}{2}\left(1 + \frac{2}{\pi}\tan^{-1}\frac{W}{2L}\right)$$

$$H_x^{outside} = \frac{M}{2}\left(1 - \frac{2}{\pi}\tan^{-1}\frac{W}{2L}\right) \qquad (2.15)$$

The first term in each expression is due to the poles at the surface of field evaluation. The second term is due to the oppositely charged poles on the opposite surface. The second term is continuous across the surface, whereas the first term provides the discontinuity of magnitude M to make B continuous across the interface.

Demagnetizing factors

Fields caused by uniformly magnetized material are frequently of interest. In such cases, as in the above example, (2.8) includes only the term containing the surface charge. For uniformly magnetized media (2.8) can be written in tensor form:

$$H(r) = -\tilde{N}(r) \cdot M \qquad (2.16)$$

where $\tilde{N}(r)$ is a tensor function of position given by:

$$\tilde{N}(r) = -\frac{1}{4\pi}\int\int\int d^3r'\nabla'\left(\nabla'\left(\frac{1}{r-r'}\right)\right) \tag{2.17}$$

This tensor is given by an integral over the object volume and can be evaluated either inside or exterior to the body. The demagnetizing field tensor $N^d(r)$, so called when evaluated in the interior, always exhibits the property that the trace or sum of the diagonal elements at all r is unity:

$$\sum_i N_{ii}^d = 1 \tag{2.18}$$

For the case of symmetric bodies with the coordinate system coinciding with the symmetry axes, then $\tilde{N}^d(r)$ is diagonal. If, in addition, the object is an ellipsoid of revolution, the tensor is not a function of position. Demagnetizing factors for ellipsoids are tabulated in the introduction to almost all books on magnetic materials (e.g. Brown, 1962; Chikazumi, 1964; Cullity, 1972).

For a sphere the diagonal elements are equal by symmetry, so that (2.17) reduces to a scalar relation. Utilizing (2.18) yields:

$$N_{sphere}^d = \frac{1}{3} \tag{2.19}$$

For a long square rod as illustrated in Fig. 2.3 the demagnetization field tensor at the center of the rod is given from (2.14) by:

$$\tilde{N}_{rod}^d = \frac{2}{\pi}\begin{pmatrix} \tan^{-1}(W/L) & 0 & 0 \\ 0 & \tan^{-1}(L/W) & 0 \\ 0 & 0 & 0 \end{pmatrix} \tag{2.20}$$

The tensor is diagonal with entries in order of the coordinate directions x,y,z. The demagnetizing field vanishes along the rod axis or z direction since, in that case, the poles are infinitely far from any evaluation point. Equation (2.18) can be shown easily to hold. If the rod is of square cross-section ($L = W$) then the non-zero factors in (2.20) are each equal to 1/2, a result that can be argued by symmetry from (2.18). By symmetry, a long rod of circular cross-section would have the same demagnetization factors as the center of a square rod, but, as in the case of the sphere, the demagnetizing fields would be independent of position.

In general, as illustrated in Fig. 2.3(c), demagnetizing fields are large if the evaluation point is near the 'poles' (or positive and nearly infinite near sharp corners). For example, from (2.20), if the rod is long in the x

direction, $L \gg W$, the demagnetizing field in the x direction is small. As the ratio W/L increases the demagnetizing field at the rod center in the x direction increases (and that in the y direction decreases). For $L \ll W$ the rectangular rod becomes a thin plate with a unity demagnetizing factor in the direction (x) perpendicular to the plane of the plate. In this case the demagnetizing field is everywhere equal in magnitude and oppositely directed to the magnetization normal to the plate plane (as long as the field is not evaluated near the essentially infinitely far plate ends).

The volume averaged demagnetization factors are of interest because they enter the expression for the energy of a uniformly magnetized object and hence are measured. For a symmetric body where the coordinate system is taken coincident with the principal axes, the volume averaged demagnetization tensor is diagonal. The diagonal terms obey (2.18) and are tabulated in standard references (Cullity, 1972; Brown, 1962) and are often referred to as the 'magnetometric' or 'ballistic' factor. For cubes, spheres, long square rods and cylinders, by symmetry, the magnetometric factors are identical to the (center) demagnetization factors discussed above.

Magnetostatic fields from flat surfaces

In general as mentioned above, (2.8) yields an incremental field that is directed away from incremental positive poles towards the evaluation point; the magnitude is given by the pole strength divided by the square of the distance between source and observation point. The complete field involves a vector integration over spatially changing charge distributions and vector orientations to observation point. An alternative view of the incremental field is that the field magnitude is given by the local charge density $(\hat{n}' \cdot M(r)$ or $-\nabla \cdot M(r)\mathrm{d}(r - r'))$ times the incremental solid angle seen from the observation point $\mathrm{d}^2 r'/|r - r'|^2$). In the case of plane surfaces with a constant surface pole density, as in the example in Fig. 2.3, the field component normal to the surface is given by the solid angle extended by the poles with respect to the observation point:

$$H_\mathrm{n} = (\hat{n} \cdot M)\Omega/4\pi \qquad (2.21)$$

where Ω is the solid angle (Fig. 2.4). The normal field direction is away from (positive) poles. Equation (2.21) can be utilized to obtain the fields in Fig. 2.3(c) as well as (2.14, 2.15). For a thin plate magnetized normal to its surfaces, (2.21) yields a demagnetizing field equal and opposite to the magnetization as long as the plate is reasonably thin, since the solid angle

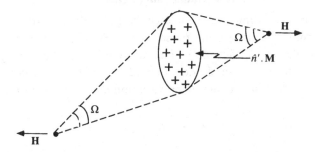

Fig. 2.4. Illustration of solid angle formula for the magnetic field component normal to a finite flat surface of constant magnetic charge. The normal field is directed away from the positive surface charge on both sides.

anywhere (far from the ends) is 2π for each surface. It is important to note that the field in this case is independent of the plate thickness: in the case of an infinite plate only the normal field component exists.

Equation (2.21), which may be derived from (2.8), gives the normal field component from a sheet of poles regardless of whether the observation point is inside or exterior to the medium. For a sheet of infinite extent, the field magnitude is everywhere equal to half the pole density $(H_n = -M/2)$, and is normal to and points away from the sheet: there is no tangential component. In fact, for any distribution of magnetization, where the complete form of (2.8) applies, the field arbitrarily close to a point on the surface is equal to half the local pole density plus the contribution from the other sources. In this sense a surface element d^2r' defines the plane at the point and the region exterior gives the field due to the other sources. (Of course care must be taken at sharp corners, and the simple picture may be complicated if volume poles exist arbitrarily close to the point at the surface.) Nonetheless, as the observation point is moved through the surface from the inside to the outside, the 'other source' fields are invariant, but, as discused above, there is a field discontinuity due to the sheet d^2r' from (2.21) of amount $(n' \cdot M(r'))$. Thus, (2.8) yields the usual boundary condition that the normal component of B is continuous across a boundary. If the tangential component of the magnetic field is considered at a surface point, then the field is due entirely to the 'other sources' and no discontinuity occurs. Thus, (2.8) also yields a second boundary condition that, for current free regions, the tangential component of H is continuous.

Review of magnetostatic fields

Two-dimensional fields

In magnetic recording the geometry is generally two-dimensional involving only the recording plane (x,y) (Fig. 1.6). Uniformity is assumed in the cross-track direction except when explicit analysis of finite track effects is desired. Equation (2.8) can be integrated in the cross-track (z) direction assuming no variation in any properties to yield:

$$H(r) = -\frac{1}{2\pi}\int_S d^2r' \nabla \cdot M(r')(r - r')/|r - r'|^2$$

$$+ \frac{1}{2\pi}\int_P dr' \,\hat{n}' \cdot M(r')(r - r')/|r - r'|^2$$

(2.22)

In two dimensions the integration over the volume poles becomes an area integration (S) over the (x,y) plane projection of the sample volume. The surface integral becomes a line integral (P) around the projected area. In (2.22) the position vector is now $r = (x,y)$. An analogous form is obtained for two-dimensional current sources using (2.7). In all subsequent discussions of 2D fields in this text the cross-section area will be drawn and the material will be assumed to extend infinitely out of the plane of illustration (as in Fig. 2.3(a)).

A useful simplification of (2.22) occurs for the fields from plane surfaces (straight lines in 2D) with constant pole density. Let the edge, without loss of generality, lie along the y direction (Fig. 2.5) and let the vector $r - r'$ be described by the complex variable $z = x + iy$. Then, letting σ denote the surface charge density and utilizing the second term of (2.22), the field can be written as:

$$H^*(z) = \frac{i\sigma}{2\pi}\int_{z_1}^{z_2} dz' \cdot (z - z')/|z - z'|^2 = \frac{i\sigma}{2\pi}\ln\frac{z - z_2}{z - z_1}$$

(2.23)

where z_1, z_2 are the end coordinates of the edge. $H^*(z)$ is the complex conjugate of the field:

$$H(z) = H_x(x, y) + iH_y(x, y)$$

(2.24)

Writing $z = re^{i\theta}$ yields:

$$H(r, \theta) = -\frac{\sigma}{2\pi}\left(\Delta\theta + i\ln\left(\frac{r_2}{r_1}\right)\right)$$

(2.25)

or

$$H_n = -\frac{\sigma \Delta\theta}{2\pi}$$

(2.26)

$$H_t = -\frac{\sigma \ln(r_2/r_1)}{2\pi}$$

where H_n, H_t are respectively the field components normal and tangential to the surface. r_1, r_2 are, respectively, the distances from the upper and lower surface end points to the observation point and $\Delta\theta$ is the subtended angle (Fig. 2.4). Thus, corresponding to (2.21), the field normal to the edge (surface) is proportional to the subtended angle. The field tangential to the edge varies logarithmically with the distance ratio. Thus, the tangential field vanishes at any point along the centerline of a surface and becomes infinite at the corners. For example at the bottom of the surface where $r_1 \rightarrow 0$ the tangential field becomes infinite and points downward. The positive sense of $\Delta\theta$ is counterclockwise from the line joining the lower edge. Thus, the field symmetrically on the opposite side of the surface changes sign only in the normal component. As illustrated in Fig. 2.5, the sense of (2.25) is that fields point away from positive surface charge. Further, after the surface charge is defined in the

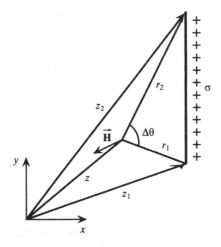

Fig. 2.5. Illustration of the derivation of the vector field from a 2D line source of constant charge.

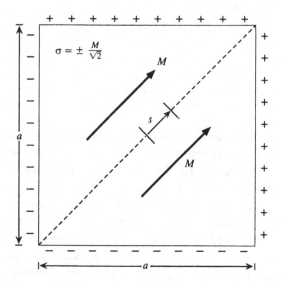

Fig. 2.6.   2D cross-section of a long rod magnetized along a diagonal uniformly to level M. a is the side length and s is the distance along the diagonal from the center. The magnetic charges reside only on the surface as shown with charge density $\sigma = \pm M/\sqrt{2}$.

application of (2.25), the field does not 'know' whether or not it is inside or exterior to the surface.

In uniformly magnetized objects with sharp corners, such as a cube, the interior magnetostatic field evaluated near the corners will become positive and infinite in magnitude as the corner is approached. An example is shown in Figs. 2.6, 2.7 of a long rod of square cross-section, uniformly magnetized to level M, with direction along a diagonal of the cross-section. The magnetic poles are solely surface charges as illustrated with surface charge density $\sigma = \pm M/\sqrt{2}$. Equation (2.25) can be readily utilized in this 2D case to obtain the field in all space. Along the diagonal corresponding to magnetization orientation, the field direction is along the direction of the magnetization and is given by:

$$H = -\frac{M}{2}\left(n - \frac{1}{\pi}\ln\frac{a^2 + 2s^2}{|a^2 - 2s^2|}\right) \qquad (2.27)$$

where $n = 1$ inside the square and $n = 0$ outside. The length of the square sides is a and s is the distance from the center along the diagonal. The corners occur at $s = \pm a/\sqrt{2}$. The field along the diagonal is plotted in Fig. 2.7. At the square center the demagnetizing field is $-M/2$, as

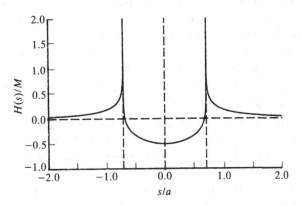

Fig. 2.7. Magnetostatic field along magnetizing diagonal of configuration of Fig. 2.6. $s = \pm a/\sqrt{2}$ denotes the sample boundaries.

expected from symmetry and (2.18). As a corner is approached from inside, the field becomes positive and eventually infinite at the precise corner. Outside the square the field decreases continuously from infinity as distance from the corner is increased.

This example gives insight into pole tip saturation of a recording head. Real materials do not remain uniformly magnetized in the presence of such a non-uniform field. In general, the magnetization will reorient toward the field direction (or demagnetize) in a complicated manner depending on the magnetic interactions and sample geometry. In particular, at corners the material will saturate, limiting the size of the large positive field. If the corners correspond to pole tips in a recording head, the external 'recording' field will also saturate. As a recording head with a given saturation magnetization is driven with higher currents in order to record on media with increased coercivity, the region of saturation expands (Bertram & Steele, 1976).

Imaging

In magnetic recording the recorded medium is close to, or in contact with, the highly permeable recording head. For any permanent magnetization distribution, the presence of high permeable material (which can include the medium itself) will alter the fields. The fields from the magnetized medium will magnetize the soft material, which then becomes a source of field itself. For a highly permeable object (keeper) of arbitrary shape, it is generally impossible to calculate the fields analytically, and the field problem must be solved by differential or integral methods (Lindholm, 1980b). In magnetic recording the permeable surface can often be approximated by an infinite flat surface (the surface of a head neglecting the gap or edge effects). In that case the method of images (Ramo, *et al.*, 1984) may be utilized to determine the fields. Imaging for infinite surfaces as well as approximations for gapped head are discussed here.

If a magnetic source is brought near an infinitely flat permeable surface demarking a semi-infinite region, the magnetization induced in the 'keeper' will, of course, be such that the boundary conditions on B,H will be satisfied. For an infinitely permeable keeper this yields from (2.5) that the tangential component of H vanishes at the interface. The magnetization pattern is in general complicated inside the permeable object; however, for linear material (2.1) and (2.4) yield no interior magnetic charges $(\nabla \cdot M(r') = 0)$ and therefore the fields arise from induced surface charges. If only fields exterior to the permeable keeper are considered, the correct boundary condition occurs if the flat infinite keeper is thought to have a mirror image of the permanent external

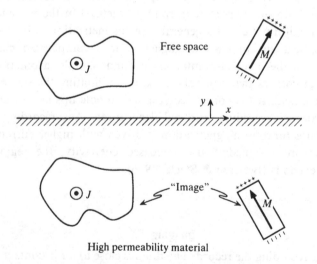

Fig. 2.8.   Imaging configurations for magnetic and current sources for an infinite flat surface of high permeability material. In both cases the configuration is imaged. In the magnetic case the image has a reversed sign of magnetization component tangential to the surface. In terms of magnetic charge all charges reverse sign in the image.

medium magnetization (Fig. 2.8). Thus, if $M(x, y)$ represents the medium magnetization with x taken parallel to the interface and the origin of y at the interface, then the mirror image magnetization is given by:

$$M_x^m(x, y) = -M_x(x, -y)$$

$$M_y^m(x, y) = M_y(x, -y)$$

(2.28)

Thus, the mirror image magnetization involves taking the magnetization at each point $x, y > 0$ in the medium and placing the same magnetization at $(x, -y)$ in the keeper but with the x component reversed. As illustrated in Fig. 2.8, (2.28) corresponds to an exact image of the equivalent charge density. Extension of (2.28) to three dimensions is immediate (Che & Bertram, 1993a). Equation (2.8) can then be used to calculate the fields everywhere external to the high permeability medium, and the boundary conditions will be satisfied: at the interface the field will be solely normal to the surface.

If the field source is due to a current distribution $J(r)$, then the appropriate image is obtained by reflecting the currents below the surface (Fig. 2.8):

$$J^m(x, y) = J(x, -y)$$

(2.29)

Equation (2.7) can then be used to calculate the fields exterior to the permeable medium including the actual currents as well as the image currents. Thus, for thin wires or current sheets lying on the high permeable medium, the effect on distant fields is to double the current.

If the high permeable medium has a finite permeability μ, then all image sources are simply multiplied by the factor $(\mu - 1)/(\mu + 1)$. It is to be emphasized that simple imaging applies *only* for the case of imaging by a flat surface of infinite extent. For a keeper of a general shape the appropriate Green's function must be determined (Jackson, 1975); numerical analysis may be required. For 2D structures, such as a gapped head or a thin film head with a finite length, approximate analytic Green's functions have been determined (Lindholm, 1977).

A simple illustrative example of imaging from a gapped head is given here (Lin, *et al.*, 1993). The image of a magnetized medium above the gap region of a head is *not* the simple image of Fig. 2.8 with that portion of the image under the gap removed. The exact solution for an infinite planar surface, which the image represents, is a magnetic charge distribution along the surface of the high permeability material. In 2D

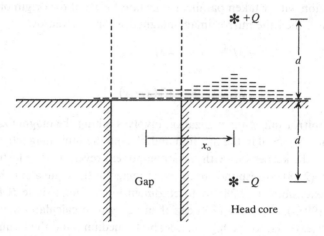

Fig. 2.9. Sketch of charges on the plane surface of a semi-infinite, high permeability body due to a 2D line source. Vertical dots above gap edge show where charge is removed to approximate the effect of a finite gap.

this charge distribution for a line charge source $Q = \hat{n} \cdot M \mathrm{d}x'$ or $-\nabla \cdot M \mathrm{d}y' \mathrm{d}x'$ is:

$$\frac{\partial Q^{\text{surface}}}{\partial x} = \sigma_{2D} = -\frac{\mathrm{d}Q}{\pi(d^2 + (x - x_0)^2)} \tag{2.30}$$

where d is the spacing and x_0 is the location along the surface. Equation (2.30) is obtained from (2.22) by evaluating the normal field at the surface. If the surface charge is set to twice the normal field, then utilization of (2.26) yields no field inside the keeper, no tangential field, and a doubling of the exterior normal field. As sketched in Fig. 2.9, this charge distribution is largest under the source charge. A good approximation for a finite gap is to simply remove the charge over the gap surface. Thus, for example, the longitudinal field component for the image term is given by:

$$H_x^{\text{im}}(x, y) = -\frac{\mathrm{d}Q}{2\pi^2}\left(\int_{-\infty}^{g/2} \mathrm{d}x' + \int_{g/2}^{\infty} \mathrm{d}x'\right)\frac{x - x'}{((x - x')^2 + y^2)(x_0 - x')^2 + d^2)} \tag{2.31}$$

Equation (2.31) can, in fact, be integrated in closed form (Lin, *et al.*, 1993). The net field versus x is plotted in Fig. 2.10 for $y = d$ and $x = -x_0$. Thus, the field is evaluated at the same spacing and at an equal distance

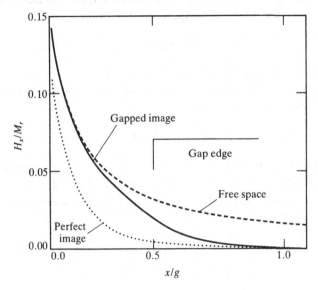

Fig. 2.10. Plot of net longitudinal field along a head surface with finite gap versus position x from the gap center due to a line source at position $-x$.

across the gap from the source. This is an appropriate example for overwrite calculations of hard–easy transition shifts, to be discussed in Chapter 7. In Fig. 2.10 the field without imaging and with perfect simple imaging is also plotted. For $x_0 < g/2$ the field approaches that of no imaging since both the source and observation points are over the gap. For $x_0 > g/2$ where the source and observation point are over the high permeable material, the field approaches that of perfect imaging, even though the source and field point separation spans the gap. For an actual medium magnetization pattern (2.31) must be integrated over all source charges. This approximation agrees well with approximate Green's functions (Fig. 5 of Lindholm, 1977).

Vector and scalar potentials

Because potential theory is useful in the analysis of magnetic recording heads, a brief review of potentials for magnetic fields will be given here.

In magnetic recording where heads and media occur with both real currents and magnetization patterns, (2.1) and (2.2) generally apply.

Since the field intensity B is always divergence free, this field may be derived from a vector potential A defined as:

$$B = \nabla \times A \tag{2.32}$$

A is a vector field that in general involves solving for *three* unknown scalar functions. Substitution of (2.32) into (2.2) and utilizing (2.12) yields Poisson's equation for A:

$$\nabla^2 A = -(\mu J + J_M) \tag{2.33}$$

where μ is the linear permeability and J_M is the total (volume and surface) effective magnetic current density due to permanently magnetized media. It is assumed that divergence free vector solutions are found and suitable boundary conditions are satisfied.

The flux Φ is defined as:

$$\Phi \equiv \int_s \int B(r') \cdot \hat{n} \, d^2 r' \tag{2.34}$$

where integration is taken over surface S. (If S is closed, Φ vanishes by (2.1) and Gauss's Theorem (e.g., Jackson, 1975).) Substituting (2.28) and using Stokes' theorem yields:

$$\Phi = \oint A(r') \cdot dl' \tag{2.35}$$

where the line integral is taken around the circumference of the surface S. For magnetic recording head analysis, solution of the fields in terms of the vector potential immediately gives the inductance, since (2.35) can be evaluated (and summed) over each S defined at each turn of the windings. This is particularly useful for complicated structures such as thin film heads. For very wide geometries where two-dimensional analysis applies, only the z or cross-track component of the vector potential is non-vanishing so that (2.33) becomes a single differential equation for a scalar. For wire loops in which the current is normal to the 2D plane (closing at infinity), the flux threading the loop per unit length from (2.35) is simply the difference in vector potential (A_z) between the two points in the plane that define the loop (Chapter 5).

For regions of space where there are no real currents, the curl of the magnetic field vanishes:

$$\nabla \times H = 0 \tag{2.36}$$

Irrotational fields can be derived from a scalar potential:

$$\boldsymbol{H} = -\nabla\phi \qquad (2.37)$$

so that a useful integral form immediately follows:

$$\int_A^B \boldsymbol{H} \cdot d\boldsymbol{l} = \phi_A - \phi_B \qquad (2.38)$$

Scalar potentials are useful, in contrast to vector potentials, since they always involve only one unknown function. They can be used in magnetic recording applications for head field analysis since often only fields away from the wires are of interest. Poisson's equation can similarly be derived for scalar fields utilizing (2.1), (2.5) and (2.37):

$$\nabla^2\phi = -\rho_m/\mu \qquad (2.39)$$

and solved utilizing suitable boundary conditions. ρ_m is the effective magnetic charge (including surface contributions) due to permanently magnetized media, and μ is the medium linear permeability.

For much of the discussion of head fields in Chapter 3, a scalar magnetic potential solution will be utilized even though the field sources are real currents. A simple way to visualize the boundary conditions is to imagine the source currents being one loop of wire with current NI (Fig. 2.11(a)). Equation (2.11) applies for any path that threads the wire. If the wire diamater is infinitesimally small, then the contribution to the line integral of the field inside the wire loop is negligible. Thus, (2.11) holds to good approximation if the line integral excludes the region in the wire. Figure 2.11(a) is an example of a loop path that goes from point A on one side of the loop around to point B just on the other side. Since the field in the region exterior to the loop may be derived from a scalar potential (2.37), we have:

$$NI = \int_A^B \boldsymbol{H} \cdot d\boldsymbol{l} = \phi_A - \phi_B \qquad (2.40)$$

Thus, A, B are any two points on either side of the wire loop face, and the line integral can be considered to go from point A to B, not passing through the loop. Since the integral is now exterior to the wire volume, the line integral is also equal to the potential difference and (2.40) holds for any chosen path from A to B. Thus, the fields exterior to the wires can be solved in the corresponding volume by assuming Dirichlet boundary

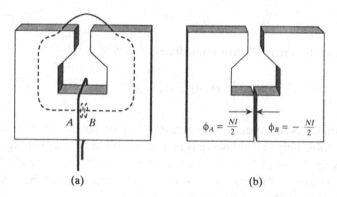

(a)   (b)

Fig. 2.11. (a) Line integral path for application of scalar potentials to field problems with current sources. (b) Equivalent scalar potential problem with scalar boundary source potentials.

conditions (Jackson, 1975) with the potential on each face of the coil area given by:

$$\phi_A = \frac{NI}{2} = -\phi_B \qquad (2.41)$$

with the potential vanishing at infinity (Fig. 2.11(b)).

Fourier and Hilbert transforms

In magnetic recording, power spectral densities of signal and noise are of interest. To obtain the frequency response, spatial Fourier transforms of recorded patterns are taken. These transforms will involve transforms of fields as well as magnetization patterns. Fourier transforms will be taken only along the head–medium motion path denoted by the spatial variable x. The transform pair utilized here will be of the form:

$$\mathscr{F}(k) = \int_{-\infty}^{+\infty} \mathrm{d}x f(x)\mathrm{e}^{-ikx}$$

$$\qquad (2.42)$$

$$f(x) = \frac{1}{2\pi} \int_{-\infty}^{+\infty} \mathrm{d}k \mathscr{F}(k)\mathrm{e}^{ikx}$$

where $k = 2\pi/\lambda$. The closure relation is:

$$\delta(x - x') = \frac{1}{2\pi} \int_{-\infty}^{+\infty} dk e^{ik(x-x')} \tag{2.43}$$

where x and k can be interchanged. Parseval's theorem assures that the total energy is the same whether computed in real space or Fourier space:

$$\int_{-\infty}^{+\infty} dx |f(x)|^2 = 2\pi \int_{-\infty}^{+\infty} dk |\mathscr{F}(k)|^2 \tag{2.44}$$

An important general consequence of the Fourier transform of fields in two dimensions can easily be shown. All potentials obey Laplace's equation in free space, which in two dimensions can be written as:

$$\nabla^2 \phi = \frac{\partial^2 \phi}{\partial x^2} + \frac{\partial^2 \phi}{\partial y^2} = 0 \tag{2.45}$$

x is the infinite-range variable along the head–medium relative motion direction and y is the direction normal to the head or medium surface. Equation (2.45) can be Fourier transformed by considering the Fourier transform as an operator on the x variable of this differential equation. We write the Fourier transform of the potential as:

$$\phi(k, y) = \int_{-\infty}^{+\infty} dx \phi(x, y) e^{-ikx} \tag{2.46}$$

where $\phi(k, y)$ denotes the Fourier transform of the potential with respect to the variable x. Thus, (2.45) transforms to a differential equation involving y only:

$$\frac{\partial^2 \phi(k, y)}{\partial y^2} + k^2 \phi(k, y) = 0 \tag{2.47}$$

which has a general solution of the form:

$$\phi(k, y) = A^+(k) e^{ky} + A^-(k) e^{-ky} \tag{2.48}$$

where suitable Fourier transforms of the boundary conditions yield the unknown coefficients. If the region of evaluation is completely to one side of the field sources, e.g. the region above a head with no permeable material present, then the boundary condition at $y = \infty$ where the potential must vanish requires that the transforms must contain only the decaying exponential term with respect to distance away from the source (Fig. 2.8):

$$\phi(k, y) = \phi(k, 0)e^{-|ky|} \tag{2.49}$$

Here, $y = 0$ can be the source plane or any plane above the source. *In two dimensions a sinusoidal variation in one direction yields an exponential decay in the other.* The fields follow a similar relation since the Fourier transform of the gradient operator is:

$$\nabla = \left(\frac{\partial}{\partial x}, \frac{\partial}{\partial y}\right) \leftrightarrow \left(ik, \frac{\partial}{\partial y}\right) \tag{2.50}$$

Thus, for this case of a simple confined source:

$$H(k, y) = (-ik\phi(k, y), |k|\phi(k, y)) \tag{2.51}$$

so that:

$$
\begin{aligned}
H_x(k, y) &= -ik\phi(k, y) = H_x(k, 0)e^{-|ky|} \\
H_y(k, y) &= \pm \mathrm{isgn}(k)H_x(k, y)
\end{aligned}
\tag{2.52}
$$

The function sgn(k) represents the sign of k and vanishes for $k = 0$. On one side of a source the Fourier transforms of the field components are identical in magnitude and shifted in phase by $\pm 90°$ depending on the sign of k. Equation (2.52) shows that each harmonic component of the vector field as a function of x is constant in magnitude and rotates with period $1/k$ in either a clockwise or counterclockwise direction (Fig. 2.12(a),(b)) (Mallinson, 1973). The sense of rotation reverses as illustrated in Fig. 2.12(b) due to the sign change of the perpendicular field component. In addition, the transformed field components follow (2.49) and decay exponentially away from the source, as does the potential.

The exponential decay in (2.49) yields the simple spacing loss when considered as a logarithmic level (dB) plotted versus linear wavenumber k or relative spacing y:

$$20 \log H_x(k, y) = H_x(k, 0) - 54.6(y/\lambda)(dB) \tag{2.53}$$

This linear form of the spacing loss is sketched in Fig. 2.13 and applies to all two-dimensional fields and potentials. The point $y = 0$ is entirely arbitrary as long as field sources are confined to only one side of the region (e.g. Fig. 2.12(a)). For example, the analysis of fields in perpendicular recording media sandwiched magnetically by a permeable

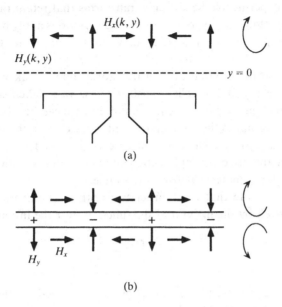

Fig. 2.12. Illustration of rotating Fourier transform components of fields exterior to a magnetized head (a) or medium (b). Note that 'above' and 'below' the medium are distinguished by a sign change in perpendicular field component resulting in an opposite sense of field rotation on either side of the medium.

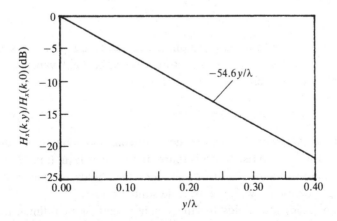

Fig. 2.13. Logarithmic plot of head-field transform versus relative spacing away from the field source.

keeper and the permeable head yields transforms that retain the general form of (2.48). Note that exponential spacing variations apply only to the harmonic components of the field: they do not apply to the fields themselves.

It is important to note that Fourier transforms may be taken of three-dimensional field patterns. The transform is always taken along the infinite x or medium motion direction, but in general can be a function of both distance normal to the medium or head surface y and the transverse or cross track direction z. This case will arise in the consideration of head side fields. In this case a simple exponential field decay with y (or z) occurs only approximately far from the source.

It is useful to note that the Fourier transforms at long wavelengths $(k \rightarrow 0)$ are given by the integral of the function over the infinite x axis:

$$\mathscr{F}(k \rightarrow 0) = \int_{-\infty}^{+\infty} dx f(x) \qquad (2.54)$$

This integral may be considered as part of the line integral in the general field integral given by (2.8). Thus, the dc Fourier component of the magnetic field is given by:

$$H_x(k \rightarrow 0) = \int_{-\infty}^{+\infty} dx H_x(x) = \int_{x axis} \boldsymbol{H} \cdot d\boldsymbol{l}$$

$$\qquad (2.55)$$

$$= NI - \int_{closure\,path} \boldsymbol{H} \cdot d\boldsymbol{l}$$

If the x axis does not thread any windings and the closure path is taken at infinity far from the magnetic structure where the field vanishes, the general relation is obtained:

$$H_x(k \rightarrow 0, y, z) = 0 \qquad (2.56)$$

This relation always holds for fields derived from mangetization sources and is true for fields outside heads (current sources). Equation (2.56) is generally relation valid in three dimensions and holds for any chosen path along x denoted by (y,z) above the source surface.

Equation (2.56) also holds for the y component of two-dimensional fields except when the field sources are not confined to one side of the region where the fields are being evaluated (e.g. heads with keepers): the relation for the y component follows from flux conservation (2.1) and (2.34). Equation (2.56) can also be shown from (2.38) so that:

$$H_x(k \to 0) = \int_{-\infty}^{+\infty} dx H_x(x) = \phi(-\infty, y) - \phi(+\infty, y) \qquad (2.57)$$

which also vanishes except in special cases.

Hilbert transforms are also useful in magnetic recording because they conveniently express the relation between field components. As discussed in Chapter 5, they relate, for example, the playback voltage from longitudinal media to that of an identical pattern but with perpendicular magnetization. The Hilbert transform, again on the variable x, is given by:

$$h(x) = \frac{1}{\pi} \int_{-\infty}^{\infty} dt' \, \frac{f(t')}{(x - t')} \qquad (2.58)$$

The Fourier transform of this relation yields:

$$H(k) = -i \, \text{sgn}(k) F(k) \qquad (2.59)$$

Thus, Fourier components of the Hilbert transform yield simply a $-$ or $+$ 90° phase shift depending on whether k is positive or negative, respectively. Equation (2.52) relating the transformed field components is exactly of this form. Thus, *all* two-dimensional field components in free space, on one-side of a source, are Hilbert transforms of each other.

Examples of Hilbert transforms that are commonly utilized in analysis of the magnetic recording process are:

$$H^{op}(\delta(x)) = \frac{1}{\pi x}$$

$$H^{op}(S(x)) = \frac{\ln|x|}{\pi} \qquad (2.60)$$

$$H^{op}\left(\frac{y}{x^2 + y^2}\right) = \frac{x}{\pi(x^2 + y^2)}$$

where H^{op} denotes the Hilbert operator (2.58) and $S(x)$ is the unit step function. These examples are plotted in Fig. 2.14. By (2.59) the Fourier transforms of each pair are identical in magnitude, but shifted 90° in phase.

Functions   Hibert transforms

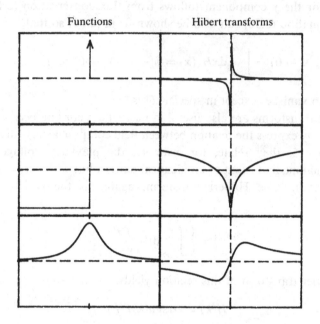

Fig. 2.14. Examples of Hilbert transforms of several functions defined in (2.60).

Integral relations for free space fields

The scalar potential at all points in a volume V defined by surface S may be written in terms of the enclosed charge ρ_m (2.3), surface potential ϕ_s, and surface normal potential derivative (normal field) (Jackson, 1975, sec. 1.8):

$$\phi(r) = \frac{1}{4\pi} \int_V \frac{\rho_m(r')\mathrm{d}^3 r'}{|r - r'|}$$

$$+ \frac{1}{4\pi} \oint_s \mathrm{d}^2 r' \left(\frac{1}{|r - r'|} \frac{\partial \phi(r')}{\partial n'} - \phi_s(r') \frac{\partial}{\partial n'} \left(\frac{1}{|r - r'|} \right) \right) \tag{2.61}$$

The geometry is illustrated in Fig. 2.15(a). Note that this is an integral relation only, not a solution to the differential equation. One is not permitted to specify *both* the potential and its normal derivative at the boundary. However, in general, to know the potential inside a closed region, knowledge of the surface potential and its normal derivative are required. If the volume were infinite, the surface term would vanish

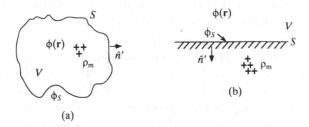

Fig. 2.15. (a) Geometry for general formulation of potential in a closed region of volume V and surface S given charge distribution inside S and the surface potential ϕ_s or its normal derivative on the boundary. (b) Simplification to semi-infinite space (2D) where the charge (field) sources are outside the region and only the surface potential need be known.

(apart from a constant average surface potential) and the gradient of the first term would give the fields in terms of charge density (2.3).

In two dimensions, with the closed surface being a half infinite space with an infinite plane surface (line) with field sources only on the outer side of the plane (e.g. Fig. 2.15(b)), (2.61) reduces to a simplified form of the Green's function solution (Jackson, 1975, sec. 1.10):

$$\phi(x, y) = \frac{y}{\pi} \int_{-\infty}^{+\infty} \frac{dx' \phi_s(x')}{((x - x')^2 + y^2)} \tag{2.62}$$

where $(x', 0)$ denotes the surface over which the surface potential $\phi_s(x')$ is given and (x, y) is the observation point. In (2.62) it is assumed that the potential vanishes at infinity far from the sources. Note that in (2.61) for a semi-infinite surface (three-dimensional) where the potential does not vanish at infinity, the first term of the surface integral is equal and opposite to that of the second term (again apart from an average constant potential). Thus, a simple relation for the potential can be written also in terms of the surface vertical field component.

The magnetic fields can be easily determined from (2.62) and (2.37) in terms of the surface longitudinal field component $H_x^s(x')$:

$$H_x(x, y) = \frac{y}{\pi} \int_{-\infty}^{+\infty} \frac{dx' H_x^s(x')}{((x - x')^2 + y^2)}$$

$$\tag{2.63}$$

$$H_y(x, y) = \frac{1}{\pi} \int_{-\infty}^{+\infty} \frac{dx' H_x^s(x')(x' - x)}{((x' - x)^2 + y^2)}$$

This result expresses the field at any point in space as the convolution of
the surface longitudinal head field $H_x^s(x')$ and an integration kernel that
is *twice* the field due to an infinitesimal wire at $(x', 0)$ (2.10 or 2.11). The
factor of two arises since specifying the field along a semi-infinite plane is
identical to placing an infinitely permeable space below that plane so that
the imaging analysis applies (Fig. 2.8). Thus, $H_x^s(x')$ can be considered a
distributed current along the integration surface. In general, $H_x^s(x')$ is not
well known since the surface, even if it coincides with the surface of a flat
head, for example, will contain regions of free space (e.g. over the gap
and beyond the ends).

The Fourier transform of (2.63) can be easily shown to be:

$$H_x(k, y) = H_x^s(k)\mathrm{e}^{-ky}$$
$$H_y(x, y) = \mathrm{i\,sgn}(k)H_x^s(k)\mathrm{e}^{-ky}$$

(2.64)

which corresponds directly to the result obtained by Fourier transforming Laplace's equation (2.52).

It is important to note that the above relations apply only to fields in
free space where the field source $H_x^s(x')$ is only *on one side* of the region
considered. Thus, the integration surface can be coincident with or
anywhere above the outer plane surface of a head or medium, but not
below that surface. A useful application of (2.62) is in the evaluation of
head fields, and in particular for head-saturation analysis. Numerical
analysis is utilized with suitable discretization to solve for the core
potential or magnetization for a range of drive currents. Only the
potential or field at the surface of the head (or a plane suitably close)
need be stored for each computer run. Subsequently, for use in analysis of
the recording process, the surface information can be used with (2.62) or
(2.63) to give the field as a continuous function in the recording region. A
practical counter-example of this analysis is a keepered pole or ring head
utilized for perpendicular recording, discussed in Chapter 3. In those
cases the recording medium where the fields are to be evaluated lies in
between two field sources: the driven head and the magnetized keeper.
Laplace's equation must be solved for the multiregion directly. The
resulting fields are not Hilbert transform pairs. Nonetheless, the general
relation (2.61) holds.

Problems

Problem 2.1   Use (2.7) to obtain a 2D form for the field (in the x, y plane) due to a current distribution $J_z(x, y)$. Find $H_x(x, y)$ for thin current sheet of width g in the x direction.

Problem 2.2   Use (2.10) to derive the field along the axis of a single loop of wire of radius a and current I. Show by direct integration that (2.11) holds.

Problem 2.3   Consider a box magnetized in a direction along one edge. Let the magnetization vary linearly in the direction of orientation and be constant in the two orthogonal directions. Determine the volume and surface magnetic charge pole densities. Show that the total charge integrated over the entire box vanishes. In addition, determine the equivalent magnetic currents using (2.12) and the equivalent magnetization pattern that would give the same surface charges and hence the same magnetic fields. Can you argue that the equivalent current distribution from this alternative pattern yields the same fields?

Problem 2.4   Consider a straight rod of length L and cross-sectional radius s uniformly magnetized along its axis with magnetization M. Show that the volume magnetic pole density vanishes and that the surface pole density resides on the ends of the rod. Use (2.8) to obtain the magnetic field everywhere along the center line interior and external to the medium. Plot B along the axis. Using (2.12) find the magnetic equivalent surface currents. Use (2.13) and find the B field everywhere. Suppose the equivalent magnetic currents are in fact real currents. Argue that external to the rod it is impossible to distinguish between a magnetic and a real current field source. Show that internal to the rod the magnetic field H does differ depending on the source.

Problem 2.5   Consider two charged sheets (2D) of height W and separation L where $L \ll W$ (approximately a thin plate). Let the charge density be M on one sheet and $-M$ on the other. Find the field H, along the centerline perpendicular to the plane of the sheets. Show two possible magnetization distributions that give the same surface charges and plot B along the centerline for each.

Problem 2.6   Use (2.26) to determine the field in a (2D) rectangle uniformly magnetized parallel to one edge. Plot the field along a line joining two opposite corners inside and outside the medium.

3

Inductive head fields

Introduction

In this section fields and Fourier transforms of a variety of inductive head configurations will be presented. The field patterns are of interest for both the recording process and, via reciprocity, the playback waveforms. First, the concept of head efficiency will be introduced. Following that, approximate and exact expressions for the fields in the gap vicinity of a very wide head will be derived and compared. The effect of finite head length as well as finite track width will be discussed in terms of approximate expressions. The results of studies of head field saturation will be included. The chapter will conclude with a discussion of the effect of keepered media on head fields.

The function of a recording head is to transfer efficiently the mmf that results from a current applied to the windings into field at the gap region where the recording medium passes. Two state-of-the-art structures are shown in Figs. 1.4 and 3.1. Figure 3.1 shows a head designed for high-data-rate recording in high-density helical scan tape recorders (Ash, *et al.*, 1990). This structure exhibits the extreme dimensional scaling typical of most head structures: the gap length is sub micron ($g \sim 0.25\mu m$), the gap depth and gap width are between one and two orders of magnitude larger ($\sim$ 10–30μm), and the head major dimensions are in the mm range. The example in Fig. 3.1 is of a structure completely laminated to allow for operation at frequencies $\sim$150MHz. An inductive head, commonly of this geometry, is either completely ferrite for medium coercivity applications or 'metal-in-gap' for use on high coercivity media for video (8mm or DAT) or data recording (Jeffers, 1986; Iizuka, *et al.*, 1988). The metal tip region of a metal-in-gap head can also be laminated. The permeability of ferrite ranges from approximately 2000 at low frequencies to a domain-

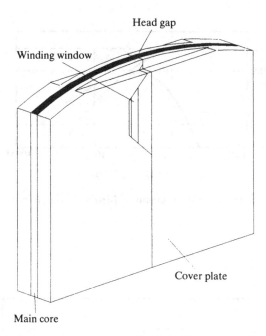

Fig. 3.1. Typical structure of a head utilized in single-track, high-density tape recording. Core structure is often Ferrite, but this illustration is of a high data-rate, completely laminated head. Taken from Ash, *et al.* (1990), © IEEE.

wall loss dominated magnitude of about 100 at 100MHz (Fig. 3.2) (Smit & Wijn, 1959). For heads composed of metal films the permeability is dominated by classical eddy current loss resulting from the high conductivity (Fig. 3.3). Loss mechanisms in soft magnetic materials are, in general, complicated: a complete analysis involves the consideration of conductivity loss coupled with domain-wall motion (Patton, *et al.*, 1966; Aharoni & Jakubovics, 1982; Yuan, 1992).

In Fig. 1.4 a thin film head is shown that is utilized in hard disk drives. For these heads the gap depth ($\sim$1–2µm) and head pole length ($\sim$2–3µm) are on the order of the gap length ($\sim$0.25µm) so that extreme scaling does not apply. Further, the distributed windings are close to the gap region, so that, in contrast to a ferrite head, the winding geometry affects the head efficiency as well as the inductance. Thin film head materials are either Permalloy, Sendust or Alfesil, so that low frequency permeabilities are in the 1000–2000 range, whereas higher frequencies are limited by eddy current losses as indicated in Fig. 3.3 (the skin depth of

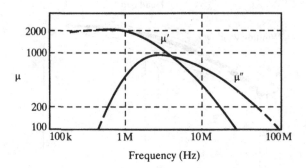

Fig. 3.2. Typical permeability spectrum of MnZn Ferrite. (Courtesy of Ampex Corporation.)

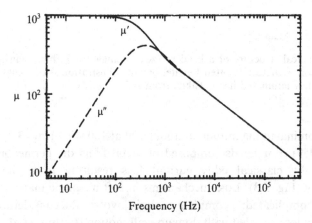

Fig. 3.3. Classical eddy current spectrum of a 2μm thin sheet with resistivity $\rho = 100$ μΩ - cm and dc permeability of $\mu_{DC} = 1000$ μ_o (e.g. see Ramo, *et al.* (1984)).

Permalloy is ~1μm at 10MHz). Ferrite and metal-in-gap heads are also utilized in hard disk drives.

The solution of magnetic-head field problems can be divided into three primary regimes, which reflect the material permeability as well as the extreme dimensional scaling of most head structures. These regimes are: (1) heads of low permeability or medium permeability where all dimensions are of the same order, (2) heads of reasonably high permeability and dimensions so that the region near the gap, which extends a distance greater than wavelengths of interest, is at constant potential, and (3) heads of any dimension where the permeability is virtually infinite.

Case 1 If the permeability is low, which generally occurs in heads at high frequencies and with heads driven to saturation, the potential varies over the core, and a complete solution is required. Low-permeability head fields and effects of saturation may be determined by finite element analysis or integral equation techniques (e.g. Kelley & Valstyn, 1980; Lindholm, 1980b, 1981, 1984; Lean & Wexler, 1982; Bertram & Steele, 1976; Rodé & Bertram, 1989 etc.). If the head structure is dimensionally on the order of the wires, such as in thin film heads and the permeability is finite, a complete solution utilizing vector potentials or integral equations may be required. A thin film head is a good example of a geometry in which the spatial distribution of the wires does not permit accurate replacement of each turn by an equivalent magnetic disc (as discussed in Chapter 2) to allow for analysis utilizing scalar potentials. However, if a two-dimensional analysis is applicable, the vector potential has only the component directed out of the plane, so that the complexity is no more than that of a scalar potential solution.

Case 2 Most heads in magnetic recording exhibit extreme dimensional scaling: the recording gap length is orders of magnitude smaller than the outer dimensions and the interior winding window. Therefore, there will be potential change from the gap face to the windings even for a core of relatively high permeability. However, a reasonably constant potential occurs near the gap. Thus, the head fields in the region of interest near the gap can be solved utilizing relatively simpler constant potential solutions and the concept of head efficiency can be utilized to relate the source potential mmf, NI, to the boundary potential in the gap region.

Case 3 If the head-core permeability is infinite, each core block on either side of the windings will be of constant potential equal to $\pm NI/2$ (Fig. 2.11(b)). For certain geometries this permits a conformal mapping solution (in 2D) by complex analysis of the entire head.

In the discussion that follows, unless explicitly stated, it is presumed that cases 2 or 3 apply.

Head efficiency and deep-gap field

Consider the scalar potentials on either side of the gap ϕ_{Ag}, ϕ_{Bg} respectively as shown in Fig. 3.4. From (2.40) we can write:

$$\phi_{Ag} - \phi_{Bg} = NI - \int_{\text{core}} \boldsymbol{H}_i \cdot \mathrm{d}\boldsymbol{l} \qquad (3.1)$$

where $\boldsymbol{H}_i$ is the internal field and the line integral is taken around the core through the windings. Sufficiently high core permeability is assumed so that the region of the core near the gap is approximately at constant potential so that ϕ_{Ag}, ϕ_{Bg} need be defined only in a region and not at specific points. For a head of finite permeability the line integrals (3.1) may be small, but do not, in general, vanish. Only if the core material has infinite permeability do the internal fields vanish, yielding core halves of constant potential and a potential drop across the gap equal to the driving mmf NI. Specific values of potentials ϕ_{Ag}, ϕ_{Bg} may be assigned arbitrarily, however, it is customary to set the potential at the gap center equal to zero so that:

$$\phi_{Ag} = -\phi_{Bg} = \frac{1}{2}\left(NI - \int_{\text{core}} \boldsymbol{H}_i \cdot \mathrm{d}\boldsymbol{l} \right) \qquad (3.2)$$

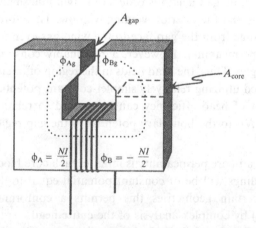

Fig. 3.4.   Schematic diagram of inductive head. A_{gap} is the gap-face area and A_{core} is a (varying) core cross-sectional area.

The head efficiency may be defined as the potential drop across the gap relative to the driving mmf (Fig. 3.5). Thus we have:

$$E \equiv \frac{\phi_{Ag} - \phi_{Bg}}{NI} = 1 - \frac{\displaystyle\int_{core} \boldsymbol{H}_i \cdot \mathrm{d}\boldsymbol{l}}{NI} \tag{3.3}$$

where (3.3) includes the total integral along the core of the internal field. It is assumed that all heads possess gap depths which are large compared to the gap length. Thus, the potential deep in the gap varies approximately linearly across the gap, yielding a constant *deep-gap field* given by:

$$H_0 = \frac{\phi_{Ag} - \phi_{Bg}}{g} \tag{3.4}$$

Combining (3.2) and (3.3) yields (Fig. 3.5):

$$H_0 = \frac{NIE}{g} \tag{3.5}$$

The head efficiency, as well as the inductance, depends on the core permeability and geometry. Equivalent circuit techniques can often be utilized to analyze 3D structures in a fashion more simplified and intuitive than complete numerical analysis (Jorgensen, 1988; McKnight, 1979; Hughes, 1983b; Katz, 1978; Jones, 1978). However, the use of such techniques requires accurate estimates of the flux paths. As an indication

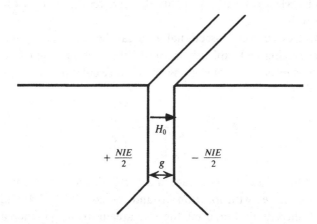

Fig. 3.5. Gap region fields and potentials. H_0 is the deep-gap field and g is the gap length so that $H_0 = NIE/g$.

of the utility of equivalent circuit methods, (3.3) and (3.5) can be combined to give the efficiency in terms of the fields along the major path circulating the core and gap region:

$$E = \frac{1}{1 + \dfrac{1}{H_0 g} \displaystyle\int_{\text{core}} H_i \cdot dl} \tag{3.6}$$

A high-efficiency head results from a small core 'potential drop' relative to that across the gap. Since the flux density is related to the field by $B = \mu H$, (3.6) can be expressed as:

$$E = \frac{1}{1 + \dfrac{1}{\mu_r g} \displaystyle\int_{\text{core}} B_i \cdot dl/(B_0 l)} \tag{3.7}$$

In (3.7) l is the length of the core, μ_r is the core permeability relative to free space (μ_0), and $B_0 = \mu_0 H_0$. High efficiency occurs if the core permeability is large compared to the ratio of core length to gap length. The internal flux density B_i integrated over the core will be small compared to the flux density across the gap $B_0 l$ if the head is designed to minimize flux fringing and maximize the core cross-sectional area to that of the gap face. As shown by the numerical example in Fig. 3.6, predominant fringing occurs across the winding window below the gap (in particular for long winding regions as in thin film heads (Fig. 1.4)). Fringing flux is in 'parallel' with the gap flux and thus will reduce the gap flux density relative to the core flux density, resulting from (3.7) in a reduction of head efficiency.

The effect of core cross-sectional area can be seen by rewriting (3.7) neglecting fringing and explicitly including the conservation of flux (2.1): BA is a constant at any point along the path circulating the head. Thus, (3.7) becomes:

$$E = \frac{1}{1 + \dfrac{l A_g}{\mu_r g} \displaystyle\int_{\text{core}} dl/(A(l))} \tag{3.8}$$

where A_g is the area of the core face and $A(l)$ is the cross sectional area of the core that varies with location around the core (Fig. 3.4). Equation (3.8) shows directly that tapering the core area near the gap increases the efficiency. Most heads have a uniform track width so that area reduction occurs by achieving a reduced gap depth. Note that (3.8) is valid only

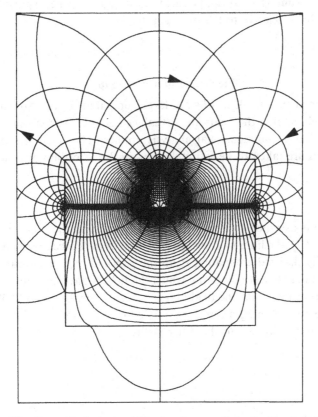

Fig. 3.6. Flux lines circulating and fringing from an inductive head. Taken from Lindholm, (1981), © IEEE.

when flux fringing from the major core path can be neglected; however, tapering the core towards the gap can increase fringing.

As exemplified in Problem 3.1, a typical video head ($\mu = 1000$, $A_g/A_c = 10$) will have a reasonably constant (to 10%) potential, spanning approximately a distance of $1000g$ on either side of the gap faces. Thus, a constant potential solution in the gap region can be found assuming boundary conditions (Fig. 3.5):

$$\phi_{Ag} = \frac{NIE}{2}, \ \phi_{Bg} = -\frac{NIE}{2} \tag{3.9}$$

satisfying (3.1) and, without loss of generality, the potential across the gap is symmetrized (3.2). The solution for the fields, given the head geometry assuming (3.9) holds, yields fields that can be expressed in a universal form relative to the deep-gap field (3.5).

Fields due to a finite gap

It is assumed that track widths are large compared to the gap length or flying height so that the fields are virtually constant over the width of the track so that two-dimensional analysis applies. The constant potential pole pieces are taken to be infinitely long in the recording direction.

Far-field approximation

If the gap is assumed to be infinitesimally small and the poles infinitely long (Fig. 3.7), the lines of constant potential in the space above the head are radially directed emanating from the gap and increase linearly with angle. The field lines (lines along which the field is directed) are therefore circular coincident with the field magnitude contour. Utilizing (2.38) and taking the path of integration to be a field line at a fixed radial distance r from one potential face to the other yields:

$$|H(r)|\pi r = NIE \qquad (3.10)$$

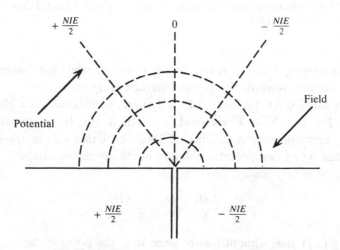

Fig. 3.7. Potential lines and field contours for far-field approximation.

so that

$$|H(r)| = NIE/\pi r \tag{3.11}$$

Thus, the far field decreases inversely with distance from the gap. Using (3.5) the field can be expressed in terms of the deep-gap field and the infinitesimal, but finite, gap length. (In reality, such a head would saturate, limiting the deep-gap field to the value of core saturation magnetization.) However, in the 'far-field' approximation, assuming linearity, the field depends only on the potential drop across the gap.

Suppose the head consisted of no permeable material, but solely of *NIE* infinitesimally thin wires placed along the gap at the surface of the head (along the z direction at the origin in Fig. 3.7). By symmetry the field lines will be unchanged, but will extend below as well as above the head surface. Application of (2.11) yields:

$$|H(r)| = NIE/2\pi r \tag{3.12}$$

which is identical to (3.11) except for a factor of 2 in the denominator. This factor arises because, in this case, the line integral must be taken along a path that completely encloses the wires. A 2π angular integration occurs instead of the π integration for the actual head with fields defined solely in the half space above the plane of fixed potential. One way to view the factor of 2 is that the far field is represented by the *NIE* wires placed above a semi-infinite space of infinite permeability. Thus, imaging yields a field equivalent to two wires of current *NIE* placed at the same point at the head surface (Fig. 2.8). Equation (3.12) with *2NIE* current yields (3.11) for the field above the head. Therefore, in general, the function of the high permeability head is, first, to translate the *NI* turns wrapped anywhere around the head core to a single infinitesimal wire placed on the head surface at the gap and, second, to double the mmf. The higher the permeability the closer the efficiency is to unity (Problem 3.1) and the more 'efficient' is this translation process.

The field components are easily derived from (3.11) and are given by:

$$H_x(x,y) = \frac{NIEy}{\pi(x^2 + y^2)}$$
$$H_y(x,y) = -\frac{NIEx}{\pi(x^2 + y^2)} \tag{3.13}$$

These fields are plotted in Fig. 3.8. The x component is 'bell' shaped and exhibits precisely a 'Lorentzian' form; the y component exhibits an antisymmetric form vanishing at the gap center. Equation (3.13) may be

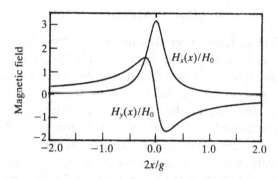

Figure 3.8.   Head field components for the far-field approximation (3.13). The fields are plotted for fixed y, but the shapes shown are invariant with separation since the x axis can be scaled as x/y.

derived from (2.63) because in this approximation the surface long-itudinal field is easily specified. Everywhere above the constant potential core material the longitudinal (x) field vanishes. Across the gap the surface field may be written as a delta function:

$$H_x^s(x') = NIE\delta(x') \tag{3.14}$$

which satisfies (2.40) applied across the gap.

Since these field components are derived from one-sided, two-dimensional sources, they are Hilbert transform pairs and (2.64) applies. The Fourier transform of the x component is readily taken and yields:

$$H_x(k, y) = NIEe^{-ky} \tag{3.15}$$

Thus, the harmonic components decrease simply following the exponential spacing loss term typical of one-sided fields. At long wavelengths $(k \to 0)$, or at the head surface $(y = 0)$, the Fourier transform yields a constant NIE. This constant corresponds to the impulse surface head field (3.14). Equation (2.56) does not hold because the unphysical example of an infinitely long pole head is being considered here. Note that the units of field Fourier transform are potential (Amps).

Medium range approximation (Karlqvist field)

The far-field expression is not accurate in the vicinity of the gap. The next level of approximation to a head field is to include a finite gap, but not to evaluate the field too close to the gap corners. It is assumed that the head is infinitely long and wide with an infinitely deep gap. The only length parameter is the gap length; Fig. 3.5 gives the geometry with appropriate boundary conditions. The field above the head ($y > 0$) can be immediately determined utilizing (2.62) if the potential along the entire surface is known or (2.63) if the surface field is known. Unfortunately, the potential variation across the gap is not known in advance.

In Fig. 3.9 lines of constant potential from an exact solution are shown. Deep inside the gap, far from the surface, the potential varies linearly: the potential lines are parallel to the gap faces and yield a deep-gap field given by (3.5). At the surface the potential will not vary linearly: the potential lines spread leaving the gap resulting in a slightly reduced field at the gap center and much increased field near the gap corners. The variation of potential across the gap at the head surface and the corresponding surface field is plotted in Figs. 3.10(a),(b).

The Karlqvist (Karlqvist, 1954) approximation assumes that the potential varies linearly across the gap. A linear potential yields a

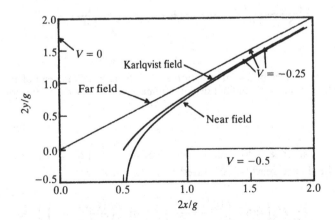

Fig. 3.9. Line of constant potential ($V/NIE = -0.25$) for far-field, medium-field (Karlqvist), and near-field (Westmijze) approximations for a 2D head with infinitely long poles and infinitely deep gap. Half the head is shown where the right core potential is $V/NIE = -0.5$ and the gap centerline potential is $V = 0.0$.

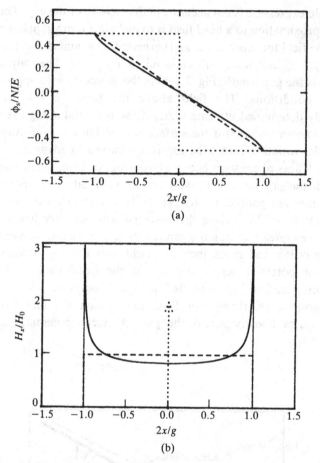

Fig. 3.10.   (a) Surface potential variation across the gap for the head of Fig. 3.9 (Fig. 3.5). (b) Surface longitudinal field component $H_x^s = -\partial V/\partial x$ corresponding to (a).

constant gap surface field equal to the deep-gap field (Fig. 3.10(b)). The resulting fields are readily calculated from (2.63) to be:

$$H_x(x,y) = \frac{1}{\pi} H_0 \left(\tan^{-1}\left(\frac{(g/2)+x}{y} \right) + \tan^{-1}\left(\frac{(g/2)-x}{y} \right) \right)$$

$$H_y(x,y) = \frac{1}{2\pi} H_0 \ln\left(\frac{((g/2)-x)^2 + y^2}{((g/2)+x)^2 + y^2} \right)$$

$$\text{(3.16)}$$

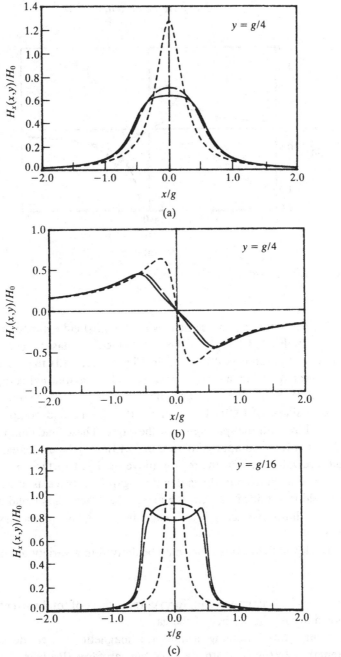

(a)

(b)

(c)

Fig. 3.11. (a) Longitudinal field component versus x/g for $y = 0.25g$. (b) Perpendicular field component versus x/g for $y = 0.25g$. (c), (d) are (a), (b) respectively for $y = g/16$. Short dash, long dash, and solid curves correspond respectively to far field, medium field (Karlqvist) and near field (Westmijze conformal map), respectively.

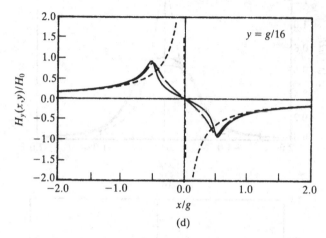

Fig. 3.11.   (*continued*)

In the limit of small gaps these fields yield the far-field expressions given by (3.13). In Figs. 3.11(a), (b) the fields (medium dashed curves) are plotted versus x for $y = 0.25g$ and in Figs. 3.11(c), (d) these plots are repeated near the head surface ($y = g/16$). The longitudinal component is symmetric and 'bell shaped,' but no longer the mathematical Lorentzian shape of (3.9). The 'half width' of this field component is determined by both the spacing y and the gap g. These field components, as well as the far-field approximation of (3.13), may be evaluated for all x,y, but are valid *only* for the region above the head surface ($y \geq 0$). (If (3.16) is evaluated deep in the gap, a deep-gap field twice as large as the correct field is obtained.) At close spacings the exact fields (solid curves) begin to exhibit the 'cusps' characteristic of fields at corners (Fig. 3.10(b)).

The Karlqvist field expressions may be derived in a variety of alternate ways:

(1) utilizing (2.7), assuming a uniform current sheet of total current $2NIE$ spread across the surface of the gap.
(2) utilizing (2.8) assuming a uniform magnetic charge density of strength $\pm 2NIE/g$ on either side of the gap faces (Problem 3.6).

The Fourier transform is easily obtained by recognizing that the surface longitudinal field is constant across that gap and vanishes on the head core surfaces (Fig. 3.10(b)). The Fourier transform of this box function is the sinc sampling function:

$$H_x^s(k) = NIE \frac{\sin(kg/2)}{kg/2} \qquad (3.17)$$

which yields with (2.52):

$$H_x(k,y) = NIEe^{-ky} \frac{\sin(kg/2)}{kg/2} \qquad (3.18)$$

$H_x^s(k)$ is plotted in Fig. 3.12. The transform approaches the zero gap limit of NIE at long wavelengths ($k \to 0$). The transform exhibits nulls at sub-multiples of the gap length. In the Karlqvist approximation, the first null occurs at a wavelength exactly equal to gap length. The envelope decreases inversely with wavenumber. The transform at a finite spacing includes the exponential spacing loss (Fig. 2.13).

Near-field expressions

The exact variation of potential across the gap surface can not be obtained in a simple analytic form. Conformal mapping techniques were utilized by Westmijze to analyze the two-dimensional configuration in Fig. 3.9 (Westmijze, 1953). The geometry has only one length scale, the gap length, since the pole faces as well as the gap depth are infinite in extent. The configuration is also simple enough to yield an analytic conformal transformation expressible by simple functions. First the basic principles of conformal transforms are summarized and then field results are discussed, including approximations for the Fourier transforms.

A complex function W is defined:

$$W = U + iV \qquad (3.19)$$

where $V(x,y)$ is taken to be the scalar potential and $U(x,y)$ is the field stream function (flux lines). It is arbitrary which function is assigned to the potential and flux. The virtue of conformal transformation is that it is often possible to find one or more simple transformations from the x,y plane to the W plane. In the W plane with coordinates (U,V), the boundaries of the original problem become parallel lines along the U axis at fixed potential V: the solution of the transformed problem is trivial. For the recording head with constant potential cores shown in Fig. 3.9, Fig. 3.13 shows the simple equivalency in the W plane where the free space region of the head lies between the core boundary potentials of

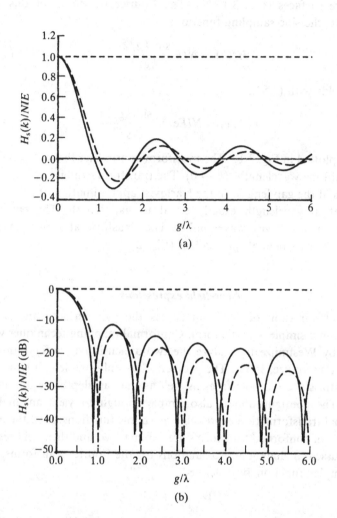

Fig. 3.12. Fourier transform of surface field versus normalized wavelength $\lambda/g = kg/2\pi$; (a) is the actual field and (b) is the log absolute value (dB).

$\pm NIE$. The disadvantage is that the back transformation yields the coordinates (x, y) as a function of the field (H_x, H_y) rather than the reverse. Generally numerical techniques are required to invert the solution.

The three functions W, U, V can be written as a function of position (x, y) or of the complex variable $z = x + iy$. W is an analytic function so

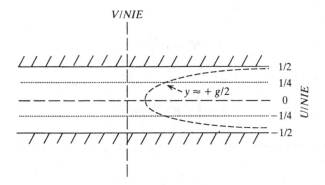

Fig. 3.13. Potential and flux lines (simply the coordinate axes) in the $W = U + iV$ plane. The medium dashed line corresponds approximately to $y = g/2$ at fixed separation from the head in the original (x,y) coordinate system.

that its derivative is independent of direction:

$$\frac{\mathrm{d}W}{\mathrm{d}z} = \frac{\partial W}{\partial x} = \frac{\partial U}{\partial x} + \mathrm{i}\,\frac{\partial V}{\partial x}$$
$$= \frac{\partial W}{\partial \mathrm{i}y} = \frac{\partial V}{\partial y} - \mathrm{i}\,\frac{\partial U}{\partial y} \qquad (3.20)$$

These two equivalent forms yield the Cauchy–Riemann relations:

$$\frac{\partial U}{\partial x} = \frac{\partial V}{\partial y}, \quad \frac{\partial U}{\partial y} = -\frac{\partial V}{\partial x} \qquad (3.21)$$

which confirm that U,V separately satisfy Laplace's equations. Note that conformal transformations maintain the orthogonality between U and V contours $(\nabla U \cdot \nabla V = 0)$.

Utilizing (3.20) and (3.21), it can be shown that:

$$\frac{\mathrm{d}W}{\mathrm{d}z} = \frac{\partial V}{\partial y} + \mathrm{i}\,\frac{\partial V}{\partial x}$$
$$= -\mathrm{i}\left(-\frac{\partial V}{\partial x} + \mathrm{i}\,\frac{\partial V}{\partial y}\right)$$
$$= -\mathrm{i}(H_x - \mathrm{i}H_y) \qquad (3.22)$$
$$= -\mathrm{i}H^*$$

where H^* is the complex conjugate of the field. Thus, the magnetic field is given by:

$$H^* = i \frac{dW}{dz} \tag{3.23}$$

The flux is given by

$$\Phi \equiv \int_s \int B(r') \cdot \hat{n} d^2 r' \tag{2.34}$$

and in two dimensions can be written as a flux per unit width in the z direction:

$$\Phi_w = \mu_o \int_A^B H \times dl \tag{3.24}$$

where dl is a path with end points A,B defining the region in the z plane over which the flux is to be determined. Using the Cauchy–Riemann relations (3.20), it follows that:

$$\begin{aligned}
\Phi_w &= \mu_o \int_A^B (-H_y dx + H_x dy) \\
&= \mu_o \int_A^B \left(\frac{\partial U}{\partial x} dx + \frac{\partial U}{\partial y} dy \right) \\
&= \mu_o \int_A^B (\nabla U \cdot dl) \\
&= \mu_o (U_A - U_B)
\end{aligned} \tag{3.25}$$

or:

$$\Phi_w = \mu_o \mathrm{Re}(W_A - W_B) \tag{3.26}$$

Once W is obtained the field as well as flux can be determined. The flux will immediately yield the inductance if (3.26) is applied near the cut defining the mmf source (Fig. 2.11(b)).

Following Westmijze, only half the head is considered by symmetry (Fig. 3.14(a), (b)) and that region is transformed first to the ζ plane by the transformation:

$$z = \frac{i}{\pi} \left(2\sqrt{\zeta} + \ln \frac{\sqrt{\zeta} - 1}{\sqrt{\zeta} + 1} \right) \tag{3.27}$$

In (3.27) all lengths are scaled to $g/2$. This transformation causes the boundary potentials to be transformed to the real ζ axis, and the space

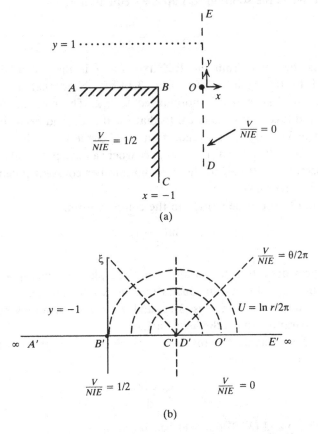

Fig. 3.14. (a) Half space for conformal head field analysis. (b) The transform ζ space. *ABC* at potential $V/NIE = 0.5$ is along the head core transform. Taken from Westmijze (1953).

where the field is to be determined becomes the upper half ζ plane (Fig. 3.14(b)). Utilizing (3.27), it can be shown that the boundary points A, B, C, D, O, E in Fig. 3.14(a) transform to A', B', C', D', O', E' in Fig. 3.14(b). B' occurs at the origin ($\zeta = (0,0)$) in the ζ plane since substitution of $\zeta = (0,0)$ into (3.27) yields $z = (-1,0)$. Similarly, the common points C', D' at $\zeta = (1.0)$ are the demarcation between the $V = 0.5NIE$ and $V = 0$ boundaries. (Substitute $\zeta = 1 + \epsilon e^{i\theta}$ into (3.27) and take the limit $\epsilon \to 0$; the variation of $0 < \theta < \pi$ causes z to traverse half the gap at $y = -\infty$.)

In the ζ plane the solution to Laplace's equation is:

$$V = \frac{\theta}{2\pi}, \quad U = \frac{\ln r}{2\pi} \qquad (3.28)$$

where θ is the angle from the Reζ axis and r is the radial distance referenced to the point C', D' (Fig. 3.14(b)). Note that V, U are normalized to NIE and r is normalized to $g/2$. The solution for the potential and flux lines is identical to that for the far-field head solution illustrated in Fig. 3.7: The boundary in the ζ plane is simplified to an infinite line bounding a semi-infinite space where a fixed potential is given on one (semi-infinite) half of the line and another constant potential is given on the other half.

Equation (3.28) can be written in the complex form:

$$W = \frac{\ln(\zeta - 1)}{2\pi} \qquad (3.29)$$

that also provides a transformation from the ζ plane to the (upper half) W plane in Fig. 3.13. As an illustration, the curves labeled $y = 1$ or ($y = g/2$ unnormalized) in Figs. 3.13, and 3.14(a), (b) denote a path of constant separation above the head surface.

If the field is normalized to the deep-gap field and all lengths to half the gap length (3.23) becomes:

$$H^* = 2\mathrm{i}\,\frac{\mathrm{d}W}{\mathrm{d}z} \qquad (3.30)$$

Utilizing (3.30) and (3.29) the field becomes:

$$H^* = 2\mathrm{i}\,\frac{\mathrm{d}W}{\mathrm{d}z} = 2\mathrm{i}\,\frac{\mathrm{d}W}{\mathrm{d}\zeta}\frac{\mathrm{d}\zeta}{z} = \frac{1}{\sqrt{\zeta}} \qquad (3.31)$$

Substitution of (3.31) into (3.27) yields the normalized form:

$$z = \frac{2\mathrm{i}}{\pi}\left(\frac{1}{H^*} + 0.5\ln\frac{1 - H^*}{1 + H^*}\right) \qquad (3.32)$$

Equation (3.32) is not generally useful because it cannot be easily inverted to yield each field component as a function of position. Numerical inversion is readily performed. In Figs. 3.11(a), (b) the field components are plotted versus x at $y = 0.25g$ in comparison with the Karlqvist approximation and the far-field approximation. The general characteristic is that the near-field longitudinal component is slightly smaller along the gap center line and larger at the pole edges. Fig. 3.15

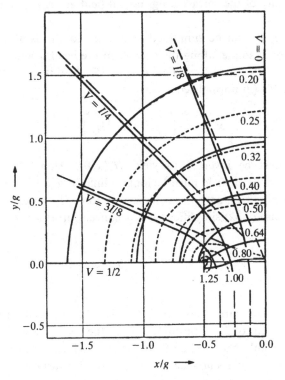

Fig. 3.15. Equipotentials, flux lines, and lines of constant field strength (dashed) for conformal map solution to finite gap head. Taken from Westmijze (1953).

gives constant potential V and flux U contours as well as lines (dashed) of constant field strength (Westmijze, 1953). Note that near the gap, flux lines that give the field direction do not coincide with lines of constant field strength. For example, the point $(-0.5, 0.25)$ in Fig. 3.15 has the field strength contour parallel to the x-axis; however, the field direction (U contour) is approximately at $60°$ to the x-axis. Far from the gap the field contours are circular and the field is directed along lines of constant strength (far 'wire' field, Fig. 3.7). In contrast to the Karlqvist approximation (3.16), the field at the top of the head at the gap center is slightly less (~ 0.8) than the deep gap field (Fig. 3.10(b)). In addition, at the gap corners both field components become infinite. In the Karlqvist approximation, only the vertical component becomes infinite (3.16). The longitudinal component as $y \to 0$ for $x = \pm g/2$ approaches half the deep-gap field or half the value of the discontinuity of the surface longitudinal field at the gap corner (Fig. 3.10(b)).

Equation (3.28) can be simplified somewhat for fields along certain directions. For example, along the surface of the head where $y = 0$ and $x < -g/2$, the field is solely vertical (Fig. 3.15). In this case $H^* = -iH_y$ and (3.32) yields (in normalized form):

$$x = -\frac{2}{\pi}\left(\frac{1}{H_y} + \tan^{-1}H_y\right) \qquad (3.33)$$

which still must be inverted to yield $H_y(x,0)$. Along the gap face $(x = -g/2, y < 0)$ the field must be longitudinal (Fig. 3.15). In addition, as seen from Fig. 3.15, as y varies from $-\infty$ to 0, H_x varies from H_0 to $+\infty$. In this case (3.32) simplifies (in normalized form) to:

$$x = \pm 1$$
$$y = \frac{2}{\pi}\left(\frac{1}{H_x} + 0.5\ln\frac{H_x - 1}{H_x + 1}\right) \qquad (3.34)$$

Plots of these fields are shown in Fig. 3.16. At the head surfaces the normal component of B is continuous. Since the head is considered to be infinitely permeable, the internal field vanishes. Thus, the external fields in Fig. 3.15 that are perpendicular to their respective surfaces are equal to the normal component of the surface magnetization or, equivalently, the

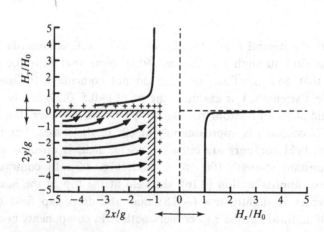

Fig. 3.16. Exact surface field plotted to reflect the surface charge on the top and gap surface of a 2D finite gap head. Schematic charge distribution is shown as well as magnetization distribution giving rise to the charge.

surface magnetic charge. Therefore, near the gap, the core magnetization is only approximately parallel to the head surface: toward the gap corners the magnetization rotates to be at an angle of ~45° to the head surface and increases in magnitude limited only by saturation.

The surface charge on the pole faces can be utilized with (2.26) to obtain the fields. As mentioned in the discussion in the previous section (and examined in Problem 3.2), the Karlqvist approximation results if a uniform magnetization parallel to the head surface, of magnitude $2NIE/g$, is assumed and used in (2.26) for $n \cdot M$ at each side of the gap. This assumption yields a deep-gap field of $2H_0$, but a field at the top of the gap ($x = y = 0$) of H_0 and a field above the head corresponding to (3.16). This dilemma is further compounded by the fact that the Karlqvist field expression yields a vertical field at the top surfaces of the head. A vertical field entails surface charges that do not occur if only poles along the gap surfaces are assumed. The discrepancy is explained by referring to Fig. 3.16. Deep in the gap the magnetization is parallel to the head surface, yielding a deep-gap field of H_0. The corresponding poles yield half the Karlqvist field. The other half arises from the additional infinite concentration of poles at the gap corners. Such a charge division is utilized in the next section to give an accurate analytic approximation to the near field.

An additional simplification of (3.32) arises when the maximum field at any fixed distance above the head surface is desired. Following any contour of constant field magnitude in Fig. 3.15, it is seen that each extends a maximum distance away from the head. These contours are concentrated around the gap corners and form a nesting, non-crossing, expanding set. The maximum field at any distance y above the head is also the field magnitude contour that has the given y as its maximum extent from the head surface. Along one of these contours only the field direction varies. A derivative of the y component of (3.32) with respect to field direction will vanish at this maximum y for a given field magnitude. From Problem 3.4 it can be shown that this condition yields:

$$\sin \alpha = \sqrt{\frac{3|H|^2 - 1}{2|H|^2}} \tag{3.35}$$

where α is the field angle at the point of furthest extent of the contour of magnitude $|H|$. Substitution of $H^* = |H|e^{i\alpha}$ into (3.32) yields the (x, y) location. In Fig. 3.17 the maximum field direction and x coordinate are plotted as a function of distance $2y/g$ above the head surface. Near the

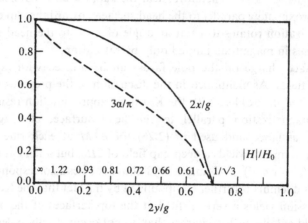

Fig. 3.17. Location ($2x/g$) and angle ($3\alpha/\pi$) of maximum field versus distance above the head surface. Beyond $y \sim 0.342g$ the maximum field occurs along the gap center line ($x = 0$) and the direction is longitudinal parallel to the gap surface.

gap corners where $|H| \to \infty$, (3.35) yields a field direction approaching 60°. (This is not the same question as asking what is the field direction at the gap corner.) Away from the head surface the angle of maximum field decreases to zero when $|H| = 1/\sqrt{3}(y = 0.342g)$. For smaller fields the maximum is along the gap centerline and is longitudinally directed.

The Fourier transform of the field at the surface of the head cannot be given in analytic form. The longitudinal Karlqvist approximation gives a constant field (as $y \to 0$) over the gap (Fig. 3.10(b)) that yields the sinc sampling function transform (3.17). The near field is not constant over the gap surface but is infinite at the gap edges and slightly less than the deep-gap field at gap center. Therefore, qualitatively, it is expected that a gap null would occur at longer wavelengths than those of the Karlqvist approximation. Westmijze gives a tabulated version of the Fourier transform, which has been fitted by Lindholm to the simple (unnormalized) form (Lindholm, 1975a):

$$H_s(k)/NIE = \frac{\sin(1.11\pi g/\lambda)}{(1.11\pi g/\lambda)}, \ g/\lambda < 0.5 = 0.326(g/\lambda)^{-2/3}\sin\pi\left(\frac{g}{\lambda}+\frac{1}{6}\right)$$

$$+ \ 0.0552(g/\lambda)^{-4/3}\sin\pi\left(\frac{g}{\lambda}-\frac{1}{6}\right), \ g/\lambda > 0.5$$

$$(3.36)$$

Equation (3.36) is plotted in Fig. 3.12 along with the transform of the Karlqvist approximation. The first gap null occurs at:

$$\lambda_1 \approx 1.136g \quad (g/\lambda_1 = 0.88) \qquad (3.37)$$

with subsequent nulls at higher, non-integrally related, wavenumbers. In addition to the shift in nulls, the envelope does not decrease as rapidly as the inverse wavenumber behavior of the Karlqvist approximation. A good approximation for this function up to this first gap null is to utilize the Karlqvist sinc function with the argument modified by a factor of 1.136:

$$H_s(k) \approx NIE \, \frac{\sin(1.136\pi g/\lambda)}{(1.136\pi g/\lambda)}, \quad \lambda > 1.136g \qquad (3.38)$$

It is to be emphasized that (3.38) is accurate only up to the first gap null; beyond that (3.37) must be utilized, especially for data analysis of measured spectra (Chapter 6). Gap nulls are shifted to even longer wavelengths if the region above the head has a finite susceptibility (Bertram & Lindholm, 1982).

Near-field analytic approximation

A simple analytic form for the near field of a finite gap head may be obtained utilizing a conformal map solution for a finite gap head that is infinitely thin (zero gap depth) (Westmijze, 1953; Ruigrok, 1990; Bertero, *et al.*, 1992). A transformation to the W plane, in normalized form (z by $g/2$ and W by NIE), is given by:

$$z = i \sinh(\pi W) \qquad (3.39)$$

Application of (3.30) yields:

$$H^* = \frac{2}{\pi\sqrt{1 - z^2}} \qquad (3.40)$$

By symmetry, the field across the gap $(-g/2 < x < g/2, y = 0)$ is longitudinal and is given simply without normalization by:

$$H_x^s(x) = \frac{2H_0}{\pi\sqrt{1 - (2x/g)^2}} \qquad (3.41)$$

The form of (3.41) exhibits the character of the exact field at the head surface plotted in Fig. 3.10(b). Ruigrok has shown that an accurate approximation to the exact surface field is given by half the Karlqvist field and half of the above thin gap field:

$$H_x^s(x) \cong \frac{H_0}{2}\left(1 + \frac{2}{\pi\sqrt{1-(2x/g)^2}}\right) \qquad (3.42)$$

Equation (3.42) yields $H_x^s(0) \cong 0.82H_0$, as does the exact field, and becomes infinite at the corners. The form of the approach to infinity varying as the square root of the distance from the corner is not exact, because for the 90° corner of a real head the variation should be as the 3/2 power (Jackson, 1975). Nonetheless, (3.42) follows the surface fields quite accurately and yields a value of $H_x^s(0) \cong H_0$ at $2x/g \sim 0.77$ as indicated in Fig. 3.10(b). This approximation clarifies the surface charge dilemma of the Karlqvist approximation (Problem 3.6). A deep gap field of H_o yields a uniform surface charge deep in the gap that results in half the Karlqvist field. The accumulation of charge at the gap corner, both in the gap and on the gap surface, as indicated in Fig. 3.16, is approximated by half the field from a thin gap head.

The Fourier transform of (3.42) may also be expressed analytically:

$$H_x^s(k) \cong \frac{H_0}{2}\left(\frac{\sin kg/2}{kg/2} + J_0(kg/2)\right) \qquad (3.43)$$

Equation (3.43) agrees well with (3.36) to wavenumbers spanning over 100 gap nulls further indicating that (3.42) is a good approximation to the surface field (Bertero, *et al.*, 1992).

Because the surface longitudinal field vanishes everywhere except over the gap for a long-pole head, (2.63) may be utilized with (3.42) to obtain the field everywhere. The integral may be performed to obtain a closed solution (Bertero, *et al.*, 1992):

$$\frac{H_x(x,y)}{H_0} = \frac{1}{2\pi}\left(\tan^{-1}\left(\frac{g/2+x}{y}\right) + \tan^{-1}\left(\frac{g/2-x}{y}\right)\right)$$

$$+ \frac{g}{2\sqrt{2\pi}}\left\{\frac{\{\sqrt{[x^2-y^2-(g/2)^2]^2 + 4x^2y^2} - x^2 + y^2 + (g/2)^2\}^{1/2}}{\sqrt{[x^2+y^2-(g/2)^2]^2 + 4y^2(g/2)^2}}\right\}$$

$$(3.44a)$$

$$\frac{H_y(x,y)}{H_0} = -\frac{1}{4\pi} \log\left\{\frac{(x+g/2)^2 + y^2}{(x-g/2)^2 + y^2}\right\}$$

$$-\,\text{Sgn}(x)\,\frac{g}{2\sqrt{2\pi}}\left\{\frac{\{\sqrt{[x^2-y^2-(g/2)^2]^2 + 4x^2y^2} + x^2 - y^2 - (g/2)^2\}^{1/2}}{\sqrt{[x^2+y^2-(g/2)^2]^2 + 4y^2(g/2)^2}}\right\}$$

$$(3.44b)$$

Equation (3.42) can be shown to give the far-field approximation (3.13) at large distance from the gap; however, the mid-range Karlqvist approximation does not occur explicitly.

Finite length heads – thin film heads

In this section general characteristics of two-dimensional, finite length heads will be discussed. So far in this chapter, idealized 2D heads with infinitely long pole lengths have been discussed. Real heads have, of course, finite dimensions: their finite length affects the fields near the head-pole edges along the medium interface surface. The general nature of fields at the pole edges in heads can be seen in Fig. 3.6. The flux flow is indicated by the arrows corresponding to a positive deep-gap field. Near the gap region the field is directed from left to right along approximately circular contours. Contours of decreasing field magnitude leave the head at increasing distances from the gap edge, and as seen in Fig. 3.6, extremely small field contours (depending on the head-pole length) leave from the side of the head circulating around the top to enter at the opposite side. These fields initially are in a direction opposite to that of the field in the gap region. For a recording medium path slightly above the head surface, the longitudinal fields beyond the head edges are therefore reversed with respect to the deep-gap field. The vertical field component direction is not changed as the point of evaluation is moved along the recording path across the head-pole edges. As discussed in Chapter 2 (Figs. 2.6, 2.7), both field components become infinite at the pole corner.

A thin film head is a complicated structure, as indicated in Fig. 1.4. For magnetic recording analysis the essential region is at the pole tips where typically the pole lengths are 2–3μm with gap lengths

(0.25–0.5µm) almost an order of magnitude smaller. In Fig. 3.6 the head scale corresponds to a video head where the pole lengths are many orders of magnitude larger than the gap ($>$ 1000:1). The fields for finite length heads, assuming infinitely deep 2D structures, have been studied by conformal transformation (Megory-Cohen & Howell, 1988; Lindholm, 1975a; Duinker & Geurst, 1964), by approximations of the surface field utilizing (2.63) (Bertero, *et al.*, 1992) and by functional fitting to scale modeling (Szczech & Iverson, 1986, 1987). The effect of finite depth has been considered for heads of elliptical cross-section (Baker, 1977; Duinker & Geurst, 1964). Fourier transforms have also been analyzed (Duinker & Geurst, 1964; Lindholm, 1975a, 1976b; Bertero, *et al.*, 1992; Arnoldussen, 1987). In this section general statements about these fields will be made and approximate relations will be derived for the pole edge field minima. The Fourier transform will also be discussed.

In Figs. 3.18(a),(b) the field components for a thin film head at various separations are plotted ($y = g/4, g/16$) utilizing (Szczech & Iverson, 1986, 1987). The head gap is $g = 0.5$ with asymmetric poles $p_1 = 1, p_2 = 2$. The pole lengths are somewhat smaller than a conventional head, and the asymmetry is exaggerated for discussion purposes. The effect of the finite pole length on the longitudinal component is to yield field minima or 'undershoots' approximately located just beyond the pole edges. The magnitude of the negative field excursion increases with decreasing separation from the head (y) and decreasing pole length (p). The vertical component does not change sign and even shows a cusp. The fields in the gap region are similar to those of the near field (or the Karlqvist approximation for y not too small), especially for the practical case of $p/g \geq 4$: neither the field peak nor the half-height width are effected by the finite length. As discussed in Chapter 5, the voltage pulse for longitudinal recording typically follows the field for $y = g/2$. Note in Figs. 3.18(a),(b) that changing the spacing y changes the field shape in the gap region much more than at the pole edges. As discussed below, the scaling is different, with y/g changing from 0.125 to 0.5 and y/L_2 ($L_2 = 2p_2 + g$) always small, varying between 0.0125 and 0.05.

A useful relation for all heads is that the line integral of the longitudinal field along the entire surface must vanish. For all heads of any shape, including finite track widths, (2.11) can be applied along the head surface with closure in the upper hemisphere at infinity. The closure integral must vanish because the fields of a finite head are dipolar and

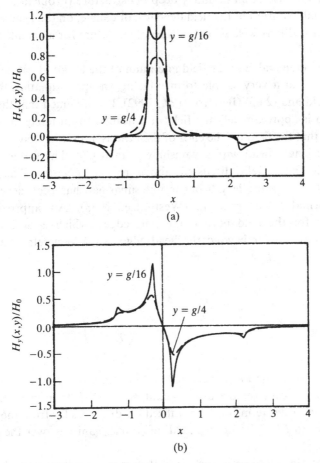

Fig. 3.18. Plots of longitudinal (a) and vertical (b) field components of an asymmetric thin film head ($p_1 = 1$, $g = 0.5$, $p_2 = 2$) for $y = g/4$, $g/16$.

decrease more rapidly with distance from the source than $1/r$. Since there is no enclosed current above the head, (2.11) reduces to:

$$\int_{-\infty}^{+\infty} H_x(x', y, z)\mathrm{d}x' = 0 \qquad (3.45)$$

where the integral can be taken at any height (y) above the head and at any transverse displacement (z) from the head center. From (2.42) this is simply the dc component ($k \rightarrow 0$) of the field Fourier transform. Thus, there is no dc component to the field from a magnetized head (2.56) except for the case of an infinitely deep 2D structure (Problem 3.7). The exact spatial variation of the field depends, of course, on the head length, depth, and width as well as on the general curvature for a rounded head surface.

The maximum value of the field minimum of the longitudinal field can be estimated in a very simple form utilizing an approximation for the surface field and (2.63) (Bertero, *et al.*, 1992). It is assumed that the field in the gap is approximately the field of a head with infinitely long pole faces and that the field off the ends of the head is approximately that due to a finite length head with a vanishingly small gap. The former is, of course, the near field discussed above in this chapter. The head configuration of finite length but infinitesimal gap has been developed by conformal transformation (Westmijze, 1953). An approximate expression for the field beyond the pole edges, which is accurate at distances not too far beyond the head edge, for a symmetric head of length L and $x \geq L/2$ is given by:

$$H_x(x,0) \simeq \frac{H_0 g}{L} \frac{3^{1/6}}{2^{4/3} \pi^{1/3}} \frac{1}{\left(\dfrac{2x}{L} - 1\right)^{1/3}} \tag{3.46}$$

This expression is for a zero gap head and therefore is independent of the gap length g except for the driving mmf $H_0 g = NIE$ that energizes the head. Higher order terms must be utilized for distances far from the head end relative to $L/2$. The surface field, of course, vanishes over the poles for $|x| < L/2$.

The field along the head surface is then given approximately by the field over the gap (3.42) and the field beyond the head pole edges (3.46). The longitudinal field of course vanishes adjacent to the head surface. Thus, the field anywhere above the head surface may be obtained analytically utilizing (2.63) (Bertero, *et al.*, 1992). The portion of the integral from $-g/2 \leq x \leq g/2$ yields the usual head field expressions (e.g. Karlqvist) whereas the portions from $-\infty \leq x \leq -L/2$ and $L/2 \leq x \leq \infty$ give the end terms. The total field is the sum of the contributions from all three portions of the integral. The extension to asymmetric heads is straightforward.

Close to the gap corner and not far above the gap surface ($y \ll L/2$) the field is dominated by the integral over the source field due to that corner. A simple approximation may be obtained for the maximum reverse field utilizing this approximation. The field near the corner $(x = L/2, y = 0)$ may be written utilizing (2.63) as:

$$H_x(x,y) \cong \frac{H_0 g y}{\pi L} \frac{3^{1/6}}{2^{4/3} \pi^{1/3}} \int_{L/2}^{+\infty} \frac{dx'}{((x-x')^2 + y^2)} \frac{1}{\left(\frac{2x}{L} - 1\right)^{1/3}} \qquad (3.47)$$

which may be rewritten using a change of variable and with suitable normalization as:

$$H_x(x,y) \cong -0.267 \frac{H_0 g}{L} \left(\frac{L}{y}\right)^{1/3} \int_0^\infty \frac{u\,du}{(1 + (u^3 - q)^2)} \qquad (3.48)$$

where $q = ((2x/L) - 1)/(2y/L)$. This term, by varying q, yields the major contribution to the field undershoot at the corner of a thin film head. The integral in (3.48) maximizes at $q \approx 1$ where the maximum value of the integral is ≈ 0.75. Thus, at any small spacing y above the head surface, the maximum reverse field (or field minimum) occurs at location $x \approx L/2 + y$ with a value given by:

$$H_x^{min} \cong -0.2 \frac{H_0 g}{L} \left(\frac{L}{y}\right)^{1/3} \qquad (3.49)$$

field minimum location : $x \sim L/2 + y$

This field minimum as well as location agrees well with Fig. 3.18(a) for different spacings.

The Fourier transforms can be obtained easily utilizing the same degree of approximation. For the field source regions beyond the head ends (3.46) is utilized in (2.42) with the integration extending over the range $-\infty < x \le -L/2$, $L/2 \le x < \infty$. The result is for a symmetric head:

$$\frac{H_x^s(k)}{NIE} \cong \frac{\sin 1.136\pi g/\lambda}{1.136\pi g/\lambda} - \frac{0.205 \cos \pi \left(\frac{L}{\lambda} + \frac{1}{3}\right)}{\left(\frac{L}{\lambda}\right)^{2/3}} \qquad (3.50)$$

The first term is simply the Fourier transform of the gap region field, accurate for wavelengths longer than the gap length. The second term is the result of a Fourier transform of (3.46) (given by Westmijze with a small numerical error). The Fourier transform is plotted in Fig. 3.19 for various pole-to-gap length ratios. Long wavelength undulations occur due to the finite head length. For a thin film head where the head length

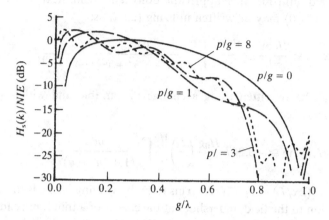

Fig. 3.19. Fourier transform of surface longitudinal field for a symmetric thin film head (p/g = 0,1,3,8).

is not extremely long with respect to the gap length, the interference between the gap and end terms produces a gap null whose location oscillates as the head length is monotonically changed. The long wavelength spectrum increases monotonically with increasing pole length.

Similar results can be obtained by superimposing the solution to a finite-gap, infinite-length head and a parallel-plate head (Lindholm, 1975a). Since the Fourier transform for a parallel-plate head can be given in closed form solution, this latter approximation is valid for all wavelengths. The approximation discussed here, with transform given by (3.50), is reasonable only for wavelengths not longer than approximately the length of the head ($\lambda < L$): it does not give the correct asymptotic result of $0.5NIE$ as $\lambda \to \infty$ (Problem 3.7).

Using the above approximations, the fields and transforms for an asymmetric head can easily be obtained (Bertero, *et al.*, 1992). The field due to a head edge (3.46) can apply to asymmetric heads if, for each edge, $L/2$ represents the distance from the gap center to the respective edge (Fig. 3.18(a),(b) where $L_1 = 2p_1 + g$, $L_2 = 2p_2 + g$). Thus, to good approximation, the reverse field maximum given by (3.49) may be utilized and agrees well with accurate conformal map solutions (Megory-Cohen & Howell, 1988, Fig. 6).

The Fourier transform may be obtained utilizing (3.46) applied to each edge. The result contains both amplitude and phase variations and is given by:

$$\frac{H_x^s(k)}{NIE} \cong \frac{\sin 1.136g/\lambda}{1.136\pi g/\lambda}$$

$$-0.1025 \left(\frac{\cos \pi \left(\frac{L_1}{\lambda} + \frac{1}{3} \right)}{\left(\frac{L_1}{\lambda} \right)^{2/3}} + \frac{\cos \pi \left(\frac{L_2}{\lambda} + \frac{1}{3} \right)}{\left(\frac{L_2}{\lambda} \right)^{2/3}} \right.$$

$$\left. + i \left(\frac{\sin \pi \left(\frac{L_1}{\lambda} + \frac{1}{3} \right)}{\left(\frac{L_1}{\lambda} \right)^{2/3}} - \left(\frac{\sin \pi \left(\frac{L_2}{\lambda} + \frac{1}{3} \right)}{\left(\frac{L_2}{\lambda} \right)^{2/3}} \right) \right) \right)$$

(3.51)

Keepered heads

The analysis of head fields presented thus far in this chapter has focused on field sources that are solely on one side of the free-space region where the fields are to be evaluated. In this case expressions (2.62, 2.63, 2.64) apply as well as the Hilbert transform relations (2.58, 2.59). In perpendicular recording it has been customary to consider use of a 'keeper' or high permeable layer that is deposited before the magnetic recording layer. Thus, even if the magnetic recording medium has essentially unit permeability, the region of field evaluation is in between two highly permeable layers: the head and the 'keeper'. Examples of such configurations, which can be considered magnetically to be a modified head structure, are shown in Fig. 3.20 for both a ring and pole head. The pole head is considered to be optimal for perpendicular recording since flux leaves the pole in a perpendicular direction and the keeper becomes the return path. In both cases fields may be determined by numerical analysis or for 2D fields by conformal mapping techniques (Minuhin, 1984).

A reasonable, simplifying approximation can be achieved by assuming that the keeper is infinitely thick and that it possesses an infinite permeability. In that case the keeper can be placed at zero potential with respect to the source potentials at $\pm NIE/2$ for the conventional inductive head and NIE for the pole head. In general (2.61) applies; however, the potential all along the head surface must be known as well as the field

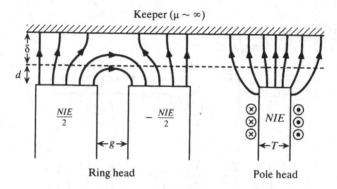

Fig. 3.20. Sketch of configuration and flux flow for keepered ring and pole heads.

gradient at the keeper interface. A clear indication that Hilbert transforms do not apply is that at the keeper interface there is a perpendicular field component while there is no longitudinal component anywhere along the keeper interface.

In general the presence of a keeper alters the potential along the head surface. In the case of the long pole head the potential variation across the gap utilizing the exact near field, the near-field approximation (3.42), or even the linear variation of the Karlqvist approximation is not correct. Figure 3.21 indicates the effect of the keeper on the surface potential for both types of head. In the case of a ring head a large keeper separation (with respect to the gap length) yields a potential close to that of the simple heads. As the keeper distance is reduced, the potential along the

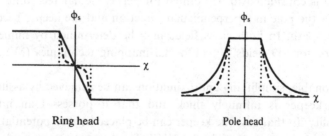

Fig. 3.21. Sketch of surface potential for ring and pole heads for various keeper separations. Solid, long dash, and short dash are for the keeper at large distances, close distances, and virtual contact, respectively.

head-core surfaces must remain at $\pm NIE/2$. However, the potential along the gap surface decreases to reflect the proximity of the zero-potential keeper. As indicated in Fig. 3.20, the consequence to the fields is that the flux flows first to the keeper and then returns to the core rather than flowing directly across the gap. This phenomenon enhances the perpendicular field component and reduces the longitudinal component. The reduction in longitudinal component is seen at the gap by a reduction in the slope of the potential variation. At very close keeper separation, the potential vanishes across the gap and the flux flow is entirely perpendicular.

In the case of a pole head the potential remains fixed at NIE along the pole surface then decreases on either side (Fig. 3.21). The decrease is gradual for large keeper separations (with respect to the pole width T) (Minuhin, 1984). As the keeper distance is reduced, the pole surface potential decreases more rapidly at distances beyond the edges of the pole. As the separation is reduced to virtual contact, the potential decreases immediately to zero beyond the pole edges. Again, as in the case of a ring head, the action of the keeper is to cause the flux flow to be perpendicular toward the keeper. At small separations the flux is essentially perpendicular, with no longitudinal component, as reflected by the vanishing potential gradient along the head surface (except at the precise corners). It has been realized that the presence of a recorded transition of magnetization caused by the pole fields yields imaging of the medium magnetic charges in the pole that considerably alter the effective recording field 'head–medium' interaction) (Nakamura & Tagawa, 1989).

General statements may be made about the Fourier transform of the fields of 2D keeper heads. The relation (2.48) applies written in terms of the transform of the surface potential $\phi_s(k)$:

$$\phi(k, y) = \phi_s(k) \frac{(Ae^{-ky} + Be^{+ky})}{(A + B)} \tag{3.52}$$

Equation (3.42) and the following in this section are written for $k > 0$ only. Because $\phi(k, t) = 0$ where t is the keeper separation, (3.52) may be expressed as:

$$\phi(k, y) = \phi_s(k) \frac{\sinh k(t - y)}{\sinh kt} \tag{3.53}$$

The field transforms, utilizing (2.50), are:

$$H_x(k,y) = H_s(k) \frac{\sinh k(t-y)}{\sinh kt}$$

$$(3.54)$$

$$H_y(k,y) = iH_s(k) \frac{\cosh k(t-y)}{\sinh kt}$$

where $H_s(k) = ik\phi_s(k)$. It is clear from (3.54) that the Hilbert relations (2.58, 2.59) do not apply in general. At short wavelengths with respect to the keeper spacing ($\lambda \ll t - y$), the keeper is not seen and the Hilbert relations hold. At long wavelengths ($\lambda \gg t - y$) the Hilbert relations do not hold:

$$H_y(k,y) = iH_x(k,y) \quad \lambda \ll (t-y)$$
$$H_y(k,y) = iH_x(k,y)/(k(t-y)) \quad \lambda \gg (t-y)$$

$$(3.55)$$

At long wavelengths the vertical component approximately 'integrates' the horizontal component. In the case of a long-pole ring head, the long wavelength transform of the longitudinal component is simply $H_x(k \to 0, y) = NIE(t-y)/t$. At the surface ($y = 0$), the field integral is the same as for no keeper. With a keeper the long wavelength component decreases linearly with distance from the head, vanishing, as required, at the keeper surface. The vertical component at long wavelengths is: $H_y(k \to 0, y) = iNIE/kt$. The vertical component increases with increasing wavelength, corresponding to the constant field between the poles and keeper at distances that extend far from the gap $H_y(x \gg g, y) = \pm NIE/2t$ (Fig. 3.20).

For a pole head the vertical field is a symmetric function similar to that of the longitudinal component of the ring head. The long wavelength component of the vertical field corresponds to the total flux (per unit head width) leaving the pole or entering the keeper and is related to the transform of the potential by: $H_y(k \to 0, y) = \phi_s(k)/t$. The longitudinal component therefore decreases at long wavelengths: $H_x(k \to 0, y) = ik(t-y)\phi_s(k)/t$. Note that the average field from the pole to keeper at the pole center is: $H_y(x = 0) = NIE/t$. The surface potential $\phi_s(k)$ decreases at all wavelengths as the keeper separation is reduced, since the decrease of potential beyond the pole edges decreases with decreasing separation (Fig. 3.21). At very close separations ($t \ll T$) the surface potential is finite only at the pole. The trends in spectral nulls are examined in Problem 3.8.

It is important to note that the relations given by (2.62) to (2.64) can be utilized considering the head and keeper as separate one-sided sources and the net field superimposed. The field due to the keeper is simply that from the image of the source fields. However, the total interactive field problem must be solved for the surface fields at the head. For an infinitely high permeability, non-saturable keeper, the net longitudinal field at the keeper interface vanishes. However, the field utilized in the expressions is that due to the sources. Thus, $H_x^s(x')$ utilized for the keeper is that surface field due to the images only. This field can be written in terms of the source surface field of the head. However, it is to be emphasized that the net field at the head surface is the sum of the head source field and that due to the keeper after the interactive problem is solved. For example, the keepered ring head is not simply the sum of a 'Karlqvist' head and its image.

Concluding remarks

Two topics have not been included in this chapter: fields of finite track-width heads and saturation phenomena. The background material presented here should enable the reader readily to comprehend and utilize the published material on these subjects. Finite track-width heads have been examined by several authors, who have included edge-field expressions (van Herk, *et al.*, 1977; Lindholm, 1975b, 1980a; Mayergoyz & Bloomberg, 1986). Simplified approximations are discussed in Chapter 5 for the estimate of off-track replay voltages and transforms. Head saturation has been studied numerically for ring heads (Monson, 1972; Bertram & Steele, 1976; Rodé & Bertram, 1989; Wachenschwanz & Bertram, 1991) and specifically for metal-in-gap heads (Kelley, 1988). There are two aspects to head saturation: the limit of the recording or deep-gap field with increasing record current and the change of field shape or head field gradients as saturation occurs. In thin film heads it is believed that saturation occurs in the narrow neck region before the gap (Fig. 1.4), limiting the recording fields without altering the recording field shape. In inductive heads saturation begins in the pole tip region as illustrated in Figs. 2.6 and 2.7. As the record current is increased, the saturation regions ($M(r) = M_s$) at the tips expand to fill first the region at the gap and eventually the entire head core. The saturating pole tips alter the field shape and cause the effective efficiency to decrease (Problem 3.10).

Problems

Problem 3.1 Consider a toroidal head of core cross-section A_c and core length L tapering to a gap face area A_g and gap length g. Let NI amp-turns be applied to the windings. Neglect fringing so that the flux is confined to a simple toroidal path. Assume that the field is constant both in the core and in the gap. Using (2.11) and the conservation of flux, calculate a simple expression for the efficiency. Write the internal field H_i in terms of the deep gap-field and thus show that infinite permeability entails vanishing internal field. Show that for infinite permeability the flux in the core (or gap) relative to the driving mmf is:

$$\frac{\Phi}{NI} = \mu_0 A_g / g$$

The right hand term is also known as the gap permeance or inverse gap reluctance.

Also calculate an expression for the change in scalar potential relative to the gap potential drop along the core centerline as a function of distance s away from the gap.

$$\text{Ans.} \quad \frac{\Delta \phi}{NIE} = \frac{s}{l}\left(1 - \frac{1}{E}\right) = \frac{sA_g}{G\mu A_c}$$

For $A_g / A_c = 0.1$, $\mu = 2000$, $g = 0.35\mu m \rightarrow s = 10\%$ drop at $s = g\mu$ $= 7$ mm.

Problem 3.2 Consider NI turns wrapped around the gap face instead of deep in the core for a head material of infinite permeability. Is the field above the gap changed?

Problem 3.3 Derive the Karlqvist field equations, assuming a uniform magnetization normal to the gap faces everywhere in the gap. If the head is energized to a deep-gap field of H_0, use boundary conditions to show that the magnetic surface charge is H_0. (The internal field vanishes due to the high permeability.) Utilize (2.26) to obtain the field and show that in this approximation the field at the gap surface is $H_0/2$ and that (2.26) yields constant head surface potentials ($y = 0$, $|x| > \pm g/2$). Thus, to obtain the correct field magnitude above the head, the deep-gap field must be artificially doubled.

Problem 3.4 Show that in the Karlqvist approximation contours of longitudinal and vertical field components are given by circles. Sketch some contours and show that longitudinal field contours pass through the gap corners with centers lying along the gap center line. Show that the circular contours of the vertical field are in pairs and have their origins on the head surface with their centers lying in the gap.

Problem 3.5 Verify by direct substitution into (3.27) that boundary points A, B, C, D, O, E in the z plane in Fig. 3.14(a) transform to A', B', C', D', O', E' in the ζ plane in Fig. 3.14(b).

Problem 3.6 Derive (3.35) for the angle of the field magnitude $|H|$ at its maximum extent from the head surface. In (3.32) let $H^* = |H|e^{i\alpha}$ where α is the field angle with respect to the x direction and find $dy/d\alpha$.

Problem 3.7 The Fourier transform of the magnetic fields of a conventional inductive head of any shape must vanish at DC; $H_{x,y}(k \to 0, y, z) = 0$. For a 2D structure of infinite pole length, the long wavelength fields at any y are, from (3.18), $H_{x,y}(k \to 0, y, z) = NIE$, because the integral for the transform or the potential difference is always taken at (infinite) distances less than the infinite pole lengths. Show that for a 2D structure with finite pole lengths and an infinitely deep core the long wavelength field is given by $H_{x,y}(k \to 0, y, z) = NIE/2$. Hint: view the head from the end, very far removed, and utilize (2.62) interchanging x and y and then (2.57).

Problem 3.8 Derive the Fourier transform of the surface potential for a keeper pole and ring head (Figs. 3.20 and 3.21) for the case of vanishingly small keeper separation. Show that for the ring head the effect of a keeper increases the gap null wavelength to a limit of $\lambda = 2g$. Show that, for the pole head the effect of a keeper decreases the gap null wavelength to a limit of $\lambda = g$.

Problem 3.9 Show that (2.11) holds for a keepered head where the path of integration compasses the entire space between the head and keeper: along the head surface ($y = 0$) and returning along the keeper surface ($y = t$), closing at $x = \pm\infty$.

Problem 3.10 Use a simple model to estimate the relation of deep-gap field to record current for a saturating inductive ring head. Assume (3.8) holds in the form: $E = 1/(1 + (1 - 1/E_0)(\mu_{dc}/\mu_r(H_i)))$, where E_0 is the low-level efficiency, μ_{dc} is the low level relative permeability, and μ_r is the general saturable permeability. Assume a simple form: $\mu_r(H_i) = (1 + \chi_0/(1 + \chi_0 H_i/M_s))$, where H_i is the internal field at the gap interface: $\mu_r H_i = H_0$ and M_s is the core saturation magnetization. Assume $M_s = 5000$, $\chi_0 = 1000$, and using $H_0 = NIE/g$ plot H_0/M_s versus $NI/M_s g$ for $E_0 = 0.5, 0.7, 0.9$.

4

Medium magnetic fields

Introduction

Magnetized recording media produce fields by virtue of divergences in the magnetization pattern. Thus, (2.8) can be utilized to obtain the magnetic fields for any specified magnetization pattern. For two-dimensional geometry it is often convenient to utilize the simple form given by (2.22), or under certain conditions (2.26). Magnetized media are particularly simple to analyze since, in general, they extend infinitely far along the x axis and possess a finite thickness or magnetization depth which does not vary along the x axis. In this section expressions are given for the fields for single magnetization transitions that are either longitudinally or vertically oriented. That discussion will be followed by a general relation for the Fourier transform of the fields. This section is concluded by a discussion of the fields from sinusoidally magnetized media. Only two-dimensional geometries will be considered.

Single transitions

We begin by deriving the fields produced by a single, perfectly sharp transition as sketched in Figs. 4.1(a) and 4.2(a) for a longitudinally and vertically directed magnetization, respectively. The coordinate system (x,y) is centered at the center of the medium at the transition center. Equation (2.26) may be utilized for both cases, since the volume charge for the case of a sharp transition of longitudinal magnetization is equivalent to a surface charge of $\sigma = 2M$ at the transition center ($x = x_0$) extending from $-\delta/2 < y < \delta/2$. The fields for a longitudinal magnetization utilizing (2.26) (r_1, r_2, θ are noted in Fig. 4.1(a)) yield:

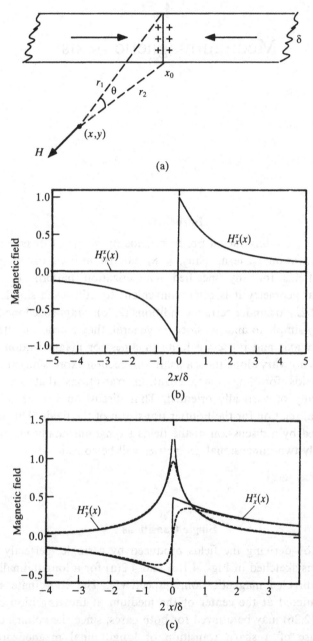

Fig. 4.1. (a) Charge distribution for a perfectly sharp transition of longitudinal magnetization. (b) Magnetic field along the medium centerline. (c) Magnetic field components versus x along the surface of the medium ($y = \delta/2$); dashed curves are just above the surface ($y = 0.55\delta$).

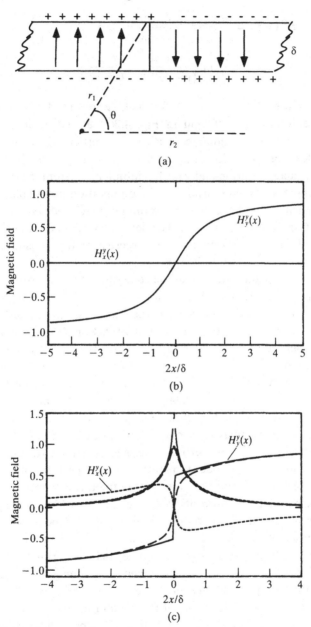

Fig. 4.2. (a) Charge distribution for a perfectly sharp transition of perpendicular magnetization. (b) Magnetic field along the medium centerline. (c) Magnetic field components versus x along the surface of the medium ($y = \delta/2$); dotted curves are just above the surface ($y = 0.55\delta$). Dashed curve is perpendicular field just below the surface ($y = 0.45\delta$).

$$H_x^x(x,y) = \frac{M}{\pi} \left(\tan^{-1} \frac{y + \delta/2}{x - x_0} - \tan^{-1} \frac{y - \delta/2}{x - x_0} \right)$$

$$H_y^x(x,y) = \frac{M}{2\pi} \ln \frac{(y + \delta/2)^2 + (x - x_0)^2}{(y - \delta/2)^2 + (x - x_0)^2} \tag{4.1}$$

These fields are plotted versus x for fixed y in Fig. 4.1(b) along the medium centerline ($y = 0$) and in Fig. 4.2(c) at the medium surface $y = \delta/2$. Along the medium centerline the vertical field component vanishes, leaving only a longitudinal field component. The field always opposes the magnetization and vanishes infinitely far from the transition. The field increases monotonically as the observation point comes closer to the transition and reaches a maximum magnitude equal to the magnetization M at the transition. The field possesses odd symmetry with respect to the transition center and is largest near the transition center and decreases toward the 'bit' edges.

At all vertical planes other than the centerline the field possesses both field components. As Fig. 4.1(c) shows for $y = \delta/2$, the longitudinal component has the same character as the field along the centerline, but increases in magnitude only to a value of $M/2$. For all y in the medium this component always increases to a value of M. For all y outside the medium the longitudinal field reaches a maximum before the transition center and then vanishes by symmetry at $x = x_0$ (shown dashed for the case $y = 1.1\delta/2$). The perpendicular field component vanishes infinitely far from the transition and reaches a maximum at the transition center. This maximum is infinitely large only for the precise location at the ends of the charged surface. The vertical field is symmetric with respect to x variations but possesses odd symmetry with respect to the medium centerline. In Fig. 4.3 field lines are plotted for this example of a single perfect sharp longitudinal transition. In Fig. 4.4 the field lines are plotted for a sharp transition of longitudinal magnetization in the presence of an infinite surface of high permeability (gapless head). Using (4.1) and the imaging rules illustrated in Fig. 2.8, the net fields exterior to the high-permeability region are readily determined. With imaging, all the flux flows into the high-permeability surface rather than being equally divided above and below the center plane of the medium.

The fields from a single, sharp transition of perpendicular magnetization can easily be expressed utilizing (2.26), but in this case there are four separate charged surfaces. These surfaces are indicated in Fig. 4.2(a). Each possesses a surface charge magnitude of $\sigma = M$, but the sign varies as indicated in the figure. Equation (2.26) must be applied separately to

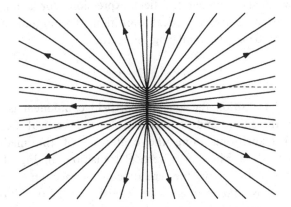

Fig. 4.3. Magnetic field lines (*H*) for a sharp transition of longitudinal magnetization corresponding to Fig. 4.1(a). A discontinuity occurs at the transition center.

each surface. In this case the *y* direction denotes the normal to the charged surface, and *x* the tangential direction. However, since the medium is assumed to extend infinitely far in either direction, r_2 should lie parallel to the *x* axis to extend infinitely far, as illustrated for the top right line of surface charges. The infinities in r_2 and r_1 for the various surfaces cancel in the total field. Thus, the net field components are:

$$H_y^y(x,y) = \frac{M}{\pi}\left(\tan^{-1}\frac{x-x_0}{y+\delta/2} - \tan^{-1}\frac{x-x_0}{y-\delta/2}\right)$$

$$H_x^y(x,y) = \frac{M}{2\pi}\ln\frac{(y+\delta/2)^2+(x-x_0)^2}{(y-\delta/2)^2+(x-x_0)^2}$$

(4.2)

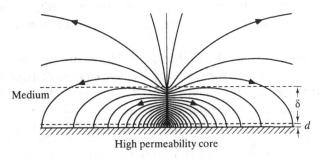

Fig. 4.4. Field lines of Fig. 4.3 for a transition close to a high-permeability surface.

Note the symmetry between the field expressions for a longitudinal transition and those for a vertical transition. The arguments of the arctangents can be inverted utilizing the relation:

$$\tan^{-1} 1/x = \pi/2 - \tan^{-1} x \qquad (4.3)$$

For a vertical transition the longitudinal field component is localized around the transition and is identical to the vertical field for a longitudinal transition: along the centerline of the medium the longitudinal field vanishes for a vertical transition. In Fig. 4.2(b) the vertical field is plotted along the centerline of the medium ($y = 0$). The field always opposes the magnetization, vanishes at the transition center and is largest far from the transition. Far from the transition the medium resembles a thin plate with maximum demagnetizing factor. Thus the field is equal in magnitude to the magnetization M so that the B field vanishes. For a transition of perpendicular magnetization, demagnetization is near the 'bit' cell edges away from the transition.

In Fig. 4.2(c) both components of the field are plotted versus x for $y = \delta/2$ and slightly below and above the top surface of the medium. The vertical component of the field just below the surface exhibits the same character as that at the center, but contains a step of magnitude M at the transition. Just above the surface the vertical field (dotted) vanishes far from the transition since there is no field exterior to a uniformly magnetized plate. At the transition this field component exhibits odd symmetry with a step discontinuity of magnitude M. Just below the surface (dashed) the vertical field exhibits a step discontinuity in M at the transition center, but increases to a value equal to M and opposite in direction far from the transition. As in Fig. 4.2(b), the net flux density B vanishes far from the transition center. The longitudinal field exhibits a symmetrical 'bulge' at the transition and varies symmetrically above and below the surface. Field lines for a vertical transition are plotted in Fig. 4.5.

If the transition is not sharp, then the field patterns are broadened accordingly. The new field expression can be derived from (2.8), (2.22) or can be written as a convolution of the step transition fields and the normalized derivative of the x variation of the transition if the transition does not change shape with depth into the medium. If

$$F(x) = \frac{1}{2M} \frac{\partial M(x)}{\partial x} \qquad (4.4)$$

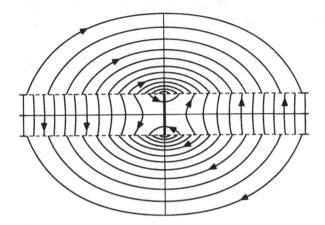

Fig. 4.5. Magnetic field lines (H) for a sharp transition of perpendicular magnetization corresponding to Fig. 4.2(a). Discontinuities occur at the medium surfaces.

where M is the saturation magnetization far from the transition, then each component of the field may be written as:

$$H(x,y) = \int_{-\infty}^{+\infty} dx' H^{\text{step}}(x - x',y)F(x')\qquad(4.5)$$

where $H^{\text{step}}(x - x',y)$ represents the field due to step transitions as given in (4.1) and (4.2). This convolution applies to both orientations of the magnetization and both components of the field. For example, if the transitions are of the arctangent form:

$$M(x) = \frac{2M_r}{\pi} \tan^{-1} \frac{x - x_0}{a}\qquad(4.6)$$

where 'a' represents the transition length, then $F(x)$ is given by

$$F(x) = \frac{a}{\pi(a^2 + (x - x_0)^2)}\qquad(4.7)$$

Note that (4.7) is the volume charge density ($\rho_m = -2M_r F(x)$ (2.3)) for the case of longitudinal magnetization.

The fields due to an arctangent transition may be obtained analytically. The longitudinal magnetostatic field for a longitudinal magnetization transition is

$$
H_x^x(x,y) = -\frac{M_r}{\pi}\left(\tan^{-1}\left(\frac{a+y+\delta/2}{x-x_0}\right) + \tan^{-1}\left(\frac{a-y+\delta/2}{x-x_0}\right)\right.
$$

$$
\left. -2\tan^{-1}\left(\frac{a}{x-x_0}\right)\right) \quad |y| \le \delta/2
$$

(4.8)

$$
H_x^x(x,y) = -\frac{M_r}{\pi}\left(\tan^{-1}\left(\frac{a+|y|+\delta/2}{x-x_0}\right)\right.
$$

$$
\left. -\tan^{-1}\left(\frac{a+|y|-\delta/2}{x-x_0}\right)\right) \quad |y| \ge \delta/2
$$

and for the vertical field:

$$
H_y^x(x,y) = \frac{M_r}{2\pi}\ln\frac{(a+\delta/2-y)^2+(x-x_0)^2}{(a+\delta/2+y)^2+(x-x_0)^2} \quad |y| \le \delta/2
$$

(4.9)

$$
H_y^x(x,y) = \frac{M_r}{2\pi}\ln\frac{(a+|y|-\delta/2)^2+(x-x_0)^2}{(a+|y|+\delta/2)^2+(x-x_0)^2} \quad |y| \ge \delta/2
$$

The vertical fields for a transition of the form of (4.6) for perpendicular orientation are given for all y by:

$$
H_y^y(x,y) = \frac{M_r}{\pi}\left(\text{sgn}(y-\delta/2)\tan^{-1}\left(\frac{x-x_0}{a+|\delta/2-y|}\right)\right.
$$

$$
\left. -\text{sgn}(y+\delta/2)\tan^{-1}\left(\frac{x-x_0}{a+|y+\delta/2|}\right)\right)
$$

(4.10)

$$
H_x^y(x,y) = \frac{M_r}{2\pi}\ln\frac{(a+|\delta/2-y|)^2+(x-x_0)^2}{(a+|\delta/2+y|)^2+(x-x_0)^2}
$$

The double script notation on the field components simply denotes the general tensor form of the field relative to a general vector magnetization. Note that all fields exterior to the medium in (4.8) and (4.9) are identical to those in (4.1) and (4.2) but with the spacing from the medium surface $d = |y| - \delta/2$ replaced by $d + a$ (Problem 4.3).

From (4.8) and (4.10) the fields along the centerline of the medium ($y = 0$) in the direction of the respective magnetization orientation (the

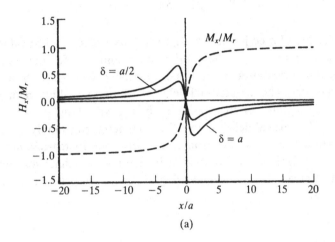

(a)

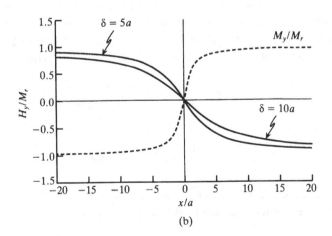

(b)

Fig. 4.6. (a) Longitudinal magnetic field along the centerline of a medium with a longitudinal magnetization transition (arctangent) of length a for relative medium thicknesses $\delta = a/2, a$. (b) Perpendicular magnetic field along the centerline of a medium with a perpendicular magnetization transition (arctangent) of length a for relative medium thicknesses $\delta = 5a, 10a$. In both cases the transition shape is shown for comparison.

diagonal tensor components) are:

$$H_x^x(x,0) = -\frac{2M_r}{\pi}\left(\tan^{-1}\left(\frac{a+\delta/2}{x-x_0}\right) - \tan^{-1}\left(\frac{a}{x-x_0}\right)\right)$$

$$H_y^y(x,0) = -\frac{2M_r}{\pi}\tan^{-1}\left(\frac{x-x_0}{a+\delta/2}\right)$$

(4.11)

These fields are plotted in Fig. 4.6(a), (b) versus x/a for typical values of $\delta = a/2$ and a for longitudinal orientation and $\delta = 5a$ and $10a$ for perpendicular orientation. Both demagnetizing fields vanish at the transition center. The longitudinal field for a longitudinal magnetization peaks in the vicinity of the transition before vanishing far from the transition. The vertical field for a vertical magnetization simply increases to a maximum of $\pm M_r$ far from the transition. Increasing medium thickness increases the demagnetizing field for longitudinal magnetization, but decreases the demagnetizing field for perpendicular media (Problem 4.6).

Field maxima and gradients for longitudinal magnetization

In the discussion of record-process modeling in Chapter 8 expressions are required for the field gradient at the transition center as well as for the maximum field for longitudinal magnetization patterns. From (4.11) the field maxima occur at $x_{max} = \pm\sqrt{a(a+\delta/2)}$. Substitution yields:

$$H_x^x(x_{max}) = \pm\frac{2M_r}{\pi}\tan^{-1}\frac{\delta}{4\sqrt{a(a+\delta/2)}}$$

(4.12)

$$M(x_{max}) = \pm\frac{2M_r}{\pi}\tan^{-1}\sqrt{1+\delta/(2a)}$$

The field gradients are obtained from a direct utilization of (2.22). With (4.6) an integral for the field that yields (4.8) is:

$$H_x^x(x,y) = -\frac{M_r}{\pi}\int_{-\delta/2}^{\delta/2} dy' \frac{x-x_0}{(x-x_0)^2 + (a+|y-y'|)^2}$$

(4.13)

If the transition length a varies with depth in the medium then (4.13) holds with $a \to a(y')$ (Problem 4.4). The field gradient at the transition center ($x = 0$, $|y| < \delta/2$) from (4.13) is:

$$\frac{\mathrm{d}H_x^x}{\mathrm{d}x}(0, y) = -\frac{M_r}{\pi} \int_{-\delta/2}^{\delta/2} \frac{\mathrm{d}y'}{(a(y') + |y - y'|)^2} \qquad (4.14)$$

for a general varying transition length. For constant transition length (4.14) yields:

$$\frac{\mathrm{d}H_x^x}{\mathrm{d}x}(0, y) = -\frac{M_r}{\pi}\left(\frac{2}{a} - \frac{1}{a + y + \delta/2} - \frac{1}{a - y + \delta/2}\right) \qquad (4.15)$$

and at the medium center is:

$$\frac{\mathrm{d}H_x^x}{\mathrm{d}x}(0, 0) = -\frac{M_r\delta}{\pi a(a + \delta/2)} \qquad (4.16)$$

Note that the field maxima (4.12) as well as the gradients increase with decreasing transition length.

It is useful to give a scaled expression for the field gradient that is not specific to the transition shape. Let the transition be denoted by:

$$M(x) = M_r f(x/a) \qquad (4.17)$$

where f can be any odd function that saturates as $x \to \pm\infty$ to $\pm M_r$ and can be written in terms of the scaled length x/a (e.g. arctangent, error function, tanh). Equation (2.22) may be written in scaled form as:

$$H_y^x(x/a, y/a) = -\frac{M_r}{2\pi} \int_{-\infty}^{\infty} \mathrm{d}s \, \frac{\mathrm{d}f(s)}{\mathrm{d}s} \int_{-\delta/2a}^{\delta/2a} \mathrm{d}t' \frac{x/a - s}{(x/a - s)^2 + (y/a - t')^2} \qquad (4.18)$$

For thin media $\delta \ll 2a$ (4.18) simplifies to give a scaled demagnetizing field gradient at the transition center:

$$\frac{\mathrm{d}H_x^x}{\mathrm{d}x}(0, 0) = -\frac{M_r\delta}{\pi a^2} \int_0^{\infty} \frac{\mathrm{d}s}{s} \frac{\mathrm{d}^2 f(s)}{\mathrm{d}s^2} \qquad (4.19)$$

The integral is simply a constant dependent on the shape as shown in Table 4.1 on page 100.

In Table 4.1 each transition shape is normalized so that the transition slope at the origin is fixed: $\mathrm{d}M/\mathrm{d}x = 2M_r/\pi a$. The demagnetizing field gradient (4.19) for thin media scales as the magnetization flux, $M_r\delta$, and inversely as the square of the transition length and depends slightly on the specific shape.

Table 4.1.

Transition shape	Integral
$f(s) = \dfrac{2}{\pi}\tan^{-1} s$	$\displaystyle\int_0^\infty \dfrac{\mathrm{d}s}{s}\dfrac{\mathrm{d}^2 f(s)}{\mathrm{d}s^2} = 1$
$f(s) = \tanh 2s/\pi$	~ 0.691
$f(s) = \dfrac{2}{\sqrt{\pi}}\displaystyle\int_0^{-s/\sqrt{\pi}} \mathrm{d}t\, \mathrm{e}^{-t^2}$	$= 2/\pi \sim 0.637$

A useful form for the magnetostatic field for the case of recording medium geometry may be derived from (2.22). For media that are bounded by flat surfaces defined by vertical location y (i.e. the medium exists over $-\infty \le x \le \infty$ and $-\delta/2 \le y \le \delta/2$):

$$H(x,y) = \frac{1}{2\pi}\int_{-\delta/2}^{\delta/2} \mathrm{d}y' \int_{-\infty}^{\infty} \mathrm{d}x' \left(M_y(\boldsymbol{r}') \frac{\partial}{\partial y'} - \left(\frac{\partial}{\partial x'} M_x(\boldsymbol{r}') \right) \right) \frac{(\boldsymbol{r} - \boldsymbol{r}')}{|\boldsymbol{r} - \boldsymbol{r}'|^2}$$

(4.20)

Fourier transforms

A general expression for the Fourier transform of the fields due to magnetized recording media can be obtained from (4.20) by separately transforming the convolved terms. Note that since the first term is an operator expression containing $\frac{\partial}{\partial y'}$, the order must be maintained and the operation performed at the end. Using contour integration, it may readily be shown that

$$\int_{-\infty}^{\infty} \mathrm{d}x\, \mathrm{e}^{-ikx} \frac{(\boldsymbol{r} - \boldsymbol{r}')}{|\boldsymbol{r} - \boldsymbol{r}'|^2} = \pi \mathrm{e}^{-ikx' - |k||y - y'|}(-i\,\mathrm{sgn}(k), \mathrm{sgn}(y - y')) \quad (4.21)$$

where the final bracket denotes the vector components for x and y directions in that order. It may also be noted that

$$\frac{\partial\, \mathrm{sgn}(y - y')}{\partial y'} = -2\delta(y - y') \qquad (4.22)$$

where $\delta(y)$ is the Kronecker delta function. Thus, the complete expression for the Fourier transform of two-dimensional fields in a planar recording medium is:

$$H(k,y) = -\frac{|k|}{2}\int_{-\delta/2}^{\delta/2} dy'(M_x(k,y')$$

$$+ i\,\text{sgn}(k)\text{sgn}(y-y')M_y(k,y'))e^{-|k||y-y'|}(1,i\,\text{sgn}(k)\text{sgn}(y-y'))$$

$$- M_y(k,y)(0,1) \quad (\text{for } y \text{ in medium only})$$

$$(4.23)$$

$M_x(k,y)$, $M_y(k,y)$ are the Fourier transforms of the x and y components of the magnetization, respectively. The second constant term involving $M_y(k,y)$ occurs only when the field is evaluated in the tape. Note that outside the medium (4.23) yields field components that are Hilbert transforms of each other as was shown in Chapter 2 for fields exterior to one side of magnetic sources. The Fourier components are simply shifted in phase by $\pi/2$ depending on the sign of k.

Utilizing (4.23), general phase conditions can be easily stated. For a first example, consider an arbitrary spatial variation of magnetization, but with the direction of the net magnetization invariant. Thus, in general

$$M_x(k,y) = Mf(k,y)\cos\theta$$

$$(4.24)$$

$$M_y(k,y) = Mf(k,y)\sin\theta$$

The angle θ is the angle that the magnetization is oriented away from the longitudinal direction ($+ \theta$ denotes orientation towards positive y). With (4.24), (4.23) becomes:

$$H(k,y) = -\frac{|k|M}{2}\int_{-\delta/2}^{\delta/2} dy'f(k,y')e^{-|k||y-y'|}e^{i\,\text{sgn}(k)\text{sgn}(y-y')\theta}$$
$$\cdot (1,i\,\text{sgn}(k)\text{sgn}(y-y'))$$

$$(4.25)$$

$$- Mf(k,y)\sin\theta(0,1) \quad \text{for } y \text{ in medium only}$$

In the evaluation of this expression note that

$$\sin(\text{sgn}(k)\text{sgn}(y-y')\theta) = \text{sgn}(k)\text{sgn}(y-y')\sin\theta \qquad (4.26)$$

since the sine function is odd. Thus, for fields outside the tape where the extra term is not included, the effect of rotating the magnetization direction simply yields a rotation of phase of the Fourier transform of the field. The sign of the phase rotation depends on the sign of k as well as the side of the medium where the field is being evaluated ($\text{sgn}(y - y') = 1$ for $y > \delta/2$ and -1 for $y < -\delta/2$). It can be shown that for each Fourier component, a rotation of the magnetization direction yields an equal counter rotation in the direction of the field external to the medium (Mallinson, 1973). For example, if the magnetization is purely long-itudinal ($\theta = 0$), a certain vector field pattern will occur. If this same magnetization pattern is now vertically oriented ($\theta = \pi/2$), (4.25) predicts that the new field is a Hilbert transform of the previous field. Since, external to the medium the field components of any given vector field are a Hilbert transform pair, the new longitudinal field is identical to the old vertical field before rotation. The new vertical field is identical to the negative of the old longitudinal field. This effect can be seen by comparing Figs. 4.1(c) and 4.2(c).

Consider next the case of magnetization patterns that are Hilbert transforms of each other at each distance y into the medium:

$$M_x(x, y) = Mf(x, y)$$
$$M_y(k, y) = \pm i \, \text{sgn}(k) M_x(k, y) \tag{4.27}$$

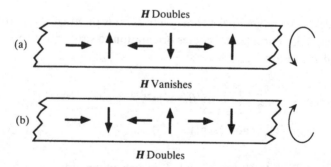

Fig. 4.7. (a) Schematic of counterclockwise rotating magnetization harmonic components that yield a doubling of the magnetic field above the medium and a vanishing of the field below. (b) Schematic of clockwise rotating magnetization harmonic components that yield a doubling of the magnetic field below the medium and a vanishing of the field above.

Each Fourier component corresponds to a spatial pattern of rotating magnetization. The patterns are illustrated in Fig. 4.7(a), (b). The plus sign yields a clockwise rotating magnetization with respect to the viewer, whereas the negative sign yields a counterclockwise rotation. The field is given in this case by:

$$H(k,y) = -\frac{|k|M}{2} \int_{-\delta/2}^{\delta/2} dy' f(k,y')(1 \pm \mathrm{sgn}(y-y'))e^{-|k||y-y'|}$$

$$\cdot (1, i\, \mathrm{sgn}(k)\mathrm{sgn}(y-y'))$$

(4.28)

$$\pm i\, \mathrm{sgn}(k) M f(k,y)(0,1) \text{ (for } y \text{ in medium only)}$$

Choice of plus $(+)$ or minus $(-)$ in evaluating this expression corresponds to the case of counterclockwise $(-i)$ or clockwise $(+i)$ magnetization in (4.27). Consider first the field exterior to the medium viewed from below ($y < -\delta/2$). In that case $\mathrm{sgn}(y-y') = -1$ for all y' in the integration. Thus, the exterior field *vanishes* for the case of a counterclockwise rotating pattern. For a clockwise rotating pattern the field is *double* that due to a single component of magnetization. On the top side of the medium the reverse is true and a doubling of the field for a counterclockwise rotating pattern and a vanishing of the field from the clockwise pattern results. If the viewer is on the top side (for example, standing on the surface of a magnetic head flying above the medium) the same conclusions hold: a counterclockwise rotating pattern viewed from one side becomes a clockwise pattern when viewed from the other, and the converse also applies. 'One-sided fluxes' have been discussed in depth (Mallinson, 1973).

Inside the medium the conclusions are slightly different and are offset by the constant term added to the vertical field component. If only the integral term of (4.28) is considered (which corresponds precisely to the longitudinal field component), then a simple rule can be stated. For a clockwise rotating pattern, at any depth y in the medium, the field arises only from the medium magnetization above it ($y' > y$). For a counterclockwise rotation pattern the field arises only from the magnetization below ($y' < y$). Thus, just inside the medium at the bottom surface for a counterclockwise rotating pattern, the longitudinal field vanishes and the vertical field equals the vertical magnetization component at that surface.

A final example is the case of magnetization that does not vary with depth into the medium. In this case the demagnetization field averaged through the depth of the medium is evaluated. Equation (4.23) is easily integrated through the medium. By symmetry it is seen that the resulting demagnetization tensor is diagonal: each component containing a single $\text{sgn}(y - y')$ term vanishes because the term is odd with respect to the medium center. These diagonal terms are:

$$N_{xx}(k) = -\frac{H_x^x(k)}{M_x(k)} = 1 - \frac{1 - e^{-k\delta}}{k\delta}$$

$$N_{yy}(k) = -\frac{H_y^y(k)}{M_y(k)} = \frac{1 - e^{-k\delta}}{k\delta}$$

(4.29)

These demagnetization factors are now wavenumber dependent but still follow the general rule that the diagonal terms sum to unity. Both terms in (4.29) are plotted in Fig. 4.8. The general wavelength dependence follows from viewing the field plots in Figs. 4.6(a), (b). For longitudinal magnetization patterns the demagnetization field is localized near the transition; thus the Fourier transform is expected to contain only high frequency components. Conversely, for a vertical magnetization pattern the field is maximum (perhaps infinitely far) away from a transition and vanishes at the transition. In this case at long wavelengths the average demagnetization field becomes equal to the magnetization and vanishes at very short wavelengths.

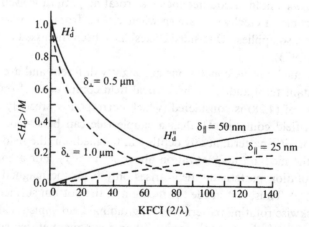

Fig. 4.8. Demagnetizing tensor components for longitudinal and perpendicular magnetization for different medium thicknesses plotted versus sinewave density $(2/\lambda)$.

Note that for an arctangent transition (4.6), the field transforms are easily evaluated since:

$$M(k) = \frac{2M_r}{ik}\, e^{-|k|a} \tag{4.30}$$

Even if the transition length varies with depth into the medium (4.23) can be easily evaluated. When a is constant, equations (4.23) and (4.30) yield fields exterior to the recording media that depend only on the sum $d + a$ or spacing plus transition length (Problem 4.3). All fields exterior to the media with recorded transitions of arctangent shape of any orientation, whether harmonic components or not, are given by expressions for the fields from perfectly sharp transitions with increased effective spacing $d + a$.

Problems

Problem 4.1 Show that the volume charge for a sharp transition of longitudinal magnetization is equivalent to a surface charge at the transition center of $\sigma = 2M$. Using (2.26) directly, show that the longitudinal field at either side of the transition ($|x| \to x_0$) at the surfaces ($y = \pm\delta/2$) away from the medium surfaces is $H_x(x_0) = \pm M/2$.

Problem 4.2 Utilize (2.8) to calculate the field components $H(x, y, z)$ for a thin medium of longitudinal magnetization or a thick medium of vertical magnetization. Assume that the magnetization is uniform over a finite width W and the transition shape is a linear ramp function ($M(x) = 2M_r x/\pi a$ in the transition region: $-\pi a/2 < x < \pi a/2$). A partial result is:

$$H_y(x, y, z) = -\frac{2M_r}{\pi a}\left((x - x')(\ln(R - (z - z')) - (x - x')\right.$$

$$+ (y - y')\tan^{-1}\left(\frac{x - x'}{y - y'}\right) + (y - y')\tan^{-1}\left(\frac{(x - x')(z - z')}{(y - y')R}\right)$$

$$\left.+ (z - z')\ln(R - (x - x'))\right)\left.\begin{array}{c}\pi a/2 \\[1mm] | \\ x'=-\pi a/2\end{array}\right.\left.\begin{array}{c}\delta/2 \\[1mm] | \\ y'=-\delta/2\end{array}\right.\left.\begin{array}{c}W/2 \\[1mm] | \\ z'=-W/2\end{array}\right.$$

$$H_z(x, y, z) = H_y(x, z, y)$$

where

$$R = \sqrt{(x - x')^2 + (y - y')^2 + (z - z')^2}$$

Problem 4.3 Show that all fields exterior to a medium with a recorded transition of arctangent shape, either longitudinal or perpendicular, are exactly the same as the fields for a perfectly sharp transition, except that the distance from the medium $(d = |y| - \delta/2)$ can be replaced by $d + a$. Use direct substitution into (4.1) and (4.2), but also use the general Fourier transform result (4.23) and (4.30).

Problem 4.4 Show that (4.13) holds for a transition whose length varies with depth into the medium: $a \to a(y')$ in (4.6). Hint: use Fourier transform analysis of (4.23).

Problem 4.5 Use (4.13) to determine the demagnetizing field gradient at the transition center for a medium with a longitudinal arctangent transition separated a distance d from an infinite planar surface demarking a semi-infinite region of infinite permeability.

Problem 4.6 Use (2.26) to give physical interpretations of the variation of the demagnetizing field versus medium thickness for longitudinal and perpendicular magnetizations as shown in Fig. 4.6(a), (b). Repeat for the harmonic components of Fig. 4.8.

Problem 4.7 Rederive (4.29) for the case of a recording medium adjacent to an infinite high-permeability keeper.

Problem 4.8 Utilizing (2.3), show that the total flux passing through a closed volume in free space equals the total magnetic charge (divided by μ_0) inside that volume. Consider an infinitely wide recording medium with a longitudinal transition of any shape, any depth variation of transition parameter or center location, and with saturation levels $\pm M_r$ at all depths. Show that the flux per unit track width leaving one side of the medium is given by:

$$\int_{-\infty}^{\infty} dx' H_y(x', y) = M_r \delta$$

If a semi-infinite keeper of infinite permeability is placed on one side of the medium (e.g. Fig. 4.4), show that the above integral doubles on the keeper side and vanishes on the opposite side.

5

Playback process:
Part 1 – General concepts and single transitions

Introduction

This chapter presents the formalism associated with the calculation of playback voltages. Expressions for both real time waveforms, such as isolated pulses, as well as spectra will be derived. The playback process involves low flux levels in the playback head; thus, linear system theory may be utilized to relate a recorded magnetization pattern to the reproduce voltage at the head terminals. The chapter begins with a simple example of the waveform obtained by direct calculation of the playback flux. However, it is generally much more convenient to utilize the formalism of reciprocity. The principle of reciprocity states that the playback flux at any instant is equal to a correlation of the recorded magnetization and the field per unit current of an energized playback head. This principle will be derived and the conditions for its usage discussed in detail. Following that, general playback formulas will be given and specific examples will be discussed for both longitudinal and vertical recording. In this chapter the playback of isolated pulses will be treated. The effects of pulse superposition, 'linear superposition', for both the 'roll-off curve' as well as linear bit shift will be analyzed in Chapter 6. The discussion will focus on playback by an inductive head. However, since reciprocity may be adapted to magnetoresistive (MR) playback heads, the results presented here apply only with slight modification. Reciprocity as applied to MR heads is discussed in Chapter 7.

In analysis of the recording process the head can be placed either above (e.g. Fig. 5.1) or below (e.g. Fig. 5.3) the medium. In the former case the medium moves to the left (or the head to the right) and in the latter the medium moves to the right. The head-below position will be utilized

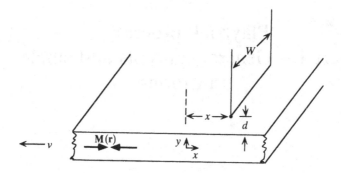

Fig. 5.1.   Sketch of a recorded medium with a single wire loop for a replay head. The wire extends a distance W in the cross-track direction. The bottom portion of the loop is a distance d from the medium surface and closes infinitely far above the medium. $x = vt$ is the distance from the coordinate system fixed in the medium to the loop plane.

primarily in this and subsequent chapters. For playback the head will sense recorded magnetizations from left to right during relative motion.

Direct calculation of playback voltage

A magnetized medium produces magnetic fields due to divergence in the magnetization. Examples were presented in Chapter 4. In the presence of a high-permeability playback head these fields are dramatically altered compared with the fringing field from the same magnetization pattern without a head present (Fig. 4.3 and Fig. 4.4). The playback head is designed to be an efficient flux collector by use of proper geometric design and utilization of core material of high permeability. The flux produced by the medium is concentrated in the core of the head and passes through the head windings. The flux in the windings is given in terms of the induction $\boldsymbol{B}$:

$$\phi \equiv \int_s \int \boldsymbol{B}(\boldsymbol{r}') \cdot \hat{\boldsymbol{n}} \mathrm{d}^2 r' \qquad (2.34)$$

where the area of integration denoted by s includes all the loops comprising the windings. The playback flux is a function of the relative position of the recorded medium and the head. An alternate form is simply to integrate the vector potential A along the total length of the coil:

$$\phi = \oint A(r') \cdot dl' \tag{2.35}$$

For a head structure that is very wide and where two-dimensional analysis applies, the only significant component of the vector potential is in the cross-track direction (A_z). In that case (2.35) is proportional to the track width times the sum over the wires of the difference in potential between the upper and lower segments of each wire (e.g. Fig. 3.4):

$$\phi = W \sum_{i=1}^{N} \Delta A_z^i \tag{5.1}$$

The reproduce voltage of an inductive head is given by

$$V(t) = -\frac{d\phi(t)}{dt} \tag{5.2}$$

which results from an enclosed coil surface integration of Maxwell's equation relating the electric field E and the flux density B:

$$\nabla \times E = -\frac{dB}{dt} \tag{5.3}$$

Utilization of (2.34) and the definition that the voltage between any two points is the line integral (here taken to be the length of the coil) of the electric field yields (5.3). In (5.2) the flux is expressed as a function of time since the relative position of the magnetization pattern with respect to the playback head is continually changing. This position changes since the medium is moving with respect to the head. In magnetic recording the relative motion is at a fixed velocity, apart from undesired speed fluctuations. Only constant motion will be considered here. Thus, (5.2) may be rewritten as

$$V(x) = -v\frac{d\phi(x)}{dx} \tag{5.4}$$

where x denotes the relative position of the head with respect to the medium and v is the relative velocity. For constant speed recording the magnetic recording process is essentially a spatial process. Time enters only through the conversion $x = vt$ and the frequency dependence of the permeability of the head. Head effects will be characterized by a constant efficiency so that the discussion will involve only spatial variables. Conversion back to time variations is therefore immediate.

The majority of the calculations are for direct on-track signal response; except in Problem 5.9 head or recorded track-edge effects will not be included. To good approximation, 2D analysis applies so that all fields have the two-dimensional forms discussed in Chapters 3 and 4. The approach taken here is to assume an infinitely wide recording and finite width playback transducer in this introductory section. Thus, the playback voltage will be 'per unit cross track distance' (as is (5.1)). If the reproduce structure has a width W, then the playback voltage can be written as:

$$V(x) = -Wv \, \frac{d\phi_w^{tot}(x)}{dx} \qquad (5.5)$$

where ϕ_w^{tot} is the total flux per unit track width threading all the turns of the playback coil. If the windings are tightly packed, or if the flux passes along the core between turn locations without fringing loss, the voltage is proportional to the number of turns:

$$V(x) = -NWv \, \frac{d\phi_w(x)}{dx} \qquad (5.6)$$

In (5.6) ϕ_w represents the flux per turn or simply the common flux in the core threading the windings. For complicated structures, such as thin film heads, the flux threading each turn decreases for turns further removed from the gap region due to flux leakage and (5.5) must be utilized (Kelley & Valstyn, 1980; Shelor, 1986; Hughes, 1983b).

The playback voltage by a single infinitesimally thin wire as illustrated in Fig. 5.1 is derived. The wire loop is rectangular of width W, separated from the surface of the medium a distance d, and extends infinitely far from the medium. The relative location of the wire to a fixed point in the medium is given by x. There are several equivalent ways to calculate $\phi_w(x)$. The flux can be calculated directly (2.34) by integrating the longitudinal or x component of the field from the medium over the depth of the wire:

$$\phi_w(x) = \mu_0 \int_{d+\delta/2}^{\infty} dy' H_x(x, y') \qquad (5.7)$$

Alternatively, the flux can also be computed by determining the net flux entering the horizontal surface to the right of the wire and leaving the horizontal surface at the left:

$$\phi_w(x) = \frac{\mu_0}{2} \left(\int_{-\infty}^{x} dx' H_y(x', d + \delta/2) - \int_{x}^{\infty} dx' H_y(x', d + \delta/2) \right) \quad (5.8)$$

Equivalently, (2.35) or (5.1) for a 2D analysis can be utilized:

$$\phi_w(x) = A_z(x, d + \delta/2) \quad (5.9)$$

The term for the wire closure is not included since the vector potential vanishes at infinity. For a finite loop the vector potential evaluated at the other part of the wire loop parallel to the z axis would be included (subtracted). For the wire sections extending away from the medium parallel to the y axis, there is no contribution to (2.35) since the vector potential is orthogonal to this part of the coil. Since, by definition of the vector potential (2.32):

$$B_y = -\frac{\partial A_z}{\partial x} \quad (5.10)$$

the playback voltage can be written as:

$$V(x) = W v \mu_0 H_y(x, d + \delta/2) \quad (5.11)$$

Thus, for playback with an infinitesimally thin wire, the reproduce voltage is proportional to the vertical field from the medium evaluated at the wire.

Voltage waveforms for specific examples will be discussed in detail following the review of reciprocity. However, for the case of a single transition, the general shape of the isolated voltage pulse can be seen immediately from the fields due to recorded transitions. For the case of longitudinal magnetization the vertical field has the 'bell' shape shown in Fig. 4.1(c). For perpendicular magnetization the vertical field outside the medium has the antisymmetric shape indicated in Fig. 4.2(c) (dotted). For (arctangent) transitions of finite length, as discussed in Problem 4.3, the distance to the wire is the actual spacing plus the transition length; the general pulse shapes do not change. These characteristic voltage pulse shapes arise also through the use of the reciprocity principle, however the interpretation is different.

Consider now a simplified (2D) reproduce structure. The head is taken to be flat, with infinitely permeable material of width W, a vanishingly small gap and infinitely long pole pieces (Fig. 5.2). It is assumed that the head efficiency is unity so that all the flux entering the top surface eventually threads the N windings. It is natural to utilize

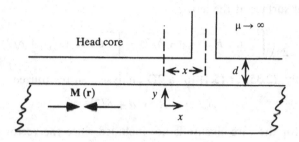

Fig. 5.2.   2D sketch of a high-permeability semi-infinite head core with a small gap above a recorded medium.

(5.8) to calculate the flux and thus the reproduce voltage. Utilizing simple imaging (Chapter 2) to account for the presence of the infinite, high-permeability surface, the effect is simply to double the vertical field at the head surface. Utilizing (5.6), an expression for the playback voltage is obtained that is identical in form to that for a single wire but of twice the magnitude:

$$V(x) = 2NWv\mu_0 H_y(x, d + \delta/2) \qquad (5.12)$$

This type of head can be thought of as a flux gathering device of two semi-infinite blocks placed on either side of the infinitesimally thin wire in Fig. 5.1. As illustrated in Fig. 4.3, the flux leaving a recorded pattern fringes both above and below the recorded medium. The effect of the high-permeability surfaces doubles the flux (Fig. 4.4), since all the flux entering the region opposite the head eventually returns through the core to the flux source.

A simple approximation that accounts for a finite gap (following the Karlqvist approximation discussed in Chapter 3) can be achieved simply by averaging (5.11) over a uniform distribution of wires or head centers of width g. However, for a realistic head with a high-permeability core, and finite gap, length and width, a determination of the fields from the recorded medium that exist in the head is extremely difficult and must be repeated for each position (time) of the head and medium.

The reciprocity principle

Reciprocity is an extremely useful tool for the computation of readback voltages. Since the playback flux can be expressed as a correlation of the recorded magnetization with the field from the playback head, the complicated problem of solving for the fields due to the magnetization in the presence of a head structure with a complicated shape and a high-permeability core is removed. For common head shapes the fields have been given in Chapter 3. Thus, with this principle, the playback response from general magnetization patterns can be quickly analyzed. Even if the head structure is not known, the playback voltage can be written as a correlation of the field at the surface of the head. For spectral analysis, the voltage then simply contains the Fourier transform of the surface head field only as a simple product.

A complication arises in the use of reciprocity in assigning specific regions to be the 'head' or the 'medium'. Consider a medium with permanent recorded magnetization as well as reversible permeability (Fig. 5.3). In addition, there can be regions of high permeability, such as keepers utilized in perpendicular recording and, of course, the permeable head core. The recorded magnetization magnetizes reversibly all regions with a reversible susceptibility: the head, keeper and the medium itself, depending on the relative position of the medium and the head. Further, the head field per unit current is utilized in the reciprocity calculation.

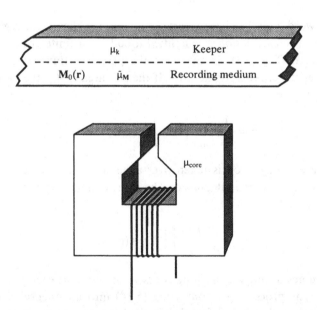

Fig. 5.3. Illustration of a complex structure of an inductive head, a permeable keeper, and a recording medium with finite permeability.

The head field will also magnetize all regions with a reversible permeability. Is the head chosen to be the head itself, or the head plus the keeper, or the head plus the keeper plus the reversible part of the medium, or perhaps only the wires? For each possible choice of head definition, what is the permissible meaning of the term 'medium magnetization'?

In this section a derivation of reciprocity will be given with specific attention to the head and medium definitions. There have been many calculations of the principle of reciprocity e.g. (Fan, 1961; Wessel-Berg & Bertram, 1978; Smith, 1987a). The analysis of Smith will be followed here. The calculation of Wessel-Berg & Bertram can be examined to understand how a direct calculation of the fields in a head can be transformed, via Green's identities, to an integration over the magnetization in the medium. The use of reciprocity for problems with regions containing a finite conductivity will not be discussed; the reader is referred to (Smith, 1987a) for that situation.

The analysis begins with the expression for the playback flux in terms of the vector potential evaluated at the coil:

$$\phi = \oint_{coil} A(r') \cdot dl' \tag{2.35}$$

where the word 'coil' has been inserted to denote explicitly the region of integration. The calculation is simplified if coils of a finite cross-section are considered (Fig. 5.4). In that case the playback flux becomes the average over the cross-sectional area. If the coil area is A_w, then the flux can be written as:

$$\Phi = \frac{1}{A_w} \int_{area} \int d^2r' \oint_{coil} A(r') \cdot dl' \tag{5.13}$$

Since the area integration is taken over a plane normal to dl' at each point along the wire, the integrals can be expressed as a volume integral over the wire:

$$\Phi = \frac{1}{A_w} \int_{wire} \int \int d^3r' A(r') \cdot \hat{l} \tag{5.14}$$

where $\hat{l}$ denotes a unit vector along the wire direction at each point.

The analysis proceeds by converting (5.14) into an integral over all space. To accomplish this a function is needed that is unity in the wire and that vanishes elsewhere. One method is to utilize a fictitious current

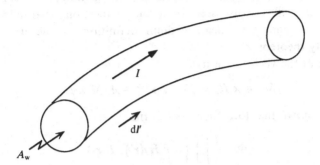

Fig. 5.4. Sketch of fictitious wire with current I for the analysis of reciprocity. A_w is the wire cross-section diameter, dl' is the vector differential integration element along the wire in (5.13). $\hat{l}$ is a unit vector along the wire or current direction in (5.14).

density. Let a current be applied to the wires, not in reality, but as a mathematical device. If the current is I, then the current density is:

$$J(r') = \frac{I\hat{l}}{A_w} \tag{5.15}$$

$J(r')$ is written as a vector function of position since, although constant in magnitude, it vanishes for all r' not in the wire. Thus, (5.14) may be written as:

$$\Phi = \frac{1}{I} \int\!\!\!\int_{\infty}\!\!\!\int d^3r' A(r') \cdot J(r') \tag{5.16}$$

where the integral is now taken over all space (∞), but of course, non-zero contributions occur only in the wire.

For time-invariant fields the current density may be written as the curl of a magnetic field (2.2):

$$\nabla \times H_J(r') = J(r') \tag{2.2}$$

Substituting (2.2) into (5.16) yields:

$$\Phi = \frac{1}{I} \int\!\!\!\int_{\infty}\!\!\!\int d^3r' A(r') \cdot \nabla \times H_J(r) \tag{5.17}$$

The field utilized in (5.17) can be *any* field with the sole requirement that its curl satisfies (2.12). Thus, the field can also include a variety of curl-free sources, such as those due to certain magnetization patterns. It is this fact, which will be discussed in detail further on, that allows for considerable choice of head–medium definition in the use of the reciprocity theorem.

By use of the vector identity:

$$\nabla \cdot A \times H_J = H_J \cdot \nabla \times A - A \cdot \nabla \times H_J \qquad (5.18)$$

it can be shown that (5.17) may be rewritten as:

$$\Phi = \frac{1}{I} \int\int_\infty \int d^3 r' H_J(r') \cdot B(r') \qquad (5.19)$$

where $B(r)$ derives from the vector potential via (2.32) and represents fields due to the magnetized recording medium. The volume integral applied to the left hand side of (5.18) may be transformed into a surface integral that vanishes at infinity. Note that H_J refers to fields that derive from the fictitious current density J.

Using the constitutive relations between B, H, and M (1.5), the integrand of (5.19) can be written as:

$$\begin{aligned} H_J \cdot B &= \mu_0 H_J \cdot (H + M) \\ &= H \cdot B_J + \mu_0 (H_J \cdot M - H \cdot M_J) \end{aligned} \qquad (5.20)$$

Thus, the flux may be expressed as:

$$\Phi(x) = \frac{1}{I} \int\int_\infty \int d^3 r' H(r') \cdot B(r', x)$$

$$\qquad (5.21)$$

$$+ \frac{\mu_0}{I} \int\int_\infty \int dr (H_J(r', x) \cdot M(r') - H(r') \cdot M_J(r', x))$$

In (5.21) the fields $H(r')$, $M(r')$, $B(r')$ all arise from the permanent magnetization recorded in the medium. $H(r')$, $B(r')$ occur everywhere in space. $M(r')$ occurs *not only* in the recording medium, but everywhere there is a non-vanishing reversible susceptibility. The permanent magnetization recorded in the medium induces fields and magnetizations in all regions with a reversible susceptibility. These fields are written in the convention taken here of a fixed coordinate system in the moving medium.

The fields $H_J(r', x)$, $M_J(r', x)$, $B_J(r', x)$ all derive from the condition (2.2). Magnetizations occur since a (fictitious) current applied to the wire will magnetize *all* regions containing a reversible suceptibility, and thereby alter $H_J(r', x)$, $B_J(r', x)$ from that due to the wires alone. The variable x represents the relative position of the head with respect to the medium and is included since these fictitious fields arise from the head. In fact the fields induced by the medium should also contain the variable x since the effect of head permeability on these fields also depends on the relative head–medium position. However, it is easy to include x later precisely where needed. The playback flux is now written as a function of relative head–medium position as well.

Dynamic effects, due to the presence of conductive materials such as an aluminum substrate for rigid-disc recording, are included in the first term of (5.21) (Smith, 1987a). The field $H(r')$ can contain eddy-current induced fields. If such effects are neglected, then the first term of (5.21) vanishes: the volume integral of the scalar product of a curl-free vector and a divergence-free vector always vanishes if taken over all space.

Thus, the reproduce flux neglecting conductivity simplifies to:

$$\Phi(x) = \frac{\mu_0}{I} \int\int_{\infty}\int d^3r' (H_J(r', x) \cdot M(r') - H(r') \cdot M_J(r', x)) \qquad (5.22)$$

Various cases for the evaluation of (5.22) according to the freedom of choice for the fields $H_J(r', x)$, $M_J(r', x)$, $B_J(r', x)$ are now discussed. All these cases produce fields that satisfy (2.2) and that are categorized by the term 'definition of the head.' The medium is always the permanent recorded magnetization plus all induced reversible magnetizations and fields with the recorded magnetization as field source.

Head definitions

Case 1: Wires only. In this case $M_J(r', x)$ does not exist and (5.22) reduces to:

$$\Phi(x) = \frac{\mu_0}{I} \int\int_M\int d^3r' H_J(r', x) \cdot M(r') \qquad (5.23)$$

This form is simple but involves an integration over all regions where magnetization occurs due to the recorded medium (the M notation under the integral sign). The field $H_J(r', x)$ is simply evaluated using (2.7); however, it is very complicated to determine $M(r', x)$ (for example in the head core). In fact, it is probably equally as difficult to evaluate the flux utilizing (5.23) as to evaluate it by a direct calculation from (2.34).

Case 2: Wires plus all permeable regions.   In this case (5.22) simplifies considerably. The 'head' consists of the wires, the head core, the permeable portion of the recording medium, and any other region with a permeability such as a keeper behind the head (Fig. 5.3). Consider first any region containing a reversible susceptibility χ. The reversible magnetization due to the medium source is:

$$M(r') = \chi H(r') \qquad (5.24)$$

and that due to the wire source is:

$$M_J(r', x) = \chi H_J(r', x) \qquad (5.25)$$

Note that the total magnetization in the recording medium is given by the sum of the permanent recorded magnetization and that due to the reversible susceptibility:

$$M_{\text{medium}}(r') = M_0(r') + \chi H(r') \qquad (5.26)$$

Substitution of these terms into (5.22) causes the integral to vanish for all regions except that of the recorded permanent magnetization. The only remaining contribution to the flux is that due to the permanent magnetization in the recording medium $M_0(r')$. Thus, the flux may be written as:

$$\Phi(x) = \frac{\mu_0}{I} \iiint_{\text{medium}} d^3r' H_J(r', x) \cdot M_0(r') \qquad (5.27)$$

The integration is now solely over the recording medium. The field $H_J(r', x)$ is evaluated with the source current density and the reversible permeabilities of all material incorporated into the head definition. Neglecting permeable media or regions that are separate from the core, (5.27) may be utilized immediately with the fields from Chapter 3.

Case 3: Wires plus permeable head core.   Here the head is considered to be the main core only, and all keepers and other regions of reversible permeability are associated with the medium. Reversible terms in (5.22) cancel, as in case 2. However, here the reversible terms occur in the head core only: $M_J(r', x)$ vanishes, by definition, in the recording medium and other permeable regions. Thus (5.22) reduces to:

$$\Phi(x) = \frac{\mu_0}{I} \int\limits_{\text{medium, keeper}} \int \int d^3 r' \boldsymbol{H}_J(\boldsymbol{r}', x) \cdot \boldsymbol{M}(\boldsymbol{r}') \tag{5.28}$$

In (5.28) the integration is over the medium and other permeable regions (denoted by 'keeper') exclusive of the head. The head field $\boldsymbol{H}_J(\boldsymbol{r}', x)$ is determined in the usual way, as in Chapter 3, exclusive of medium and keeper permeability effects. The reversible components of the magnetization due to the medium source magnetization are evaluated in the medium and the keeper in order to obtain $\boldsymbol{M}(\boldsymbol{r})$; however, *all* permeable material, including the head core, is utilized to determine the associated magnetostatic fields or scalar potentials in the evaluation of $\boldsymbol{M}(\boldsymbol{r})$.

There are, of course, other head definitions that can be employed. It is a matter of personal choice which form to utilize for any given circumstance.

Summary comments

1. The fields and magnetizations due to the fictitious wire current density are to be evaluated with a sufficiently small current so that the head is in its linear regime.
2. If the magnetic material possesses a reversible tensor susceptibility, (5.22) must be modified accordingly (e.g. (Wessel-Berg & Bertram, 1978)).
3. For any fixed instant during head–medium relative motion, reciprocity represents a space integration or correlation of the product of fields and magnetizations (Mallinson & Minuhin, 1984).

Case 2, which gives the most inclusive head definition, is frequently employed, for example in the analysis of 'pole-keeper' heads for perpendicular recording (Tagawa, *et al.*, 1992) and in the analysis of recording media with an anisotropic permeability (Bertram, 1978).

Generalized playback voltage expressions

The reciprocity formalism derived in the last section will be utilized here to develop expressions for the replay voltage and spectrum. Simple magnetization patterns will be assumed, but both longitudinal and perpendicular magnetization orientations will be considered. In the analysis presented here the form of the reciprocity relation given by case 2 (5.27) will be utilized. The integration is therefore only over the permanent recorded magnetization in the medium since the head comprises the wire and all permeable material. In virtually all the expressions analyzed, the only permeable material will be the head core. In that case (5.27) is simplified since the head expressions developed in Chapter 3 can be utilized directly. The case of perpendicular recording with a high permeable keeper layer will be treated in Chapter 6. Not only is this a case of technological importance but it provides an example of the use of reciprocity for a more generalized head definition.

In Chapter 3 all head field expressions were shown to scale naturally with the mmf across the gap: NIE. Thus, the fields in the replay expressions derived will be normalized by defining a lower case field:

$$h(r') \equiv \frac{H(r')}{NIE} \qquad (5.29)$$

Thus, the playback flux may be written, utilizing (5.27), as:

$$\Phi(x) = NE\mu_0 \iiint d^3r' h(r' + x) \cdot M(r') \qquad (5.30)$$

where the subscripts 'J' for the wire field, 'medium' denoting the region of integration, and '0' denoting the zero field medium remanent magnetization have been deleted. The explicit dependence on x is given referenced to a coordinate system fixed on the medium (Fig. 5.5). x is a vector in the head–medium motion direction defined to be from the gap center to the coordinate system origin. Thus, $r' + x$ can be written as

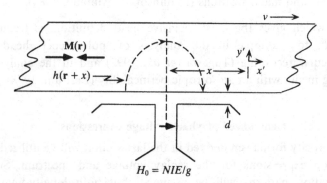

Fig. 5.5. Configuration for reciprocity integral in (5.30) illustrated for the head 'below' the medium. The coordinate system (x', y') is fixed in the medium and x is the distance along the motion direction from the coordinate origin to the head center.

$(x' + x, y', z')$. With this convention for x, as the medium moves by the head, x varies from $-\infty$ to $+\infty$ and time is related directly to x by $t = +x/v$.

For most cases where the track width is wide, a two-dimensional approximation suffices and the flux per unit track width can be utilized:

$$\Phi_w(x) = NE\mu_0 \int_{-\infty}^{\infty} dx' \int_{-\delta/2}^{\delta/2} dy h(x' + x, y') \cdot M(x', y') \qquad (5.31)$$

In (5.31) the recorded magnetization is assumed to be invariant across the track width (z' direction) and the fields have only a two-dimensional form. The Fourier transform of the flux is given immediately by (2.42):

$$\Phi_w(k) = NE\mu_0 \int_{-\delta/2}^{\delta/2} dy h^*(k, y') \cdot M(k, y') \qquad (5.32)$$

where $h^*(k, y')$ is the complex conjugate of the Fourier transform (on x) of the field and $M(k, y')$ is the transform of the recorded magnetization at each y' in the medium.

The voltage for a two-dimensional approximation is:

$$V(x) = -Wv \frac{d\Phi_w(x)}{dx} \qquad (5.5)$$

The Fourier transform relating the voltage spectrum to that of the flux is, therefore:

$$V(k) = -ikWv\Phi_w(k) \qquad (5.33)$$

The differentiating behavior of an inductive head causes the voltage spectrum to be rotated by '6dB per octave' as compared to the flux. Differentiation requires that the voltage spectrum must vanish at dc, but gives a high frequency boost. Incorporating (5.31) into (5.5) and (5.32) into (5.33) yields:

$$V(x) = NWEv\mu_0 \int_{-\infty}^{\infty} dx' \int_{-\delta/2}^{\delta/2} dy' h(x' + x, y') \cdot \frac{dM(x', y')}{dx'} \qquad (5.34)$$

for the voltage and:

$$V(k) = ikNWEv\mu_0 \int_{-\delta/2}^{\delta/2} dy' h^*(k, y') \cdot M(k, y') \qquad (5.35)$$

for the spectrum. Note that an integration by parts has been performed to obtain (5.34) as a derivative on the magnetization. The playback voltage can also be written as a derivative of the head field. For completeness, the general three-dimensional form for the replay voltage may be generalized from (5.30) and is given by:

$$V(x) = NEv\mu_0 \iiint_{\text{medium}} d^3r' h(r' + x) \cdot \frac{dM(r')}{dx'} \tag{5.36}$$

with corresponding Fourier transform. General forms in terms of the potential and charge are found in Problems 5.12, 5.13.

Utilizing (2.63) the (2D) voltage expressions can be written in a form that explicitly shows the dependence on the head field at the surface of the medium. Equation (2.63) gives the field from a head as twice the convolution of the field from a wire and the longitudinal field along the surface of the head. Thus, substitution into (5.36) yields a voltage that is the convolution of the longitudinal field at the surface of the head (per *NIE*) and the playback voltage for a playback from a head with vanishingly small gap and infinitely long pole faces (a wire with imaging). Because the latter is given by (5.12), the playback voltage can be written as:

$$V(x) = 2NWEv\mu_0 \int_{-\infty}^{\infty} dx' h_x^s(x') H_y(x - x', d + \delta/2) \tag{5.37}$$

where $h_x^s(x')$ is the normalized field at the surface of the head. (Recall that $H_y(x - x', d + \delta/2)$ is the free-space vertical field component due to the magnetization pattern evaluated at the head surface (e.g. (4.20)) with the coordinate y' origin at the medium center). Equation (5.12) is recovered when the surface field in (5.37) is represented by a delta function corresponding to a head with an infinitesimal gap and long poles. Note that (5.37) applies for a medium completely above a head structure, not for one sandwiched by a keeper.

The Fourier transform of the voltage can be obtained from (5.37) utilizing (4.23). Alternatively, (5.35) may be utilized, noting that the Fourier transform of (2.63) is:

$$h^*(k, y') = h_s^*(k)e^{-|k||y'+d+\delta/2|}(1, -i\,\text{sgn}(k)) \tag{5.38}$$

Substitution of (5.38) into (5.34) yields:

$$V(k) = ikNWEv\mu_0 h_s^*(k) \int_{-\delta/2}^{\delta/2} dy' e^{-|k||y'+d+\delta/2|}(M_x(k, y') \\ - i\ \text{sgn}(k)M_y(k, y'))$$

(5.39)

Note that these expressions are for a geometry where Hilbert transforms of the head fields are valid. In addition, it is assumed that the configuration of Fig. 5.5 applies where the head is 'below' the medium. The transformation or proper use of y' for the opposite configuration of the head above the medium is immediate: the physics is that the fields decrease away from the head surface. The Fourier transform of the voltage greatly simplifies analysis of the recording process, since the effect of various head shapes arises only in the multiplicative term $h_s^*(k)$. All the information about the recorded magnetization is contained in the integral where the magnetization transforms are weighted by the exponential spacing loss. It is also seen immediately, as was discussed in Chapter 4, that if the vertical magnetization is simply a Hilbert transform of the longitudinal magnetization (4.27), the voltage is double that for a single component.

Equation (5.39) may be put in a form that clearly indicates the role of spacing during replay. The term 'exp$(-|k|d)$' can be factored and, in addition, the vertical original of the coordinate system can be placed at the surface of the medium closest to the head:

$$V(k) = ikNWEv\mu_0 h_s^*(k)e^{-|k|d} \int_0^{\delta} dy' e^{-|k|y'}(M_x(k, y') \\ - i\ \text{sgn}(k)M_y(k, y'))$$

(5.40)

All the information about the recording process is contained in the integral *including the effect of spacing during the recording process.* The effect of spacing during playback is contained simply in the term 'exp$(-|k|d)$.' As discussed in Chapter 2 this variation arises from the general property of Fourier transforms of solutions of Laplace's equation in two dimensions. An exponential spacing variation yields a linear variation when plotted log voltage versus linear wave number, versus spacing d or versus the product $|k|d$ (Fig. 2.13). This is a common method of plotting, because voltage is often measured in decibels relative to a reference level. Thus, (5.40) can be expressed as:

$$V(dB) = V_0(dB) - 20\log_{10}(e^{-kd})$$
$$\approx V_0(dB) - 8.686kd \qquad (5.41)$$
$$\approx V_0(dB) - 54.6(d/\lambda)$$

where $V_0(dB)$ is the voltage level referred to zero playback spacing or perfect head to medium contact. Figure (2.13) applies to (5.41) as well with the abscissa replaced by $V(dB) - V_0(dB)$. for example, at a 1μm wavelength, an increase in spacing of 0.05μm yields a decrease in playback voltage of almost 3dB. It is to be emphasized that the spacing loss given in (5.41) applies *only* to spectral analysis. The voltage waveforms given, by example in (5.34) or Problem 4.5, depend in a complicated manner on playback spacing through the correlation integral. It is important to re-emphasize that the replay spacing loss given by (5.41) occurs only if the medium does not possess a reversible permeability, nor does any permeable material, such as a keeper, occur beyond the surface of the head.

Isolated transitions

In this section the reciprocity expressions derived above are utilized to examine the playback voltage and spectra of a single isolated transition. Rather than continuing to write the reciprocity relations in completely general form, simplified restricted examples that clarify the playback process will be considered. The replay voltage due to a perfectly sharp transition is evaluated first, since the result for any arbitrary magnetization variation can be computed by convolution with the normalized magnetization derivative. Corresponding to (4.4) and (4.5) for the fields from a distributed magnetization transition, the voltage due to a distributed magnetization pattern may be written:

$$V(x) = \int_{-\infty}^{\infty} dx' F(x - x') V^{\text{step}}(x') \qquad (5.42)$$

and

$$V(k) = F(k) V^{\text{step}}(k) \qquad (5.43)$$

where $V^{\text{step}}(x)$ is the playback response from a sharp, step transition. $F(x - x')$ may, in fact, be different for each magnetization component so that the convolutions must be performed separately. Equations (5.42) and (5.43) assume that the magnetization pattern does not change with depth into the medium. For depth-varying transitions, (5.34) and (5.40) can be used directly. In general, for each component the relations:

$$F(x) = \frac{1}{2M_r}\frac{dM}{dx}, \; F(k) = \frac{ik}{2M_r}M(k) \tag{5.44}$$

hold where M_r denotes the saturation remanent magnetization far from the transition.

Sharp transitions

A perfect transition of magnetization with an arbitrary orientation is shown in Fig. 5.6. It is assumed that the transition center of each component is at the same location and that no variation in the vertical direction occurs. The magnetization is a step function whose derivatives required for (5.34) are:

$$\frac{dM_x(x',y')}{dx'} = 2M_r\delta(x'-x_0)\cos\theta$$
$$\frac{dM_y(x',y')}{dx'} = -2M_r\delta(x'-x_0)\sin\theta \tag{5.45}$$

θ is the angle the magnetization makes with the x axis. M_r denotes the saturation remanent magnetization far from the transition and x_0 represents the location of the transition along the x axis. Substitution of (5.45) into (5.34) yields:

$$V(x) = 2NWEv\mu_0 M_r \int_{-\delta/2}^{\delta/2} dy' \begin{pmatrix} h_x(x+x_0,y'+d+\delta/2)\cos\theta \\ -h_y(x+x_0,y'+d+\delta/2)\sin\theta \end{pmatrix} \tag{5.46}$$

In Problem 5.5, it is shown how this form can be written as a single convolution with the surface head field. For purposes of discussion, the

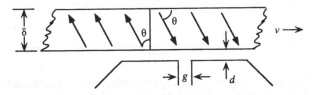

Fig. 5.6. A perfect transition of magnetization oriented at an arbitrary angle θ.

limit of very thin media is considered. In that case (5.46) becomes:

$$V(x) = 2NWEv\mu_0 M_r \delta \big(h_x(x + x_0, d + \delta/2)$$
$$\cos\theta - h_y(x + x_0, d + \delta/2)\sin\theta \big) \qquad (5.47)$$

This expression yields a different interpretation of the readback voltage than that obtained from the direct calculation. *For a sharp transition the playback process simply samples the head field at the transition center. A longitudinal transition samples the longitudinal field and a vertical transition samples the vertical head field.* Of course, for thick media and for broad transitions suitable averaging occurs.

In Fig. 5.7 the playback isolated pulse voltage (5.47) is plotted for four different cases: (a) longitudinal magnetization ($\theta = 0$), (b) magnetization at $\theta = \pi/6$, (c) magnetization at $\theta = \pi/3$, and (d) perpendicular magnetization ($\theta = \pi/2$). The voltage pulse for a longitudinal magnetization transition shows the symmetric 'bell' shape corresponding to the longitudinal head field component (Fig. 3.11(a)). As discussed for the direct calculation, this pulse shape also follows the shape of the vertical field outside of and produced by a transition of longitudinal magnetization (Fig. 4.1(c)). The voltage pulse shape for a perpendicular magnetization transition is shown in Fig. 5.7(d). The shape has the characteristic asymmetry of the vertical head field (Fig. 3.11(b)). Note that simple rotation of the magnetization direction from longitudinal to perpendicular yields a decrease in the maximum pulse voltage for

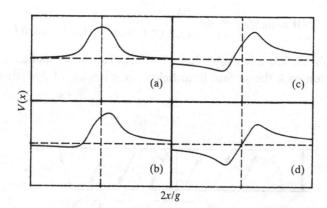

Fig. 5.7. Playback voltage for a sharp transition with magnetization oriented with respect to the longitudinal direction (Fig. 5.6) for $\theta = 0°$ (a), $30°$ (b), $60°$ (c), and $90°$ (d). Head fields evaluated at $y = g/2$ were utilized.

playback corresponding to the reduced amplitude of the vertical field component when not precisely in contact. Due to the sense of transition definition in Fig. 5.6 the leading edge of the transition is negative opposite to that of the head field. This orientation corresponds to recording with a ring head at the gap edge last seen by the medium as it moves past the head.

For a transition with an arbitrarily directed magnetization, the resulting voltage pulse is asymmetric due to a weighted combination of the longitudinal and vertical pulses (Fig. 5.7(b), (c)). For a magnetization that is primarily longitudinal (Fig. 5.7(b)) the effect of a small vertical component is to give an asymmetric pulse that rises more quickly than it decreases. Such a pulse shape is commonly measured in longitudinal tape recording (Bertram *et al.*, 1992), however, as discussed later on, the effect is dominated by phase shifts with depth of the longitudinal magnetization component. In Fig. 5.7(c) a pulse shape is shown resulting from predominately perpendicular magnetization. A small longitudinal component causes the initial pulse minimum to be smaller than the subsequent positive maximum. A whole family of pulse shapes could be given for the range of orientations varying from longitudinal to vertical (Potter, 1970).

[Historical note: during the period of intensive investigations of perpendicular recording from 1976 to 1980, pulses similar to that of Fig. 5.7(c) were measured. It was concluded that the medium was not sufficiently vertically oriented since the two peaks did not have the same magnitude. Great effort was undertaken to improve the orientation and many publications quoted the X-ray rocking curve measurement of the dispersion angle $\Delta\theta_{50}$ (Iwasaki & Ouchi, 1978). Reducing $\Delta\theta_{50}$ did not alter the shape of the pulse; as will be discussed in Chapter 8, the pulse shape arises from the nature of the recorded magnetization pattern, which does not possess the symmetry of a sharp transition.]

Broad transitions – arctangent example

Real transitions in magnetic recording media are not only broad but can have quite complicated asymmetric shapes. The following is a discussion of the effect of transition broadening utilizing the arctangent approximation:

$$M(x) = \frac{2M_r}{\pi} \tan^{-1}\left(\frac{x}{a}\right) \qquad (4.6)$$

where a is the transition length. This form is utilized because it is extremely simple and correlation with head fields may be readily performed analytically. The Fourier transform (4.30) is given in normalized form (5.44) by:

$$F_{at}(k) = e^{-|k|a} \tag{5.48}$$

According to (5.43) the Fourier transform of the pulse is obtained simply by multiplying (5.40) by (5.48). The result is the same equation as (5.40) *except that the spacing d now is replaced by $a + d$*. Because the voltage waveform for the isolated pulse may be obtained by back Fourier transforming the pulse spectrum (5.42), it is clear that *for an arctangent transition the isolated pulse waveform, after correlation, is simply the sharp transition waveform given by (5.46) or (5.47) except that d is replaced by $d + a$*. The effect of transition broadening is to broaden the pulse, and for an arctangent transition that broadening is exactly as though an increase in replay head-to-medium spacing of distance a occurred. This argument holds for all possible magnetization orientations and for thick media, as long as the transition length is a constant (Potter, 1970). If the transition shape is not arctangent, e.g. error function or tanh (Middleton & Miles, 1991), the concept of effective spacing is only approximate. However, for a finite transition width the results plotted in Figs. 5.7(a), (b), (c), (d) apply where the spacing y to the head surface from the medium center (5.47) denotes $d + a + \delta/2$. The examples plotted for $y = g/2$ correspond to typical disk drives where $a \sim d$ and $d \sim g/4$.

Explicit accurate formulas for thin media with a longitudinal magnetization transition that corresponds to typical disk drives are readily given. Assuming an arctangent magnetization transition a given by (4.6), then (5.47) yields a replay form of:

$$V(x) = 2NWEv\mu_0 M_r \delta h_x(x + x_0, d + a) \tag{5.49}$$

where it is assumed that $\delta \ll d + a$. Because the field is not evaluated too close to the surface $((d + a)/g$ is not too small), for a long pole head the Karlqvist field approximation (3.16) suffices. If we set $x_0 = 0$ then:

$$V(x) = \frac{2}{\pi g} NWEv\mu_0 M_r \delta \left(\tan^{-1}\left(\frac{g/2 + x}{d + a}\right) + \tan^{-1}\left(\frac{g/2 - x}{d + a}\right) \right) \tag{5.50}$$

The peak voltage of the isolated pulse occurs at the transition center $(x = 0)$:

$$V^{peak} = \frac{4}{\pi g} NWE \, v\mu_0 M_r \delta \, \tan^{-1} \left(\frac{g}{2(d+a)} \right) \tag{5.51}$$

The voltage pulse (5.50) is plotted in Fig. 5.8 normalized to its peak value (5.51) for typical values of $d + a = 0.5g$. In addition, the playback voltage

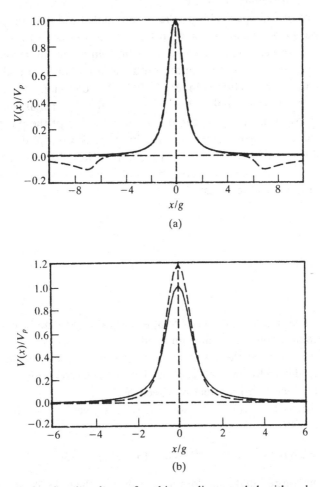

Fig. 5.8. Isolated pulse shapes for thin media recorded with a longitudinal transition of (a) arctangent shape with long pole (solid) and thin film (dashed) playback heads and (b) playback with a long pole head comparing an arctangent (solid) and tanh (dashed) transition shape. Constants are $a = d = g/2$ and $p/g = 6$, and pulses are normalized to the peak voltage given by (5.51). The arctangent and tanh shapes are defined to have a common center slope with a given a parameter (Table 4.1).

utilizing a thin film head of $p_1 = p_2 = 6g$ is plotted (dashed). The long pole head exhibits the characteristic 'bell' shaped curve whereas the thin film head exhibits undershoots corresponding to the finite pole length. The thin film pulse was normalized to (5.51) so that near the center of the pulse the effect of finite poles of realistic length hardly changes the pulse amplitude until the pulse is reduced to perhaps 30% of its peak value.

In Fig. 5.8 the replay voltage is also plotted, using (5.37) directly, for a transition of *tanh* shape (Table 4.1) for a long pole head, thin media, and longitudinal magnetization. The tanh function is a closer representation of the actual transition shape of typical thin film recording media since the approach to saturation is not as gradual as an arctangent and not as abrupt as an error function (Lin, *et al.*, 1992; Middleton & Miles, 1991). Note that the playback pulse shape hardly differs from that due to an arctangent except in the more rapid approach to zero of the pulse tails.

It is common to characterize the pulse by its peak value and the width of the pulse at half amplitude. The half-amplitude pulse width, termed PW_{50}, may easily be determined by finding that x that makes (5.50) equal to half of (5.51). The result is:

$$x_{50} = \pm\sqrt{g^2 + (a + d)^2/2} \qquad (5.52)$$

so that:

$$PW_{50} = \sqrt{g^2 + 4(d + a)^2} \qquad (5.53)$$

Equation (5.53) is exact for thin media, a long pole head, and clearly is a good approximation for a thin film head if the poles are not too short. Equation (5.53) gives the pulse width in terms of distance along the medium. If the width in time is desired, it is necessary only to divide (5.53) by the head–medium relative speed, v.

The Fourier transform of the isolated pulse is given by the product of (5.40) and (5.48)

$$V(k) = 2NWEv\mu_0 M_r \delta e^{-|k|(d+a)} \sin(1.136kg/2)/(1.136kg/2) \qquad (5.54)$$

In (5.54) the approximate gap loss for a long pole head (3.38) is utilized for $h_s(k)$. This transform is plotted in Fig. 5.9(a) (solid) for $a = d = g/4$. The voltage plotted in dB is normalized to $2NWEv\mu_0 M_r \delta$. At long wavelengths all wavelength dependent terms in (5.54) approach unity. As the wavenumber is increased the spacing losses and eventually the gap loss decrease the voltage. Usually system noise during measurement is

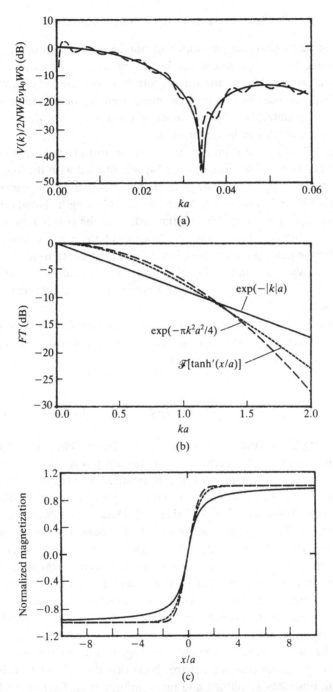

Fig. 5.9. (a) Isolated pulse Fourier transform corresponding to Fig. 5.8(a) normalized to $2NWEv\mu_0M_r\delta$. (b) Isolated pulse Fourier transform corresponding to Fig. 5.8 (corrected by spacing loss and head surface field) normalized to $2NWEv\mu_0M_r\delta$ for transition shapes of arctangent (solid), tanh (dashed), and error function (dotted). (c) Transition shapes of arctangent (solid), tanh (dashed), and error function (dotted) with a common center slope.

large enough so that the gap null can not be seen. However, at long wavelengths, this measurement can be accurately utilized to obtain the constant $2NWEv\mu_0 M_r\delta$. If the track width has been measured optically and $\mu_0 M_r\delta$ has been determined by magnetometer measurements, the product NE is obtained. If the number of turns is known, then the low frequency efficiency can be determined.

In Fig. 5.9(a) the pulse Fourier transform is also plotted for a typical thin film head ($p/g = 6$). Equation (5.54) was utilized with the long pole gap loss replaced by (3.50). The general character is the same except for the additional undulations due to the finite pole length. However, the long wavelength limit may still be estimated, since the pole length is large compared to the gap. For a long pole head this limit is equivalent to the area under the pulse (Problem 5.6). For a finite pole length head the pulse integral vanishes so that a Fourier transform measurement at long wavelengths probably yields the greatest accuracy for a simple estimate of $2NWEv\mu_0 M_r\delta$.

Equation (5.54) can be written generally, utilizing (5.40) in terms of the head surface transform $H_s(k)$ (or strictly its complex conjugate) and the generalized transform of the normalized transition derivative $F(k)$ (5.44):

$$V(k) = 2NWEv\mu_0 M_r\delta e^{-|k|d}F(k)h_s^*(k) \qquad (5.55)$$

If the spacing d is known, as well as the head surface field transform, the measurement can be corrected to leave the long wavelength constant times the Fourier transform of the normalized transition derivative. Equation (5.55), corrected by these terms, is plotted in Fig. 5.9(b) for three different transition shapes (Table 4.1). Data for a typical thin film disk agree well with the transform of the *tanh* functional shape (Middleton & Miles, 1991). On a log voltage, linear wavenumber plot, a simple linear form occurs for the spacing loss form of the arctangent transform. Both the error function and the *tanh* transition shapes show a plateau at long wavelengths, so that a determination of the long wavelength limit, even without correction for a thin film head transform, is somewhat easier for realistic thin film recording media. In general, with a calibrated replay head and known spacing, this measurement can be utilized to obtain the transition shape. Note that the Fourier transform, in general, possesses amplitude and phase information. Even if the head is symmetric, an asymmetric transition shape will yield a complex $F(k)$ and $V(k)$.

This chapter concludes with a brief discussion of the isolated pulse and transform for transitions recorded in thick media. In general, (5.34) and (5.39) must be utilized in a 2D approximation. If the transition is arctangent in shape and the magnetization is longitudinal in orientation, (5.34) (or an integration of (5.49) versus depth) yields:

$$V(k) = 2NWEv\mu_0 M_r \int_0^\delta dy' h_x(x + x_0(y'), d + a(y')) \tag{5.56}$$

In (5.56) the transition length $a(y')$ and the transition center $x_0(y')$ are allowed to vary arbitrarily with depth into the medium. The y' origin is taken at the medium surface closest to the head. It has been found by numerical simulation that the transition center changes monotonically with depth into the medium (Bertram, *et al.*, 1992). Such a change yields a measured pulse shape of the form illustrated in Fig. 5.7(b) (Problem 5.10). Even though the pulse shape can be explained by several mechanisms, analysis of the phase of the Fourier transform at long wavelengths shows that a rotated magnetization direction does not occur.

For thick media with an arctangent transition of constant transition length and no phase change with depth, (5.56) yields (utilizing long pole heads) an approximated expression for the PW_{50} (Middleton, 1966):

$$PW_{50} \cong \sqrt{g^2 + 4(d+a)(d+a+\delta)} \tag{5.57}$$

However, it must be emphasized that thick tape media will not follow (5.57) in a simple fashion because of the changing transition shape and position with depth. In addition, in both (5.56) and (5.57) the depth δ is not the medium thickness but a much smaller depth that is determined by the recording current, the medium coercivity and the head parameters (Armstrong, *et al.*, 1991).

Problems

Problem 5.1 In the formulation of the reciprocity expression show that if the head is defined by the wires, the head core, and the reversible permeability of the recording medium, then (5.22) reduces to:

$$\Phi(x) = \frac{\mu_0}{I} \underbrace{\iiint d^3r' H_J(r', x) \cdot M_0(r')}_{medium} + \frac{\mu_0 \chi_k}{I} \underbrace{\iiint d^3r' H_J(r', x) \cdot H(r')}_{keeper}$$

Explain how all the fields in this expression are determined.

Problem 5.2 Show that the reciprocity relations may be written as a correlation of the scalar head field potential and the magnetic charge. Use the Case 2 form for simplicity. Recall that for regions not including the wires the field may be derived from a scalar potential ϕ : $H = -\nabla\phi$; the charge density is written as $\rho = -\nabla \cdot M$. Thus, show that the replay voltage may be written as a correlation of the longitudinal head field and the magnetic charge. Find a form for the replay voltage in terms of the vector potential for the head field and equivalent currents for the magnetization.

Problem 5.3 Derive (5.34).

Problem 5.4 Derive (5.37) utilizing (2.63) and (4.20).

Problem 5.5 Show that (5.42) can be written as a single integral over the surface head field by utilizing (5.37) and (4.1, 4.2):

$$V(x) = \frac{2NWEv\mu_0 M_r}{\pi} \int_{-\infty}^{\infty} dx' h_x^s(x')$$

$$\left(\begin{array}{c} \dfrac{\cos\theta}{2} \ln \dfrac{(x + x_0 - x')^2 + (d + \delta)^2}{(x + x_0 - x')^2 + d^2} \\[2mm] + \sin\theta \left(\tan^{-1}\left(\dfrac{d + \delta}{x + x_0 - x'}\right) - \tan^{-1}\left(\dfrac{d}{x + x_0 - x'}\right) \right) \end{array} \right)$$

What $h_x^s(x)$ is required to give the published analytic result for a finite arctangent transition of length a where in the above $d \to d + a$ (Potter, 1970)?

Problem 5.6 Utilizing (5.36) show that the integral of the voltage pulse for an arbitrary isolated magnetization transition always vanishes. Consider a ring head with infinitely long pole pieces. Argue that the integral of the pulse depends only on the saturation remanence of the longitudinal component M_r^x and can be written simply as:

$$\int_{-\infty}^{\infty} V(x)dx = 2NWEv\mu_0 M_r^x \delta$$

or, a time integral independent of speed:

$$\int_{-\infty}^{\infty} V(t)dt = 2NWE\mu_0 M_r^x \delta$$

Discuss why this is a good approximation for very long pole pieces. What result applies for pole heads, keepered or not? Note that the temporal integral under the pulse for a long pole ring head (apart from the replay constants $NWE\mu_0$) is the total charge per unit track width in the medium. Alternatively, the pulse integral per NWE is twice the flux in the saturation region. Show that these results hold for thick media with arbitrary transition shape and depth variation.

Problem 5.7 Consider the Fourier transform given by (5.35b) at very long wavelengths. Show that

$$\lim_{k \to 0} ikM_x(k, y') = 2M_r^x$$

$$\lim_{k \to 0} ikM_y(k, y') = 2M_r^y$$

and therefore

$$\lim_{k \to 0} V(k) = 2NWEv\mu_0 h_s^*(k \to 0)\delta(M_r^x - i \, \text{sgn}(k)M_r^y)$$

Consider a measurement of the rms spectrum (suppose you averaged many pulses, digitized and performed an FFT on the pulse). Explain why this measurement gives a different result (approach to zero at long wavelengths) from that given by an integral of the pulse shape (Problem 5.6). Under what conditions (especially over what range of wavelengths) does it give the same result as that of the long pole approximation above?

Problem 5.8 A simple relation frequently utilized (Comstock & Moore, 1974) between the peak voltage and the pulse width is:

$$V^{peak}PW_{50} = \frac{4}{\pi} NWEv\mu_0 M_r \delta$$

Show that this relation holds exactly for a Lorentzian pulse shape:

$$V(x) = \frac{V^{peak}(PW_{50}/2)^2}{(x^2 + (PW_{50}/2)^2)}$$

by integrating the pulse and comparing the result with that in Problem 5.6. Show that a Lorentzian pulse shape occurs only for the case of thin media, zero replay gap, and an arctangent transition of longitudinal magnetization.

Problem 5.9   This is an example of the use of the general 3D form of reciprocity. Fourier analysis yields the result that for fields off the edge of a head a distance greater than the wavelength under consideration and zero spacing $(y = 0)$, the side field is approximately (Lindholm, 1978b, eq.(12)):

$$h_x(k, o, z) \approx \frac{e^{-kz}}{2}$$

Give a geometric interpretation of this result and thus write an approximate expression for the longitudinal field off the side of a head. Consider a track of width W off-set from the record head by a distance (guard band) H. Using the general form of reciprocity, calculate the read voltage pulse for side reading. Assume an arctangent recorded longitudinal magnetization with transition parameter a, and assume thin media. Neglect any spacing effect (y). Write an expression for the side reading ratio, defined as the ratio of the peak voltage from side reading to the peak voltage on-track. For the on-track voltage include the spacing term d. Evaluate the ratio for $W = 10\mu m$, $H = 2\mu m$, $g = 0.5\mu m$, $a = d = 0.125\mu m$.

Problem 5.10   Assume for thick media a longitudinal arctangent transition shape where the transition length increases linearly with depth into the medium $a(y) = a_s + (a_b - a_s)y/\delta$ and the transition center increases linearly with depth $x_0(y) = \beta y$. Use (5.56) to evaluate numerically the isolated pulse, and show that a finite β gives the pulse shape of Fig. 5.7(b). Fit $a(y)$ and $x_0(y)$ to results in (Bertram, *et al.*, 1992). Use (5.40) to obtain an analytic form for $V(k)$ and discuss data analysis to determine a_s, a_b, δ, and β.

Problem 5.11   Show that (5.11) holds but with a factor of two increase for a long pole head of infinitesimal gap length, where the winding is a small wire lying along the gap at the head surface. Show that in this case the head efficiency is unity and the replay structure captures the total flux from the transition (see Problem 4.8).

Problem 5.12 Show that the general expression for the replay voltage (5.36) may be written as:

$$V(x) = NEv\mu_0 \int\int\int d^3r' \phi(r' + x') \partial \rho(r')/\partial x'$$

$$+ NEv\mu_0 \int\int dx' dz' \phi(x + x', d + \delta, z') \partial M_y(x', d + \delta, z')/\partial x'$$

$$- NEv\mu_0 \int\int dx' dz' \phi(x + x', d, z') \partial M_y(x', d, z')/\partial x'$$

where ϕ is the potential normalized to NIE of the head field including all permeable material (e.g. keepers) when Case 2 (5.27) is utilized to define the head field. ρ is the charge in the medium and the additional terms represent surface charge $\sigma = \hat{n} \cdot M = \pm M_y$ at the two surfaces $y = d + \delta, d$, respectively with $y = 0$ at the head surface.

Problem 5.13 Show that integration by parts transforms the result in Problem 5.12 (apart from an overall sign change) to:

$$V(x) = NEv\mu_0 \int\int\int d^3r' h_x(r' + x')\rho(r')$$

$$+ NEv\mu_0 \int\int dx' dz' h_x(x + x', d + \delta, z') M_y(x', d + \delta, z')$$

$$- NEv\mu_0 \int\int dx' dz' h_x(x + x', d, z') M_y(x', d, z')$$

Problem 5.14 Show that the reciprocity integral for Case 3 (5.28) applies to a head defined by the wires plus all permeable material except for possible reversible permeability of the recording medium. In that case the integration is solely over the recording medium. Utilizing (5.18) show that (5.36) transforms to:

$$V(x) = NEv \iint\int d^3r' A(r' + x') \cdot J_{\text{vol}}(r')$$

$$+ NEv \iint dx'dz' A(x + x', d + \delta, z') \cdot J_s(x', d + \delta, z')$$

$$- NEv \iint dx'dz' A(x + x', d, z') \cdot J_s(x', d, z')$$

where A is the vector potential for the field (per NIE) and J denotes the equivalent current distribution (2.12) due to the magnetized medium including reversible magnetization.

6

Playback process:
Part 2 – Multiple transitions

Introduction

In this chapter analysis of the playback process is extended to consider the effects of multiple transitions. First the concept of linear superposition in magnetic recording is introduced. Two examples are discussed in detail: (1) square wave recording of alternating transitions with a fixed bit spacing and (2) dibit recording of a pair of transitions. In the former both the 'roll-off' curve of the peak voltage versus density will be derived as well as the spectrum or Fourier transform. The utility of spectral measurements for analysis of the recording process will be emphasized as in Chapter 5. The relation of the square wave spectrum to an 'effective' channel transfer function will be given. In this chapter, as in the previous one, a comparison will be made between longitudinal and perpendicular recording. The effects that relate solely to differences in a change in magnetization orientation will be emphasized. In this way the playback effects and record phenomena discussed in Chapter 8 can be distinguished.

Linear superposition

The magnetic recording process is not strictly linear. A linear integral relation does not occur where a change in input amplitude yields a proportional change in output voltage. In fact an impulse does not exist because the 'input' to the recording channel consists of step functions of voltage or current that produce transitions of magnetization. However, a restricted linearity applies as long as transitions are not written too closely. A transfer function can be defined whose product with the input spectrum yields the output spectrum. This 'quasi' linearity is called linear superposition. The principle of linear superposition states that the replay voltage from a sequence of step current changes is given by the sum of the playback responses of the individual transition playback voltages with regard to the spacing and polarity that comprise the current sequence:

$$V(x) = \sum_n (-1)^n V_{sp}(x - x_n) \qquad (6.1)$$

where V_{sp} is the playback isolated transition response discussed in Chapter 5 and x_n is the location of the n transitions written in the sequence. In digital recording x_n is an integral multiple of the fundamental bit spacing B. The polarity will always alternate from transition to transition independent of the spacing. For superposition to hold the individual transitions must be identical. As discussed in Chapter 8 this phenomenon will occur for fixed current amplitude (in reality fixed deep-gap field amplitude) applied to the head.

The Fourier transform of a given sequence is simply:

$$V(k) = V_{sp}(k) \sum_n (-1)^n e^{ikx_n} \qquad (6.2)$$

For example, a dibit consisting of two transitions written at a spacing B, has a voltage playback and transform given by:

$$V(x) = V_{sp}(x - B/2) - V_{sp}(x + B/2)$$
$$V(k) = 2iV_{sp}(k) \sin kB/2 \qquad (6.3)$$

Even when linear superposition holds, the playback waveform can be complicated. In Fig. 1.7 an example of the NRZI sequence [1011011100] is shown where $B \sim 0.9PW_{50}$. Only the first transition retains the character of the isolated transition. The waveform of the interior transitions is complicated, and some peak shift of the individual transition maxima occurs. The patterns can become quite complicated, as the example in Fig. 6.1 shows, where the individual transition width is broader than that in Fig. 1.7 ($B \sim 0.5PW_{50}$). Here the overlap is so great that voltage peaks occur outside the bit cell in which they were written. In addition, there is significant 'pulse droop' as well as amplitude loss.

A system transfer function in terms of an equivalent impulse response can be defined and is given after the following discussion of square wave recording. Non-linearities due to the recording process at high densities, which cause linear superposition to fail, are discussed in Chapter 9.

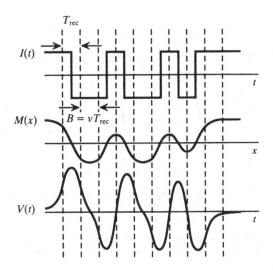

Fig. 6.1. Voltage waveform and magnetization pattern for an NRZI current pattern '101101110' as in Fig 1.7, but with a larger transition length a relative to the bit spacing B.

Square wave recording

Square wave recording refers to an alternating series of step record current changes at fixed amplitude with a fixed bit spacing. As an NRZI data pattern, it is referred to as an all '111s' pattern. In Fig. 6.2 two patterns of recorded magnetization are shown alternating in polarity at bit interval B between fixed normalized amplitude; the dotted curve represents perfectly sharp transitions that follow the current waveform, whereas the broadened curve represents possible transition broadening that reduces the amplitude and decreases the slope at the transition centers. The density D is defined as $D = 1/B$. Assuming linear superposition, the net replay voltage can be written in terms of the voltage from a single transition as:

$$V(x) = \sum_{n=-\infty}^{\infty} (-1)^n V_{sp}(x - nB) \qquad (6.4)$$

where $V_{sp}(x - nB)$ now denotes the playback voltage pulse from a single transition located at positions $x_n = nB$. For a single transition voltage the general form for a sharp transition will be utilized (5.46). For a broad transition the final result can be convolved with the normalized derivative

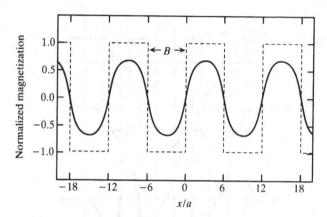

Fig. 6.2.   Illustration of magnetization patterns for square wave recording for two cases of a perfectly sharp transition and transition length.

$F(x)$ (4.4) of the broadened transition. To visualize the effect of transition broadening, the results derived are identical in form for an arctangent transition, except for the utilization of an increased effective spacing: $d \to d + a$. Rather than utilizing an angle to represent orientation, the saturation remanence of each component is given (5.46):

$$V_{\mathrm{sp}}(x - nB) = 2NWEv\mu_0 \int_{-\delta/2}^{\delta/2} \mathrm{d}y' \left(\begin{array}{c} h_x(x - nB, y' + d + \delta/2)M_r^x \\ -h_y(x - nB, y' + d + \delta/2)M_r^y \end{array} \right)$$

(6.5)

Here, it is assumed that the remanent magnetization does not vary with depth, nor does the phase of each component. Utilizing the general form for the fields in terms of the head surface field, (2.63), (6.5) can be written as:

$$V_{\mathrm{sp}}(x - nB) = \frac{2NWEv\mu_0}{\pi} \int_{-\infty}^{\infty} \mathrm{d}x' h_s^x(x') \int_d^{d+\delta}$$
$$\mathrm{d}y' \left(\frac{y'M_r^x + (x - nB - x')M_r^y}{(y')^2 + (x - nB - x')^2} \right)$$

(6.6)

To obtain the net voltage (6.4) it is necessary to perform the sum:

$$S = \sum_{n=-\infty}^{\infty} (-1)^n \left(\frac{y'M_r^x + (x - nB - x')M_r^y}{(y')^2 + (x - nB - x')^2} \right)$$

(6.7)

Following Williams and Comstock who evaluated (6.7) for a longitudinal pattern (Williams & Comstock, 1971), the complex variable expression is utilized:

$$\sum_{n=-\infty}^{\infty} (-1)^n f(n) = -\pi \sum_{z=z_{res}} \frac{res(z)}{\sin \pi z} \qquad (6.8)$$

where $f(z)$ is written as:

$$f(z) = \frac{(y'/B)M_r^x + (z - ((x - x')/B))M_r^y}{(y'/B)^2 + (z - ((x - x')/B))^2} \qquad (6.9)$$

The residues of $f(z)$ occur at:

$$z_{res} = ((x - x')/B) \pm i(y'/B) \qquad (6.10)$$

so that (6.8) can be evaluated using (6.9) with (6.10) and the identity:

$$\frac{1}{\sin(A \pm iC)} = \frac{\sin A \cosh C \pm \cos A \sinh C}{\sin^2 A + \sinh^2 C} \qquad (6.11)$$

to yield:

$$V(x) = \frac{2NWEv\mu_0}{B} \int_{-\infty}^{\infty} dx' h_s^x(x') \int_d^{d+\delta}$$

$$dy' \left(\begin{array}{c} \dfrac{M_r^x \sinh(\pi y'/B) \cos(\pi(x - x')/B)}{\sinh^2(\pi y'/B) + \sin^2(\pi(x - x')/B)} \\[2mm] + \dfrac{M_r^y \cosh(\pi y'/B) \sin(\pi(x - x')/B)}{\sinh^2(\pi y'/B) + \sin^2(\pi(x - x')/B)} \end{array} \right) \qquad (6.12)$$

Equation (6.12) gives the voltage waveform resulting from square wave recording at bit separation B. The first evaluation is for thin media ($\delta \ll d$), a long pole head with a vanishingly small replay gap (3.14), and an arctangent magnetization transition of parameter a for both components:

$$V(x) = \frac{2NWEv\mu_0\delta}{B} \left(\begin{array}{c} \dfrac{M_r^x \sinh(\pi(d + a)/B) \cos(\pi x/B)}{\sinh^2(\pi(d + a)/B) + \sin^2(\pi x/B)} \\[2mm] + \dfrac{M_r^y \cosh(\pi(d + a)/B) \sin(\pi x/B)}{\sinh^2(\pi(d + a)/B) + \sin^2(\pi x/B)} \end{array} \right) \qquad (6.13)$$

Equation (6.13) is plotted in Fig. 6.3 for the two cases of (a) longitudinal and (b) vertical magnetization at a low–medium density of $B = 6(d + a)$. Both plots are normalized to the peak voltage for longitudinal recording

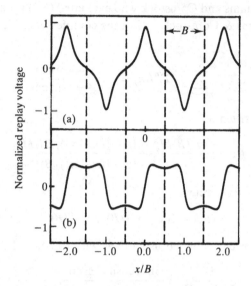

Fig. 6.3. Replay voltage for square-wave recording at low–medium densities where the individual transition playback pulses are beginning to overlap. (a) longitudinal magnetization (b) perpendicular magnetization.

((5.51) with $g \to 0$). For longitudinal recording the waveform indicates overlap of somewhat isolated pulses with the peak of each pulse decreasing due to adjacent voltage pulses of the opposite polarity. For perpendicular recording the waveform of each bit cell center indicates the asymmetric perpendicular field shape (Fig. 3.8). At the cell edges the overlap of asymmetric transitions of opposite polarity increases the voltage above that of the long tail of the isolated pulse. The amplitude of vertical recording at low–medium density is about half that of longitudinal, because the amplitude of the vertical head field component for a zero-gap long-pole head is half that of the corresponding longitudinal field.

Equation (6.13) can be utilized to obtain the 'roll-off' curve or the peak voltage at any density. For a longitudinal magnetization the peak voltage can always be determined by evaluating (6.13) at $x = 0$:

$$V_{long}^{peak} = \frac{2NWEv\mu_0 M_r^x \delta}{B \sinh(\pi(d+a)/B)} \tag{6.14}$$

For perpendicular magnetization the peak voltage occurs for low densities at x given by:

$$\sin \frac{\pi x}{B} = \sinh \frac{\pi(d+a)}{B} \tag{6.15}$$

When the density increases (B decreases) so that:

$$\sinh \frac{\pi(d+a)}{B} = 1 \tag{6.16}$$

then the peak voltage occurs for higher densities at:

$$x = B/2 \tag{6.17}$$

Thus, the density variation of the peak voltage for perpendicular recording can be expressed as:

$$V_{perp}^{peak} = \frac{NWEv\mu_0 M_r^y \delta \cosh(\pi(d+a)/B)}{B \sinh(\pi(d+a)/B)}, \quad B > \pi(d+a)/\sinh^{-1} 1 \tag{6.18}$$

$$V_{perp}^{peak}(B) = \frac{2NWEv\mu_0 M_r^y \delta}{B \cosh(\pi(d+a)/B)}, \quad B < \pi(d+a)/\sinh^{-1} 1$$

The peak voltages given in (6.14) and (6.18) are plotted versus density (1/B) in Fig. 6.4. They are normalized to the low density peak voltage for longitudinal recording (5.51), for the same saturation magnetization level $M_r^x = M_r^y = M$ and a small gap:

$$V_{long}^{peak}(B \rightarrow \infty) = \frac{2NWEv\mu_0 M_r \delta}{\pi(d+a)} \tag{6.19}$$

In Fig. 6.4 the peak voltage for longitudinal recording decreases monotonically with increasing density. For perpendicular recording at low densities the voltage peak increases to a maximum at $B \sim 2(d+a)$ and then becomes identical to the longitudinal curve at high densities. The initial increase and subsequent decrease can be visualized from Fig. 6.3. The coincidence of the two curves at high densities can be seen from harmonic analysis. At high densities the square voltage response has overlapped to yield a sine wave voltage pattern. As discussed in the following section on harmonic analysis, the higher harmonics of each waveform have vanished and all that remains is the fundamental. As discussed in Chapter 5, a change in orientation alters only the phase of a single spectral component. The waveforms are now purely sinusoidal, identical in amplitude and (as can be surmised from Figs. 6.3(a), (b))

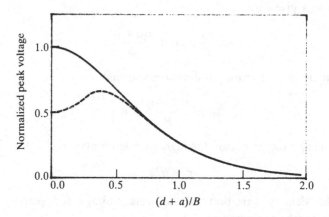

Fig. 6.4.   Peak voltage versus density for square wave recording ('roll-off' curve) for longitudinal (solid) and perpendicular (dashed) recording. Plots are for thin media with a long-pole, zero-gap replay head and are normalized to the peak voltage of the isolated pulse for longitudinal recording.

shifted in phase by $\pi/2$. At high densities the peak voltage becomes from (6.14) and (6.18):

$$V^{peak}(B) = \frac{4NWEv\mu_0 M_r \delta e^{-\pi(d+a)/B}}{B} \tag{6.20}$$

where the same saturation magnetization level is assumed for both orientations.

'Roll-off' curve and D_{50}

The variation of the peak voltage versus density of the all 111s pattern for longitudinal recording is referred to as the 'roll-off curve'. A density, D_{50}, where the amplitude has reduced to half that of the isolated pulse, gives a reasonable estimate of a maximum system operating density. D_{50} relations for a finite gap head including the influence of the finite pole length of a thin film head are discussed later in this chapter. In general, (6.12) can be utilized for any magnetization pattern with appropriate convolution and any given head structure. However, a simple relation in the zero-gap, thin media approximation can be given. Using (6.14), the half-voltage density occurs when:

$$B\sinh(\pi(d+a)/B) = 2 \tag{6.21}$$

or:

$$D_{50} = \frac{1}{B_{50}} \cong \frac{2.1773}{\pi(d+a)} \qquad (6.22)$$

From (5.53) for thin media and a small gap head, the pulse width of the isolated pulse is:

$$PW_{50} = 2(d+a) \qquad (6.23)$$

Thus, the relation holds (Comstock & Moore, 1974):

$$D_{50}PW_{50} \approx 1.39 \qquad (6.24)$$

This relation holds for the linear superposition of isolated Lorentzian pulses. This is the case for longitudinal arctangent magnetization transitions recorded on thin media and played back with a head with a vanishingly small gap (6.6). However, as will be shown following the section on spectral analysis, (6.24) is a good approximation over a wide range of practical values of g and $d+a$.

Thin film head

The effect of finite pole lengths on the 'roll-off' curve for longitudinal recording is of interest for replay with a thin film head. Equation (6.12) is utilized for thin media with a longitudinal arctangent transition, but the integral over the surface field is retained:

$$V(x) = \frac{2NWEv\mu_0 M_r\delta}{B} \int_{-\infty}^{\infty} dx' h_s^x(x') \, \frac{M_r^x \sinh(\pi(d+a)/B)\cos(\pi x'/B)}{\sinh^2(\pi(d+a)/B) + \sin^2(\pi x'/B)}$$

$$(6.25)$$

This expression is plotted normalized by (5.51) in Fig. 6.5(a). A finite playback gap with $g = 2(d+a)$ is utilized in field expressions for a symmetric thin film head with p/g values of $2,4,8,\infty$ (Szczech & Iverson, 1987). The effect of the finite pole length is to cause oscillations in the 'roll-off' curve that are set by the ratio of pole to gap length. The D_{50} density (or any definition of a density limit) will oscillate as a function of the ratio p/g (Bertero, et al., 1992). In Fig. 6.5(b) a logarithmic plot in dB of Fig. 6.5(a) is shown. This figure clearly shows that the 'roll-off' curve vanishes at various frequencies and exhibits a first gap null, dependent on the p/g ratio, in the vicinity of the gap null of the Fourier transform of the head field $B = 2g$ (Figs. 3.12, 3.19).

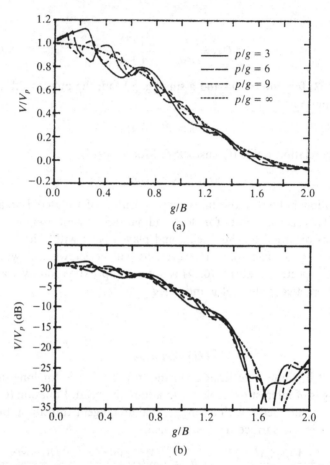

Fig. 6.5.   Peak voltage versus density for square wave recording for a finite gap, finite pole-length symmetric thin film head for longitudinal arctangent magnetization transitions with $g = 2(d + a)$. The curves are normalized to the isolated pulse peak voltage and are for various ratios of pole length to gap length $p/g = 3, 6, 9\,\infty$. (b) is a logarithmic (dB) plot of (a) that clearly shows the zeros of the 'roll-off' curve.

Spectral analysis

Equation (6.4) describes the linear superposition of alternating identical transitions into an all '111s' square wave pattern at transition separation B. The continuous infinite Fourier transform from (6.2) is:

$$V(k) = V_{sp}(k) \sum_{n=-\infty}^{\infty} e^{-in(kB-\pi)} \tag{6.26}$$

Using the relation:

$$\sum_{n=-\infty}^{\infty} e^{i2\pi nx} = \sum_{m=-\infty}^{\infty} \delta(x-m) \tag{6.27}$$

(6.26) can be written as:

$$V(k) = V_{sp}(k) \sum_{m=-\infty}^{\infty} \delta\left(\frac{kB}{2\pi} - \frac{1}{2} - m\right) \tag{6.28}$$

Since the wavenumber of the fundamental component is related to the bit length by:

$$k_0 = \frac{\pi}{B} \tag{6.29}$$

and delta functions possess the property that:

$$\delta(ax) = \frac{\delta(x)}{|a|} \tag{6.30}$$

(6.28) can be expressed as:

$$V(k) = 2k_0 V_{sp}(k) \sum_{m=-\infty}^{\infty} \delta(k - (2m+1)k_0) \tag{6.31}$$

Equation (6.31) yields equally weighted pulses at all odd harmonics for both positive and negative k. At each harmonic the amplitude changes by $V_{sp}(k)$. Note that the pulse weights do not decrease inversely with the harmonic number. That occurs in the Fourier analysis of a square wave. Here the Fourier transform is obtained of a series of alternating polarity pulses that corresponds to the derivative of a square wave. Equation (6.31) is plotted in Fig. 6.6 utilizing (5.54) for the spectrum of the isolated pulse for thin film media (for any given magnetization orientation) with $d + a = 2g$ as would be seen using spectral analysis ($f > 0$). The fundamental was chosen to correspond to Fig. 6.3 so that $k_0 = \pi/6(a+d)$. The height of each arrow denotes the odd harmonic components at multiples of k_0. The envelope of the spectrum can be utilized to analyze data for a and g as well as the long wavelength constants, a process similar to that discussed in Chapter 5 for the Fourier transform of the isolated pulse.

It is useful to give (6.31) in a form corresponding to quantitative spectral measurements. A spectrum analyzer measures the *rms* power in a bandwidth Δf centered about frequency f. It presents the power at each

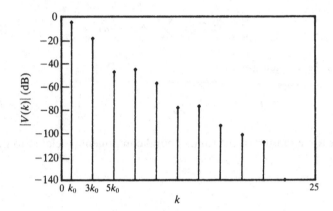

Fig. 6.6. Square wave voltage spectrum versus $k > 0$ for $a + d = 2g$ and $k_0 = \pi/6(a + d)$ corresponding to Fig. 6.3. The height of the arrows at each odd harmonic is proportional to the Fourier transform of the isolated pulse.

positive frequency f as the sum of the power at f and the negative frequency $-f$. The frequency bandwidth is related to the wavenumber bandwidth by:

$$\Delta f = v\Delta\left(\frac{1}{\lambda}\right) = \frac{v\Delta k}{2\pi} \tag{6.32}$$

Utilizing (6.32) and (6.31) yields:

$$V_{rms}(k = mk_0) = \frac{2k_0}{\sqrt{2\pi}} |V_{sp}(k)|, \ m = 1, 3, 5, 7, \dots \tag{6.33}$$

Equation (6.32) represents the power in each harmonic, as would be measured on a spectrum analyzer, so only positive frequency components are given. For real functions $V(x)$ the power is the same for $+k$ and $-k$ even though $V(k) \neq V(-k)$. Since the transform is given by a series of infinitely narrow pulses, the power is independent of the measuring bandwidth. The factor of k_0 in (6.33) shows that the result of recording an infinite sequence of transitions at fixed spacing and alternating polarity is equivalent to a differentiation.

The Fourier transform of the isolated pulse (5.40) combined with (6.33) yields:

$$V_{rms}(k = mk_0) = \frac{2k_0 kNWEv\mu_0}{\sqrt{2\pi}} \, |h_s^*(k)|e^{-|k|d}$$
$$\left| \int_0^\delta dy' e^{-|k|y'}(M_x(k, y') - i\,\text{sgn}(k)M_y(k, y')) \right|$$

(6.34)

In (5.34) the phase information of the transition center location is part of $M(k, y')$. An important case is discussed first: a transition of magnetization described by the normalized derivative $F(k)$ (from 5.44) but with arbitrary fixed angle at each depth with respect to the longitudinal direction (as in (5.45)). Equation (6.34) simplifies to:

$$V_{rms}(k = mk_0) = \frac{4k_0 NWEv\mu_0}{\sqrt{2\pi}} \, |h_s^*(k)|e^{-|k|d}$$
$$\left| \int_0^\delta dy' e^{-|k|y'} F(k, y') e^{i(\text{sgn}(k)\theta(y') - kx_0(y'))} \right|$$

(6.35)

In (6.35), $F(k, y')$ denotes the Fourier transform of the transition at each depth with center location at $x_0 = 0$ at all y'. The phase shift due to a changing transition center is in the exponential. Whatever the depth variation of the magnetization orientation, $\theta(y')$, if all the magnetizations are rotated by a fixed angle $(\theta(y') \to \theta(y') + \theta_0)$, then the spectrum does not change. If the magnetization orientation is not a function of depth, then (6.35) becomes:

$$V_{rms}(k = mk_0) = \frac{4k_0 NWEv\mu_0 M_r}{\sqrt{2\pi}} \, |h_s^*(k)|e^{-|k|d} \left| \int_0^\delta dy' e^{-|k|y'} F(k, y') \right|$$

(6.36)

Even though a rotation of the magnetization changes dramatically the shape of the isolated pulse as discussed in Chapter 5, *the amplitude of the spectrum does not change.* Only the phase changes.

Phase information can be examined by utilization of (6.31) without the absolute value signs appropriate to an amplitude (rms) measurement. The phase associated with (6.34) is simply (without the absolute value signs) the arctangent of the imaginary part divided by the real part. In tape recording the phase is complicated due to a changing phase of the transition center with depth as well as a contribution of vertical magnetization to a largely longitudinal magnetization (e.g. Haynes, 1976, 1977; Yeh & Niedermeyer, 1990; Bertram, *et al.*, 1992), but at long wavelengths the phase is solely due to the orientation of the saturated regions far from the transition (assuming a symmetric replay head). Neglecting variations with depth, the phase is $\theta(k \to 0)$ $= \tan^{-1}(M_y^r/M_x^r)$. A measurement of low frequency phase yields the angle of remanent magnetization. This discussion of phase applies generally for Fourier transforms and therefore as well to the isolated pulse analyzed in Chapter 5.

The spectrum is simplified further to the case of a magnetization pattern that does not vary with depth into the medium. In that case the spectrum becomes:

$$V_{rms}(k = mk_0) = \frac{4NWEv\mu_0 M_r}{\sqrt{2\pi m}} \; (1 - e^{-mk_0\delta})|h_s^*(mk_0)|e^{-mk_0 d}|F(mk_0)|$$

$$(6.37)$$

Note that the harmonics are weighted by spectral components of the head field, the transition shape, spacing loss, and a term dependent on the medium thickness from $m = 1, 3, 5, \ldots$.

The 'spectrum'

Square wave recording is often performed over a wide range of bit separations B or fundamental wavenumber k_0. The fundamental component is measured at each frequency on a spectrum analyzer and plotted as an envelope versus recording frequency. Such a plot is often referred to as the 'spectrum,' and with the simplifications leading to (6.37) is given by:

$$V_{rms}^{\text{Fund}}(k_0) = \frac{4NWEv\mu_0 M_r}{\sqrt{2\pi}} \; (1 - e^{-k_0\delta})|h_s^*(k_0)|e^{-k_0 d}|F(k_0)| \qquad (6.38)$$

or for thin media:

$$V_{rms}^{\text{Fund}}(k_0) = \frac{4NWEv\mu_0 M_r k_0\delta}{\sqrt{2\pi}} \; |h_s^*(k_0)|e^{-k_0 d}|F(k_0)| \qquad (6.39)$$

It is to be emphasized that to obtain the 'spectrum' a virtual continuum of k_0 must be utilized for recording and the fundamental component measured at each one. For the spectrum to have meaning $F(k_0)$ must be constant in functional form: the transition shape must not change as a function of k_0 so that linear superposition holds. As discussed in Chapter 8, this occurs if a constant recording field is utilized to write the transition. A constant deep-gap field is obtained by utilizing a constant record current amplitude. If the head efficiency is frequency dependent, the record current must be adjusted appropriately.

The spectrum differs from the Fourier transform of the isolated pulse by the factor $\sqrt{2k_0}/\pi$ as given in (6.33) for any general magnetization transition. For the special case of thin media the spectrum and the Fourier transform of the isolated pulse are given respectively by (6.39) and (5.55). A comparison of the 'spectrum' of square wave recording and the Fourier transform of the isolated pulse is shown in Fig. 6.7 for a simple long-pole head of gap length g and an arctangent transition of length a and head medium spacing d. As in Fig. 5.9 the Fourier transform of the isolated pulse (except for the modification due to thin film heads) is largest at low frequencies and decreases at short wavelengths due to the effective spacing loss and gap loss. The spectrum (6.33) due to the factor k_0 increases linearly at low frequencies (curved on a logarithmic scale) and reaches a maximum, subsequently decreasing due to the above losses. The constants $NWEv\mu_0 M_r \delta$ may be obtained from the long wavelength measurements of either the asymptote of the isolated pulse or from the linear slope of the 'spectrum.' In actual measurements the presence of system noise makes analysis of the isolated pulse transform more accurate for the playback proportionality constants. However, a determination of the gap length from the first gap null is more accurate with the 'spectrum' measurement.

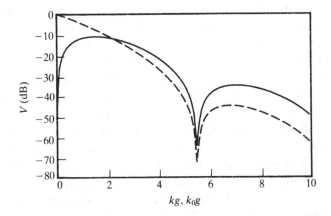

Fig. 6.7. Plot of the square wave 'spectrum' of the fundamental component (solid) and Fourier transform of the isolated pulse versus $k_0 g$, kg respectively. Equations (5.55) and (6.39) were utilized with an arctangent transition with $a + d = g/2$. The logarithmic plot is normalized to 0dB $\rightarrow 2NWEv\mu_0 M_r \delta$.

In Fig. 6.7 or (5.44) the Fourier transform of the isolated pulse results from one measurement (suitably averaged) of isolated pulses recorded by square wave recording at a single frequency. The isolated pulse is independent of the frequency as long as the bit separation is large compared to the extent of a single transition. The 'k' in (5.44) or Fig. 6.7 is simply the transform wavenumber that spans an infinite range and is not at all related to the bit separation of recording. The 'spectrum' is an envelope of many measurements over a range of wavenumbers k_0 (each single measurement resembling Fig. 6.6) and represents the fundamental component of each measurement.

Wallace factor or 'thickness loss'

The integration through the medium thickness results in a spectral term that is equivalent to the weighting of each lamina at each separation from the head by a spacing loss $\exp(-ky')$. This integration can be complicated if the magnetization pattern is depth dependent; however, if the magnetization is invariant with depth the effect is the difference between (6.38) and (6.39). The effective depth, which applies to all Fourier transforms and spectra, is therefore:

$$\frac{\delta_{\text{eff}}}{\delta} = \frac{1 - e^{-k\delta}}{k\delta} \qquad (6.40)$$

This thickness factor is termed the Wallace factor or occasionally the 'thickness' loss (Wallace, 1951). This factor is plotted on a log–log scale in Fig. 6.8. For long wavelengths $k\delta \ll 1(\lambda \gg 2\pi\delta)$, the factor is unity and the spectrum varies as the thickness of the film. This factor continually decreases with increasing wavenumber: hence the term 'thickness loss.' For short wavelengths $k\delta \gg 1(\lambda \ll 2\pi\delta)$, (6.40) decreases inversely with wavenumber, yielding an effective thickness less than the recording thickness: $\delta_{\text{eff}} = \lambda/2\pi$. At these short wavelengths the exponential weighting causes the reproduce voltage spectrum to 'see' only a fraction of the coating thickness. The term 'thickness loss' is in fact a misnomer because increasing the thickness, for example in (6.38), never reduces the output: at short wavelengths it simply makes no difference (apart from the effect of thickness on the transition shape $F(x)$ during the record process). Modification of the thickness term arises, for example, from a finite medium permeability (Westmijze, 1953; Bertram, 1978) and a transition length that increases with depth into the medium (Middleton & Wisely, 1976).

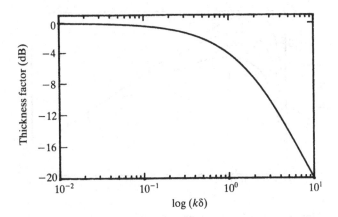

Fig. 6.8. Log–log plot of spectral thickness factor versus normalized $(1 - e^{-k\delta})/k\delta$ wavenumber $k\delta$.

It is important to note that the thickness term results from the playback process and does not arise from the recording process. For thick particulate media the depth of recording at high density optimization (~ 0.25–$1.0\ \mu$m) is less than the coating thickness (3–8μm) and results from an optimization of many factors, as discussed in Chapter 8. It is not simply equal to the effective depth given by (6.40).

Analysis of spectra

Experimental measurements of 'spectra' of square wave recording can be utilized to determine recording parameters, as was discussed in Chapter 5 for a measurement of the Fourier transform of the isolated pulse. Several aspects of analysis will be discussed in terms of simple depth-independent transitions. For an arctangent transition (6.38) becomes:

$$V_{rms}^{\text{Fund}}(k_0) = \frac{4NWEv\mu_0 M_r}{\sqrt{2}\pi}\ (1 - e^{-k_0\delta})|h_s^*(k_0)|e^{-k_0(d+a)} \qquad (6.41)$$

The long wavelength linear range is given (independent of the transition shape) by:

$$V_{rms}^{\text{Fund}}(k_0) = \frac{4NWEv\mu_0 M_r k_0\delta}{\sqrt{2}\pi} = \frac{8NWEf\mu_0 M_r\delta}{\sqrt{2}} \qquad (6.42)$$

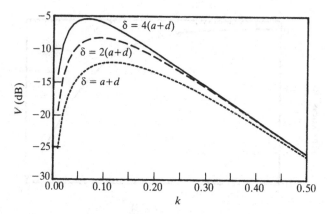

Fig. 6.9. Plot of replay voltage spectra (dB) versus linear wavenumber for longitudinal media where the recording layer thickness δ varies compared to the effective spacing $(a + d)$.

In (6.42) playback with a long-pole head is assumed, or that the spectral corrections for a thin film head have been accurately made. In addition, it is assumed that the wavelength is not too long so that edge track effects occur. For analysis of the transition shape (6.41) is generally plotted on a log–linear scale after correction by $|h_s(k)|$ of either the gap loss for a long-pole head or a general thin film head surface transform. The result is plotted in Fig. 6.9, assuming an arctangent transition with $d + a = 3$ and $\delta = 1, 6, 12$ (arbitrary units). For thick recording layers ($\delta = 12$) the thickness term $1 - e^{-k_0\delta}$ becomes unity at frequencies low enough so that the good linear range of $e^{-k_0(d+a)}$ on a log–linear plot can be analyzed for $d + a$. If the recording layer is not sufficiently large, as the examples of $\delta = 1, 6$ show, the thickness term variation overlaps into the short wavelength region. The short wavelength can look approximately linear, and analysis in terms of $e^{-k_0(d+a)}$ will give an erroneously small effective spacing. Such an error will arise with films of medium thickness, such as metal evaporated tape ($d \sim 0.15\mu\text{m}$) when measured over practical frequency ranges (to $\lambda \sim 0.5\mu\text{m}$). For very thin films it is simply necessary to divide the spectra by k independent of the thickness for a determination of the effective spacing. In general, the thickness term in any measurement should be carefully removed even for thin films, in order to determine an effective spacing or, accurately, the transition shape via $F(k)$.

'Roll-off' curve and spectrum compared – simplified D_{50} analysis

The waveform for square wave recording shows an alternating polarity of isolated pulses at low density (Fig. 6.2). As the density is increased, the pulses overlap so that successively, beginning with the highest harmonics, the amplitudes are attenuated, until at very high densities only the fundamental component remains. At these high densities the roll-off curve becomes proportional to the rms fundamental voltage spectrum separated only by $1/\sqrt{2}$ relating peak to rms. An example is shown in Figs. 6.10(a), (b) for the case of thin longitudinal media played back with a long-pole head. Fig. 6.10(b) is the identical curve seen in Fig. 6.10(a), except that the voltage is plotted in a logarithmic (dB) scale. In Fig. 6.10(b) the proportionality of $1/\sqrt{2}$ or -3dB can easily be seen at high densities.

At low densities the roll-off curve levels asymptotically at the voltage for an isolated pulse, given in (5.51) for the case of an arctangent transition and playback with a simple long-pole head:

$$V^{\text{peak}} = \frac{4}{\pi g} \, NWEv\mu_0 M_r \delta \tan^{-1}\left(\frac{g}{2(d+a)}\right) \qquad (5.51)$$

The fundamental voltage component, the 'spectrum', at any recording density $D = k_0/\pi$ is given by (6.39):

$$V^{\text{fund}}_{\text{rms}} = \frac{4NWEv\mu_0 M_r \delta k_0}{\sqrt{2\pi}} \, e^{-k_0(d+a)} \, \frac{\sin k_0 g/2}{k_0 g/2} \qquad (6.43)$$

where a Karlqvist approximation for the head surface field (3.17) has been utilized.

A simple relation for the D_{50} density can be obtained from (5.51) and (6.43) if the assumption is made that at D_{50} the waveform is approximately sinusoidal (Problem 6.9). In that case the D_{50} density occurs at that k_{50} where:

$$\sqrt{2}V^{\text{Fund}}_{\text{rms}}(k_{50}) = V^{\text{max}}/2 \qquad (6.44)$$

At D_{50} half the peak voltage of the isolated pulse is simply equal to the fundamental of the spectrum increased by $\sqrt{2}$ to account for the rms level. Of course, viewing Fig. 6.10, the solution of (6.44) involves utilization of the high density side of the spectrum. Using (6.39) and (5.51) in (6.44), the simplified condition:

$$\tan^{-1}\left(\frac{g}{2(d+a)}\right) = 4e^{-\pi\left(\frac{d+a}{g}\right)D_{50}g} \sin\frac{\pi D_{50}g}{2} \qquad (6.45)$$

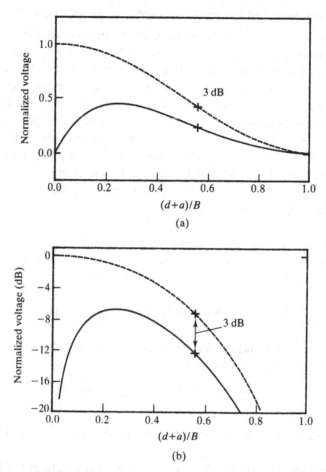

Fig. 6.10.   Plot of 'roll-off' curve (as in Fig. 6.4) and corresponding spectrum for square wave recording versus recording density on thin longitudinal media. Voltage is normalized to peak voltage of the isolated pulse. Plot (b) is (a) with the voltage scale plotted logarithmically. The D_{50} density is shown, where the 'roll-off' curve decreases to half its maximum.

is obtained. Equation (6.45) may be written in a simple normalized form of $D_{50}g$ versus $(d + a)/g$ for the assumption of an arctangent transition. This expression can easily be solved by simple iteration and is plotted in Fig. 6.11. Thus, given the transition parameter, the flying height, and the playback gap, D_{50} can be read off this curve. PW_{50} is also given utilizing the thin medium expression given in (5.53). The product $D_{50}PW_{50}$ is also shown dashed where it approaches the relation (6.24) for a Lorentzian pulse at small gaps. Typically, in high density recording on thin film

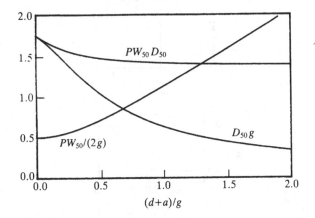

Fig. 6.11. Plot of $D_{50}g$, $PW_{50}/2g$ and the product $PW_{50}D_{50}$ versus the scaled ratio $(a + d)/g$ for thin longitudinal media assuming an arctangent transition.

media, $a + d \sim 0.5g$ so that $D_{50}g \sim 1.0$ and $PW_{50} \sim 1.44g$. For $g = 0.5\mu m$, Fig. 6.11 yields $D_{50} \sim 50$ KFCI and $PW_{50} \sim 72nm$. Fig. 6.11 is accurate when the transition is closely approximated by an arctangent. However, extension for any transition shape, such as tanh, is straightforward. Utilization of thin film replay heads with finite pole length may be treated by the approximate method above; the finite pole lengths yield about 10% undulations in D_{50} versus the ratio p/g (Bertero, *et al.*, 1992).

Transfer function

Signal processing techniques commonly utilize the transfer function for linear system analysis. As discussed, the magnetic recording channel possesses the 'quasi' linearity of linear superposition and only under certain conditions. When linear superposition is applicable, a transfer function can be defined. In general, the transfer function of a linear system is the Fourier transform of the impulse response. In magnetic recording an impulse cannot be written; the fundamental process is the writing of a step or a single transition (Fig. 6.12): the medium responds naturally to the writing of a step transition. The writing of an impulse would involve writing a dibit or two transitions arbitrarily closely together. In that case, as discussed in Chapter 8, the recording process becomes completely non-linear and the two transitions comprising the impulse demagnetize each other.

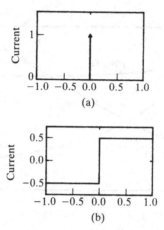

Fig. 6.12. (a) Impulse for specification of a transfer function. (b) Step function input characteristic of the magnetic recording channel.

Nevertheless, a transfer function can be formally written and utilized for linear signal processing if it is realized that the amplitude is fixed: only linear superposition holds. The impulse response is related to the step response by a simple derivative:

$$V_{imp}(x) = \frac{1}{2} \frac{\mathrm{d}V_{step}(x)}{\mathrm{d}t} \tag{6.46}$$

The Fourier transform of (6.46) yields the transfer function:

$$T(k) = \frac{ikV_{sp}(k)}{2} \tag{6.47}$$

Thus, the transfer function can be obtained from the Fourier transform of the isolated pulse. However, the fundamental component of square wave recording at wavenumber $k_0 = \pi/B > 0$ is related to the step response by (6.31, 6.33):

$$V^{\mathrm{Fund}}(k_0) = \frac{2k_0}{\pi} V_{sp}(k_0) \tag{6.48}$$

which has both amplitude (6.34) and phase information. Thus the amplitude of the transfer function is directly proportional to the fundamental component of the spectrum:

$$T(k > 0) = \frac{i\pi}{4} V^{\mathrm{Fund}}(k) \tag{6.49}$$

with amplitude:

$$|T(k > 0)| = \frac{\sqrt{2}\pi}{4} V_{\text{rms}}^{\text{Fund}}(k) \qquad (6.50)$$

A measurement of the spectrum yields the transfer function. The proportionality constant is simply the difference between the amplitude of a sine wave (transfer function) and the fundamental component of a square wave (spectrum).

Linear bit shift

Linear superposition of the transitions that comprise an all 111s or square wave pattern yields a playback voltage consisting of a sequence of transitions of alternating polarity where the peak voltage decreases monotonically with density. However, peak shift does not occur: the peaks of alternating polarity occur equispaced at bit spacing B. For all other patterns, which will not have the symmetry of a square wave, the peak locations shift affecting detection accuracy in a digital channel. An example of linear peak shift is given here for the case of a dibit or the playback of two bits of opposite polarity. For simplicity a Lorentzian transition shape is considered:

$$V_{sp}(x) = \frac{V_{sp}^{peak}(PW_{50}/2)^2}{(x^2 + (PW_{50}/2)^2)} \qquad (6.51)$$

Consider two bits of opposite polarity written at $x = -B/2$ and $x = B/2$. The net voltage by linear superposition is simply:

$$V(x) = \frac{V_{sp}^{peak} PW_{50}^{\ 2}}{4}\left(\frac{1}{(x+B/2)^2 + (PW_{50}/2)^2} - \frac{1}{(x-B/2)^2 + (PW_{50}/2)^2}\right) \qquad (6.52)$$

Equation (6.52) is plotted in Fig. 6.13 for various bit separations B relative to the pulse width PW_{50}. The addition of the two pulses causes the amplitudes to decrease and the peak voltages to shift apart. Linear bit shift in longitudinal recording tends to 'push bits away from each other'. In general, in Figs. 6.13(a)–(e) the dibit amplitude continually decreases, and the peak separation is reduced as the transition separation is reduced.

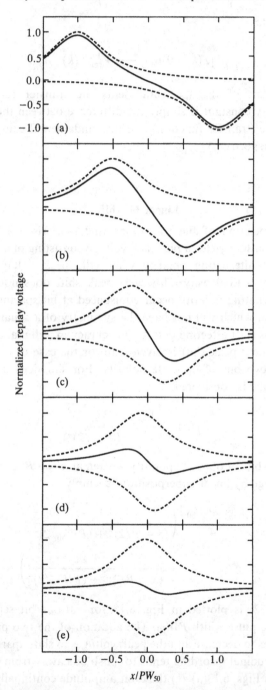

Fig. 6.13. Dibit response versus density utilizing the linear superposition of Lorentzian pulses. (a) $B = 2PW_{50}$, (b) $B = PW_{50}$, (c) $B = 0.5PW_{50}$, (d) $B = 0.2PW_{50}$, (e) $B = 0.1PW_{50}$. Individual pulses are shown dashed.

At high densities $B \ll PW_{50}$ (e.g. Figs. 6.13(d), (e)), the dibit peaks remain at a fixed location and only the amplitude decreases.

For Lorentzian pulses the bit shift can be written from (6.52) as:

$$\frac{\Delta x}{B/2} = \sqrt{\frac{1 - (PW_{50}/B)^2 + 2\sqrt{1 + (PW_{50}/B)^2 + (PW_{50}/B)^4}}{3}} - 1 \quad (6.53)$$

At high densities this expression reduces to:

$$\Delta x = \frac{PW_{50}}{2\sqrt{3}} - \frac{B}{2} \quad (6.54)$$

so that the bit locations occur at:

$$x_{peak} = \pm \frac{PW_{50}}{2\sqrt{3}} \quad (6.55)$$

independent of the bit length and proportional to the pulse width.

A simple, general, view of high-density dibit recording is that the dibit pulse is proportional to the derivative of the single pulse:

$$V_{dibit}(x) \approx B \frac{\partial V_{sp}(x)}{\partial x} \quad (6.56)$$

Equation (6.56) immediately shows that the high density dibit pulse shape (peak shift and peak location) are fixed and that only the amplitude decreases, proportional to the dibit separation. In Fig. 6.14 dibit peak location, peak shift, and dibit amplitude are plotted versus density. The

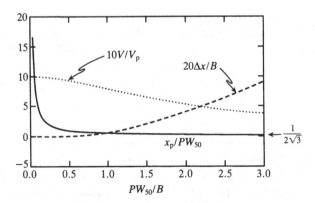

Fig. 6.14. Dibit peak location (solid), peak shift relative to half bit cell size (dashed), and dibit amplitude (dotted) versus density.

amplitude decreases continuously, the peak shift *relative* to the bit separation increases continuously, but the absolute dibit peak position decreases to a fixed location as density is increased. In dibit (multi-transition) recording the decrease in amplitude with increasing density is increased profoundly by non-linearities in the recording process, as discussed in Chapter 9.

Problems

Problem 6.1 Numerically evaluate D_{50} and PW_{50} for the transitions in Table 4.1 for thin film media with longitudinal magnetization transitions and a finite-gap, long-pole head. Plot versus $(d + a)/g$ even though such scaling applies only to an arctangent transition. Show where scaled plots are in error.

Problem 6.2 Show that the 'roll-off' curve should exhibit a gap null, as does the spectrum. For a long-pole head show that the first null of the 'roll-off' curve occurs at a slightly shorter wavelength than that of the head–field transform ($\lambda_1 < 1.136g$). In the Karlqvist approximation show that both nulls occur at the same wavelength ($\lambda = g$).

Problem 6.3 Using (6.38), show that at short wavelengths, for a thick medium recorded uniformly with depth, 90% of the output arises from a surface layer a distance $\lambda/3$ into the medium and that 50% of the output arises from a surface layer of thickness $\lambda/6$.

Problem 6.4 Consider a disk drive utilizing thin longitudinal media and long-pole, finite-gap heads with $a = d = 5\mu''$, $g = 0.25\mu\text{m}$, $M_r\delta = 2$ memu/cm^2, $N = 14$, $E = 0.6$, $v = 400$ in/s, and $W = 0.5$ mil. Find the peak voltage of the isolated pulse, PW_{50}, and D_{50}. Note that all formulae are in MKS units.

Problem 6.5 Assume thin media recorded with a transition of longitudinal and vertical magnetization both of the same arctangent transition shape but with different transition locations x_0^x, x_0^y. Derive an expression for the amplitude and phase of square wave recording of this fundamental transition. Sketch the variation of the fundamental component ($k = k_0$) amplitude and phase with frequency (k_0).

Problem 6.6 A simple model of a transition in longitudinally oriented tape (Beardsley, 1982a), is longitudinal magnetization except in the center of the transition where magnetization vector rotation produces vertical magnetization at the transition center. Model such a transition by a distribution of sharp transitions whose transition center depends on the angle of magnetization. Let the transitions have the form of (5.45) where θ ranges uniformly from $-\theta_0$ to $+\theta_0$. Let the transition centers follow $x_0 = t\theta/\theta_0$ where t is a constant. Derive expressions for the isolated pulse shape and sketch the pulse using $\theta_0 = \pi/4$ and $t = 0.5$. Derive and sketch the spectrum and phase, assuming no depth variation.

Problem 6.7 At D_{50} density given by (6.44), show that the waveform is approximately sinusoidal by evaluating the ratio of the third harmonic to the fundamental. Obtain the harmonics from (6.37). Use the typical value of $a + d = g/2$.

Problem 6.8 Using a simple approximation for the playback with a thin film head of a single transition of longitudinal magnetization recorded in thin media, plot dibit waveforms versus density. Plot the peak shift and dibit amplitude versus density and compare with that from a long-pole head.

Problem 6.9 Utilize a simple playback pulse form for perpendicular recording and show that linear superposition in dibit recording causes the bits to shift closer than the written separation.

7

Magnetoresistive heads

Introduction

Ferromagnetic materials exhibit a variety of phenomena associated with changes in resistivity due to changes of the state of magnetization. The magnetization can be altered in magnitude by changes in temperature or application of an applied field. A magnetic field can also rotate the direction of magnetization. The most useful effect is the anisotropic change in resistivity as the magnetization direction of a saturated specimen is rotated with respect to the direction of an applied current. A review of anisotropic magnetoresistance that covers both phenomenological as well as microscopic effects has been given by (McGuire & Potter, 1975). Field sensors using the anisotropic MR effect can be fabricated into extremely small devices using thin film technology. In general the anisotropic magnetoresistance of a ferromagnetic material is characterized by the two tensor components of resistivity parallel to the magnetization, $\rho_{\parallel}$, and perpendicular to the magnetization $\rho_{\perp}$. If a current is applied to an MR material (e.g. Fig. 7.1(a)), then the resistance as monitored by the voltage across material in the current direction, depends on the angle of the magnetization θ with respect to the current direction:

$$\rho = \rho_{\perp} + \Delta\rho \cos^2 \theta \qquad (7.1)$$

where ρ is the film resistivity and $\Delta\rho = \rho_{\parallel} - \rho_{\perp}$ represents the anisotropic MR effect. The magnetization angle is altered from an initial configuration by the application of fields transverse to the film (H_{t}). In general, $\rho_{\parallel} > \rho_{\perp}$ so that the resistance decreases as the magnetization rotates away from the current direction. Equation (7.1) represents only a first order expansion of a general even symmetric variation. The most

166

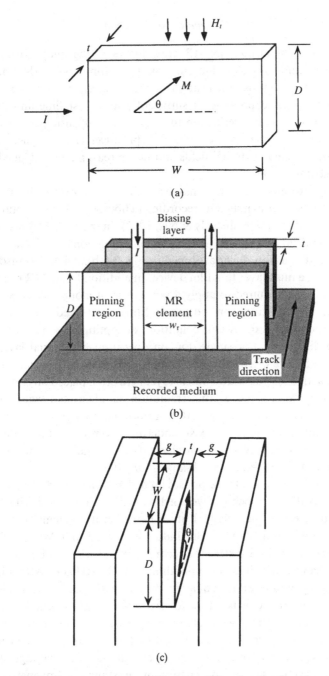

Fig. 7.1. (a) Basic geometry of an MR element showing the direction of current (*I*) across the element width *W*, the thickness *t*, and the depth *D*. θ is the angle of the magnetization with respect to the current direction. H_t is the field direction, termed the transverse field when uniform for testing. (b) Schematic of biasing and stabilization of an MR element for use as a playback head. (c) Illustration of a shielded element; *g* is the distance from the element to the shield for a centered element (Additional layers are not shown).

common material utilized in MR transducers is the alloy NiFe with $\Delta\rho/\rho \sim 2\%$ and $\rho \sim 20$ μohm-cm for thin films with thicknesses $t \sim 200\text{Å}$. At a composition near (81%Ni, 19%Fe), permalloy is non-magnetostrictive and possesses a sufficiently low crystalline anisotropy and high saturation magnetization to yield a high intrinsic permeability. Thus, a transducer may be designed that produces large magnetization rotation due to small external fields and hence reasonable MR playback voltages (Hunt, 1971).

The magnetoresistance effect may be utilized as an extremely efficient playback transducer in magnetic recording (Thompson, 1975; Thompson, *et al.*, 1975; Jeffers, 1986; Shelledy & Nix, 1992). In the basic MR process a current is applied to a thin film along its long direction (Fig. 7.1(a)). The film is placed in a transducer as shown in Fig. 7.1(b). An MR transducer is, in general, a multilayer thin film device. In addition to the MR element, current leads are required to apply a sense current along the cross-track direction. In addition, an adjacent film is utilized to bias the magnetization to a quiescent condition of optimum sensitivity and minimum distortion. For single domain operation, additional layers of 'exchange tabs' are often utilized to stabilize the film magnetization and define the active sensing width. Shield layers are often placed on either side of the element to enhance spatial resolution, as illustrated (without the additional layers shown) in Fig. 7.1(c). Additional layers for insulation as well as for interlayer adhesion exist rendering a complicated manufacturing process, especially when small elements are required for ultra high density recording (Jeffers, 1986; Tsang, *et al.*, 1990).

Multilayer films with MR responses an order of magnitude larger than that of permalloy are being developed as well as high sensitivity 'spin valve' structures (White, 1992). Future high density systems will likely utilize a combination of thin film inductive record and MR playback heads. In such a hybrid system the record gap and track width can both be larger than those values corresponding to MR playback. With a large record gap (generally about equal to that of a shielded MR shield to shield spacing) the effects of head saturation and subsequent poor overwrite can be reduced without compromising high density output. If the MR element is narrower than the width of the record head, then edge-track error signals may be minimized. Although the output voltage of the MR transducer caused by the change in resistance by magnetization rotation is limited by the requirement of linearity, large sense-current densities can be utilized ($J \geq 10^7$ A/cm^2) without overheating or film degradation. Thus, an advantage of MR heads is their large replay voltages compared to inductive heads, rendering the system signal to medium noise limited (Jeffers & Wachenshwanz, 1987).

The essential characteristic of MR heads is that the replay voltage is fundamentally independent of head–medium relative speed. The playback voltage depends only on the magnetic field in the element due to the presence of a recorded medium. The playback voltage utilizing (7.1) may be written approximately (Smith, 1987b) as:

$$V_{MR}(x) = \Delta\rho JW(\langle \cos^2 \theta(x,y,z) \rangle - \langle \cos^2 \theta_0 \rangle) \qquad (7.2)$$

where x denotes the relative position of the medium and the transducer; $\langle \rangle$ denotes an average over the film plane (y, z), because the MR magnetization is in general not completely uniform. The current density J is assumed to be uniform across the film. θ_0 represents the magnetization orientation distribution in the film without a signal present and $\theta(x, y, z)$ represents the orientation distribution in the presence of the medium with relative head–medium position x. Note that the speed v does not enter the expression, and time enters simply by $x = vt$. For a track width of $W = 5\mu m$, $J = 10^{11}$ A/m^2, $\Delta\rho = 4 \times 10^{-9}$ ohm-m, and a maximum change in the angular averages of about 0.25 to maintain linearity, a typical zero–peak voltage of $V_{MR} = 500$ µV is obtained.

The purpose of this chapter is to develop expressions that yield the angular variation in MR magnetization due to the presence of recorded transitions. The chapter is organized as follows. First a discussion of magnetization biasing for uniformly magnetized films is presented for both single and double layer films. Realistic magnetization patterns from numerical micromagnetic calculations are shown. Next a review of reciprocity principles for MR heads is given; the analysis is for elements where the magnetization does not vary across the track width, but a discussion of modification for truly finite track widths is given. The reciprocity principle is applied to obtain the MR playback voltage for shielded heads, including a quantitative comparison with inductive heads. Fourier transforms are presented, including a comparison of accurate and approximate voltage pulse shapes. The chapter concludes with a discussion of asymmetric side reading, as well as an SNR relation based on Johnson thermal noise.

Magnetization configurations

Due to large planar demagnetization fields the magnetization of MR films is restricted largely to the film plane. In general the magnetization is not uniform because of a micromagnetic configuration dominated by magnetostatic fields: θ varies somewhat from a desired equilibrium, reducing the sensitivity. Magnetization non-uniformity arises from demagnetization fields at the top and bottom edges as well as magnetizing pinning at either side of the defined track. In Fig. 7.2(a) the vertical magnetization component versus depth in the element along with the adjacent biasing film (SAL) magnetization is plotted for zero external field. A numerical calculation for an essentially infinitely wide track (2D) was utilized (Smith, 1987b). The MR element depth and film thickness were 4 μm and 250Å, respectively. Both elements had a saturation flux density of 1T, and the SAL thickness and spacing to the MR element were 200Å and 100Å, respectively. A current density of $J = 10^7$A/cm^2 was applied only to the MR element. In both films the magnetization is approximately uniform. The configuration is such that the SAL layer is saturated and the MR element is magnetized to approximately 45° to the current direction. At the top and bottom surface in both films, the magnetization rotates to be parallel to the surfaces reducing the magnetostatic energy.

In Fig. 7.2(b) the change in MR resistance or device transfer characteristic is plotted. The solid curve is for a SAL layer well into saturation for near zero external fields and the long dash curve is for a SAL layer not into saturation for small fields. For the saturated SAL, parameters corresponding to those for Fig. 7.2(a) were utilized, whereas for the unsaturated SAL the sole difference was a current density of 0.5×10^7 A/cm^2. Equation (7.1) was utilized with the cos$^2\theta$ averaged over the MR element area (y,z plane) for each value of uniform external field. An effective bias field from the SAL layer of $H_{\text{bias}} \sim -53$ Oe can be inferred from Fig. 7.2(b), because an external field of $H_t \sim 53$ Oe is required to bring the MR magnetization along the current direction ($\theta = 0°$ uniformly over the film). At zero field, $H_t = 0.0$, the MR magnetization is approximately at 45° in the center for the solid curve example (Fig. 7.2(a)) with a slightly smaller transfer average ($\langle\cos^2\theta\rangle \sim 0.6$). Increasing the field (in the negative direction for this example) brings the MR magnetization to saturation in the vertical direction, reducing the resistance. Both transfer curves in Fig. 7.2(b) (solid and long dash) exhibit a gradual approach to saturation due to the surface demagnetizing fields. Figure 7.2(b) is one example of a transducer; however, it is clear that magnetization patterns as well as the transfer curve depend on the geometry, applied current and film magnetization of the structure. In general the configuration is set so that

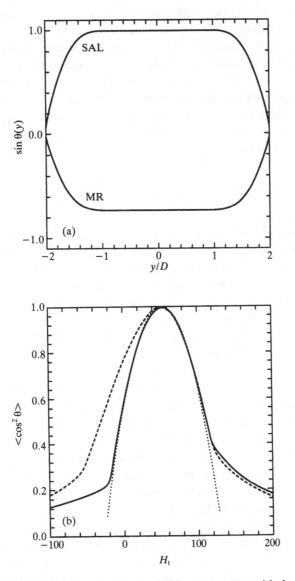

Fig. 7.2. (a) Variation of vertical magnetization component with depth into the film for the MR and SAL films in a double film structure. (b) Variation of MR response due to a uniform transverse (vertical) external field. The solid and long dashed curves are for a double layer film where the SAL bias layer is in and out of saturation respectively. The short dash curve is the result of a simple model. These numerical results are courtesy of Neil Smith.

the quiescent magnetization of the MR element is approximately 45°. As viewed in Fig. 7.2(b), such a condition yields a transfer characteristic of maximum sensitivity and minimum distortion.

The transfer curve may be estimated using simple approximations. The variation of magnetization angle with external field for a long (2D) film with no cross-track variation is calculated first for a uniform magnetization over the element depth. A single film is considered first. In general, neglecting exchange coupling, the magnetic energy per film area in the (x, y) plane for a very thin film $(t \ll D \ll W)$ may be written (e.g. Cullity, 1972) as:

$$E/M_{\mathrm{s}} = -\langle H_{\mathrm{t}} \sin\theta(y)\rangle + \frac{1}{2} H_{\mathrm{K}} \langle \sin^2\theta(y)\rangle - \frac{1}{2} \langle H_{\mathrm{D}}(y) \sin\theta(y)\rangle \quad (7.3)$$

In (7.3) $\theta(y)$ is the variation of magnetization angle versus depth into the element $(M_y(y) = M_{\mathrm{s}} \sin\theta(y))$. $\langle\ \rangle$ denotes an average over the film depth. The first term yields a reduction in energy if the magnetization rotates toward the external field H_{t} (increasing θ) in the transverse (y) direction. The second term results from a presumed (uniform) uniaxial anisotropy field H_{K} in the long direction that attempts to keep the mangetization along the track direction (decreasing θ). The last term is the film magnetostatic energy, where H_{D} is the demagnetizing field that depends on position as well as the magnetization configuration $\theta(y)$. In general, the magnetostatic field opposes the magnetization and is proportional to the vertical magnetization component $(\sin\theta)$. Thus the magnetostatic energy varies quadratically with $\sin\theta(y)$ and adds to the net anisotropy energy. The difficulty in solving (7.3) is due to the spatial variation of $H_{\mathrm{D}}(y)$ (as well as possible spatial variations of H_{t}). The concentration of magnetic poles at the surface causes H_{D} to be largest at the top and bottom surface (Problem 7.1), so that the magnitude of θ is largest in the film center (Fig. 7.2(a)).

A simplified solution of (7.3) is found, first, by assuming that the magnetization does not vary throughout the depth of the film $(\theta(y) = \theta)$. As indicated by the numerical example in Fig. 7.2(a), for realistic structures, this is a good approximation. In this case the demagnetizing field is still non-uniform (2.14); however, (7.3) dictates that average fields be utilized for uniform magnetization. Averaging (2.14) yields:

$$\langle H_{\mathrm{D}}(y)\rangle = -\langle N\rangle M_y \approx -\frac{tM_y}{\pi D} \ln\frac{D-\Delta}{\Delta} \quad (7.4)$$

In (7.4) Δ is a distance from each edge where the averaging does not occur, to account, approximately, for the reduction in magnetostatic energy due to magnetization rotation (Problem 7.2). In (7.4) $\Delta^2 \gg (t/2)^2$ is assumed where t is the film thickness. (The center value ($y = 0$) of the demagnetizing field is: $H_D(0) = -2tM_y/\pi D$). Assuming a uniform magnetization and finding the magnetization angle that minimizes the energy in (7.3) yields:

$$\sin\theta = \frac{H_{ext} + \langle H_{bias}\rangle}{H_K + \langle N\rangle M_s} \qquad (7.5)$$

where the applied field has been separated into an external field and an element averaged bias field: $H_t = H_{ext} + \langle H_{bias}\rangle$. The anisotropy field is generally much smaller than the magnetostatic field: $H_K \sim 0.4\text{kA/m}$ (5 Oe) and $\langle N\rangle M_s \sim 6.4\text{ kA/m}$ (80 Oe) for $t \sim 200$ Å, $D \sim 1\mu\text{m}$, $M_s \sim 800\text{kA/m}$, and $\Delta = 750$ Å. Thus, H_K in the denominator in (7.5) may be neglected. (The intrinsic permeability of the film, $\mu_r = 1 + M_s/H_K \sim 2000$, is quite large.) The bias field is set so that the quiescent magnetization angle is $\theta_0 = 45°$ in (7.5): $\langle H_{bias}\rangle = \langle N\rangle M_s/\sqrt{2}$. Thus, the variation of resistance with field is simply:

$$\rho = \rho_0 + \Delta\rho\left(1 - \left(\frac{H_{sig} + \langle H_{bias}\rangle}{\langle N\rangle M_s}\right)^2\right) \qquad (7.6)$$

The quadratic variation with field is sketched in Fig. 7.2(b) (short dash) to compare with the simulation. The zero field position of (7.6) was shifted to match the peak position of the simulation. The agreement is quite good except in the high field region where a slow approach to saturation occurs. The zero field sensitivity, is:

$$\frac{\partial\rho}{\partial H_{sig}} = \frac{\sqrt{2}\Delta\rho}{\langle N\rangle M_s}, \text{ or } \frac{\partial\cos^2\theta}{\partial H_{sig}} = \frac{\sqrt{2}}{\langle N\rangle M_s} \approx \frac{\sqrt{2}\pi D}{M_s t \ln(D/\Delta)} \qquad (7.7)$$

The above discussion applies to a bias field generated by a fixed bias film, for example as produced by a permanent magnet. If a SAL bias is utilized, it is important to configure the structure so that the SAL is approximately saturated at the operating condition (Fig. 7.2(a)). A simplified analysis of SAL biasing is given here. It is assumed, as shown in Fig. 7.3, that an MR element of thickness t_{MR} and uniform magnetization at angle θ_{MR} is adjacent to a SAL element of thickness t_{SAL}, also uniformly magnetized. Positive angle θ_{SAL} of the SAL magnetization is taken opposite to that of the MR element. The SAL

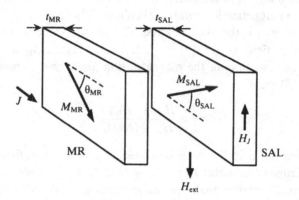

Fig. 7.3. Magnetization orientation definitions for analysis of a coupled film structure.

element experiences a field H_J due to current in the MR element. Both elements experience demagnetizing fields and magnetostatic coupling fields directed along the respective magnetizations. The assumption of uniform magnetization is reasonable from Fig. 7.2(a) over most of the film depth for both the MR and SAL.

Equation (7.3) applies to both films. Including all fields, the coupled equations for the vertical component of the magnetizations for each film are:

$$\sin \theta_{MR} = \frac{H_{ext}}{\langle N_{MR} \rangle M_s^{MR}} + \left(\frac{M_s^{SAL} \langle N_{SAL} \rangle}{M_s^{MR} \langle N_{MR} \rangle} \right) (1 - \alpha) \sin \theta_{SAL}$$

$$\sin \theta_{SAL} = -\frac{H_{ext}}{\langle N_{SAL} \rangle M_s^{SAL}} + \frac{H_J}{\langle N_{SAL} \rangle M_s^{SAL} } \qquad (7.8)$$

$$+ \left(\frac{M_s^{MR} \langle N_{MR} \rangle}{M_s^{SAL} \langle N_{SAL} \rangle} \right) (1 - \alpha) \sin \theta_{MR}$$

The last terms in each expression are the magnetostatic coupling terms. It is assumed that the coupling field is similar in form, but slightly smaller by $(1 - \alpha)$ than the demagnetization term. The term α will depend on film thicknesses t_{MR}, t_{SAL} as well as the inter-film spacing. The coupled equations are, of course, readily solved. Energy minimization dictates that, in either equation, if the right hand side exceeds ± 1, the magnetizations will be saturated in the respective direction ($\theta = \pm 90°$).

In zero external field, the MR magnetization is set at $\theta_{MR} = 45°$. Two cases are considered corresponding to Fig. 7.2(b) of a saturated and unsaturated SAL layer. In the first case sufficient current, and, or magnetostatic coupling, is utilized so that the right hand side of the second equation exceeds unity at least for fields near $H_{ext} = 0$. If the SAL is saturated, $\theta_{SAL} = 90°$, then $\theta_{MR} = 45°$ is achieved by the condition:

$$\frac{M_s^{SAL} \langle N_{SAL} \rangle}{M_s^{MR} \langle N_{MR} \rangle} \approx \frac{M_s^{SAL} t_{SAL}}{M_s^{MR} t_{MR}} = \frac{1}{\sqrt{2}} \qquad (7.9)$$

This approximation occurs because the logarithmic factors in the average demagnetizing field are relatively insensitive to film thickness. In essence the saturation flux of the SAL should be less than that of the MR element, approximately in the ratio of $1/\sqrt{2}$, in order to properly bias the MR element. If, in the operating range, the SAL stays approximately in saturation, then (7.5) applies with H_{bias} replaced by the field due to the SAL acting on the MR element (the numerator of the last term of the first equation in (7.8) with $\theta_{SAL} = 90°$). The short dashed curve in Fig. 7.2(b) is (7.6) where H_{bias} has been adjusted for horizontal alignment and a best fit of $\Delta = 1.5 t_{MR}$ has been utilized in (7.4) to account for magnetization non-uniformities at the surfaces.

If, however, the SAL comes out of saturation, then the transfer curve is determined by simultaneous solution of the coupled equations (7.8). The vertical component of the MR magnetization is given (with $\alpha \ll 1$) by:

$$\sin \theta_{MR} \approx \frac{H_{ext}}{2 \langle N_{MR} \rangle M_s^{MR}} + \frac{(1 - \alpha) H_J}{2 \alpha M_s^{MR} \langle N_{MR} \rangle} \qquad (7.10)$$

This expression is in the same form as the first term of (7.8) for the case of a saturated SAL except for the factor of two reduction in the sensitivity to external field. Substitution of (7.10) into (7.1) yields a transfer curve with the same form as (7.6), but with a reduced sensitivity at zero field by a factor of two that is in excellent agreement with the numerical solution plotted (long dash) in Fig. 7.2(b). The factor of two reduction in sensitivity occurs independent of SAL magnetization orientation as long as the SAL is not saturated. For the numerical solution in Fig. 7.2(b), positive fields saturate the SAL so that the two cases (solid and long dash) coincide.

The equilibrium magnetization in the MR element due to a constant bias field that does not vary with depth may be expressed analytically by a very accurate approximation. A simple solution is indicated here to show the variation of magnetization near the film surfaces due to demagnetization fields. The film is assumed to be extremely thin with infinite permeability. Assuming a constant bias field H_{bias} the vertical component of bias magnetization is given (in MKS) by:

$$\frac{M_y^0(y)}{M_s} = \frac{H_{bias}D}{t}\sqrt{1-(2y/D)^2} \qquad (7.11)$$

Conformal transformation of the strip into a circle that has a uniform magnetization has been utilized to obtain (7.11), which agrees well with numerical micromagnetic analysis (Smith, *et al.*, 1992b). An alternative derivation that deals directly with the fields is given here. If the film is assumed to have an infinitely large permeability, then the internal field must vanish at all depths y in the element:

$$H_{bias} + H_d(y) = 0 \qquad (7.12)$$

Utilizing (2.22) for an infinitely thin film yields:

$$H_{bias} = \frac{t}{2\pi}\int_{-D/2}^{D/2} dy' \frac{\frac{\partial M_y^0(y')}{\partial y'}}{y-y'} \qquad (7.13)$$

A convenient variable change: $y' = D\sin\theta/2$, $y = aD/2$ yields:

$$H_{bias} = \frac{t}{\pi D}\int_{-\pi/2}^{\pi/2} d\theta \frac{\partial M_y^0(\theta)/\partial\theta}{a-\sin\theta} \qquad (7.14)$$

The vertical component of the bias magnetization is assumed to vanish at the element surfaces ($\theta = \pm\pi/2$) and since, by symmetry, the magnetization derivative is an odd function, the derivative may be expanded in a Fourier sine series that comprises only odd terms:

$$\frac{\partial M_y^0(\theta)}{\partial\theta} = \sum_{n=1,3,5,\dots}^{\infty} A_n \sin n\theta \qquad (7.15)$$

Subsitution of (7.15) into (7.14) and term by term integration yields a constant only for $n = 1$. The integration of the higher harmonics ($n \geq 3$) yields terms that depend on the depth y. Thus, all A_n must vanish except for A_1. Substitution of the first term of (7.15) in (7.13) gives

$A_1 = -H_{bias}D/t$. Resubstitution of the variable change, $\sin\theta = 2y/D$, yields (7.11).

The expression given by (7.11) approximately describes the magnetization variation near the film surfaces as seen in Fig. 7.2(a). In films biased by a permanent magnetization or a saturated SAL, the magnetostatic field from the SAL that biases the MR element is not constant, but decreases at the film center. This variation causes the film magnetization to be approximately constant over a large region at the film center.

Reciprocity for MR heads

If the signal field from recorded media is sufficiently small, the MR response will be linear. In that case a useful reciprocity relation similar to that derived in Chapter 5 can be developed (Potter, 1974; Smith & Wachenschwanz, 1987). Expanding (7.2) to first order, yields:

$$V_{MR}(x) = -2\Delta\rho JW\langle\sin\theta_0\Delta\sin\theta(x)\rangle \qquad (7.16)$$

where $\sin\theta_0$ is the equilibrium orientation distribution and $\Delta\sin\theta(x)$ represents variations from equilibrium due to a signal field. Since the vertical component of magnetization is $M_y = M_s\sin\theta$, (7.16) may be expressed as:

$$V_{MR}(x) = -\frac{2\Delta\rho JW}{DM_s^2}\int_0^D M_y^0(y')M_y^{sig}(x,y')\mathrm{d}y' \qquad (7.17)$$

In (7.17) the averaging is taken over the depth of the element. To include the effect of a finite track width an integral over the element width should replace the constant factor W. However, the analysis here is only for equilbrium or bias magnetizations that do not vary over the element width; nevertheless, the induced magnetization variation may be non-uniform over the width. The signal magnetization is related to the flux in the element by:

$$M_y^{sig}(x,y') = \frac{\Phi^{sig}(x,y')}{\mu_0 tW} \qquad (7.18)$$

Equation (7.18) holds for a high-permeability element (Problem 7.4). Subsition of (7.18) into (7.17) yields:

$$V_{MR}(x) = -\frac{2\Delta\rho J}{\mu_0 DtM_s^2}\oint_0^D M_y^0(y')\Phi^{sig}(x,y')\mathrm{d}y' \qquad (7.19)$$

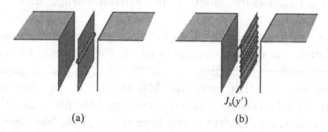

$J_s(y')$

(a)   (b)

Fig. 7.4. Illustration of (a) path of vector potential integration for direct calculation of replay voltage and (b) effective current distribution for reciprocity calculation.

Following the derivation of reciprocity for an inductive head in Chapter 5, the flux in the element may be expressed in terms of the vector potential (2.35).

$$\Phi^{\mathrm{sig}}(x, y') = \int_{\substack{\mathrm{MR} \\ \mathrm{area}}} A^{\mathrm{sig}}(x, y') \cdot \mathrm{d}l' \qquad (7.20)$$

The integral in (7.20) is taken around the MR element at a given depth y' (Fig. 7.4(a)). In general A^{sig} varies in direction and magnitude across the track as well as somewhat over the element thickness (A^{sig} is an odd function in x with respect to the element center).

The final result in this section applies to a finite track width as long as the equilibrium magnetization does not vary across the track. A 2D approximation that assumes no variation of signal flux over the track width direction (z) will be utilized for simplicity; in that case A^{sig} is independent of z and is directed along the track width direction. Treating A^{sig} as a scalar, (7.20) is simply $\Phi^{\mathrm{sig}}(x, y') = 2WA^{\mathrm{sig}}(x, y')$ where $A^{\mathrm{sig}}(x, y')$ is the vector potential at one edge of the film ($x' = t/2$ or $-t/2$). Subsituting (7.20) into (7.19) yields:

$$V_{\mathrm{MR}}(x) = -\frac{4\Delta\rho JW}{\mu_0 DtM_s^2} \int_0^D M_y^0(y') A^{\mathrm{sig}}(x, y') \mathrm{d}y' \qquad (7.21)$$

An equivalent surface current in terms of the equilibrium magnetization can be defined via (2.12):

$$J_s^0(y') = M_y^0(y') \qquad (7.22)$$

$J_s^0(y')$ can be thought of as a coil wrapped non-uniformly around the element (Fig. 7.4(b)), where the concentration of turns is proportional to the vertical component of the equilibrium magnetization. Substitution of (7.22) into (7.21) yields:

$$V_{\mathrm{MR}}(x) = -\frac{4\Delta\rho J W}{\mu_0 D t M_{\mathrm{s}}^2} \int_0^D J_{\mathrm{s}}^0(y') A^{\mathrm{sig}}(x, y') \mathrm{d}y' \qquad (7.23)$$

The integral in (7.23) is in the same form as (5.16) except that it is not per unit current I_{w}. In (5.16) the integral is over all space where a current density occurs. For the MR head expressed by (7.23) this integration is equivalent to an integration of the surface current density down the depth of the element times the length of the 'coil':

$$\iiint_{\mathrm{Vol}} J(r') \mathrm{d}^3 r' \to 2W \int_0^D J_{\mathrm{s}}(y') \mathrm{d}y' \qquad (7.24)$$

Using the equivalence of (5.16) with the form of (5.27) yields:

$$V_{\mathrm{MR}}(x) = \frac{2\Delta\rho J W}{D t M_{\mathrm{s}}^2} \int_{\mathrm{Area}} H(r' + x\hat{x}) \cdot M^{\mathrm{rec}}(r') \mathrm{d}^2 r' \qquad (7.25)$$

where the 2D formulation is assumed because the linear variation of track width is explicitly given and the integral (Area) is over the cross-section of the recording medium in the (x, y) plane. In (7.25) the explicit relative position of the head relative to a coordinate system in the medium, $x\hat{x}$, is shown. A general form for 3D structures, in order to include track edge effects, is

$$V_{\mathrm{MR}}(x) = \frac{2\Delta\rho J}{D t M_{\mathrm{s}}^2} \int_{\mathrm{Vol}} H(r' + x\hat{x}) \cdot M^{\mathrm{rec}}(r') \mathrm{d}^3 r' \qquad (7.26)$$

where the integration is over the entire volume of recorded medium.

In (7.25) and (7.26) $M^{\mathrm{rec}}(r')$ is the vector distribution of the permanent recorded magnetization, neglecting reversible effects due to a finite medium permeability. The magnetic field $H(r')$ is the field generated by the surface current $J_{\mathrm{s}}^0(y')$ (7.22), including the effects of all permeable regions: the MR element, the shields, reversible medium permeability, and possible keepers. In (7.24) and (7.25), as in subsequent expressions, the minus sign in front of the voltage expression will be neglected. The sign of the voltage depends on the direction of the equilibrium magnetization $M_0(y')$. The correct sign, if desired, can be deduced physically as discussed in the next section. Note that (7.25) and (7.26) apply for finite track width, where the effective current is wrapped around the element; however, a bias magnetization invariant across the track is assumed.

The expression given by (7.26) may be put in a form similar to those in Chapter 5 for an inductive head. Beyond the head surface, in the region of the recording medium, there is no source current. The effective magnetic reciprocal field therefore may be written in terms of a scalar potential: $H = -\nabla\Phi_{MR}$. Substitution into (7.26) with integration by parts yields:

$$V_{MR}(x) = \frac{2\Delta\rho J}{DtM_s^2}\int_{Vol} \Phi_{MR}(r' + x\hat{x})\nabla \cdot M^{rec}(r')d^3r' \qquad (7.27)$$

Thus, the MR replay voltage depends on the occurrence of magnetization variations or 'poles' in the recorded medium. As for playback with an inductive head, no voltage occurs for DC or uniformly magnetized media.

For a longitudinally recorded medium (7.27) simplifies to a form comparable to (5.36):

$$V_{MR}(x) = \frac{2\Delta\rho J}{DtM_s^2}\int_{Vol} \Phi_{MR}(x' + x, y', z') \frac{\partial M_x^{rec}(x', y', z')}{\partial x'} d^3r' \qquad (7.28)$$

A sketch of the potential variation of a typical (unshielded) single MR element is shown in Fig. 7.5. For a longitudinal recorded transition that is represented by localized charge in the medium, the replay voltage will have the symmetric shape versus time or position x of Fig. 7.5, which has the same symmetry as the replay voltage with an inductive head. The rule for converting the effect of a finite transition width of an arctangent shape to an effective spacing applies to all 2D fields. Thus,

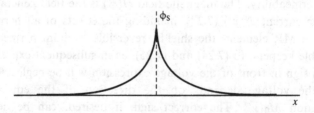

Fig. 7.5. Sketch of surface potential variation for a single, unshielded MR element.

for a wide track MR head, with a longitudinal magnetization transition of arctangent shape (parameter a) and thin media, (7.28) may be written as:

$$V_{MR}(x) = \frac{4\Delta\rho JWM_r\delta}{DtM_s^2}\,\Phi_{MR}(x+x_0, d+a) \qquad (7.29)$$

The voltage is simply the potential evaluated at effective spacing $d+a$. Similar approximations apply to magnetization patterns recorded at any orientation for wide-track (2D) replay expressions. Note that in (7.27–7.29), Φ_{MR} denotes an un-normalized scalar potential resulting from the equivalent current of the vertical component of the bias magnetization; it is not the flux as utilized in Chapter 3. Normalized potential is introduced in the section on shielded heads.

The voltage of an MR head utilizing reciprocity, in contrast to that for an inductive head, is not given in terms of a head field per unit current, nor is the correlation of effective field with recorded magnetization pattern in terms of the longitudinal derivative of magnetization. The MR voltage does not contain a speed term v, since there is no time derivative. The voltage expression (7.26) has the form of a reproduce flux (5.27), as for inductive heads, but as discussed below, the MR reciprocity field is of different form, approximately like a derivative, compared to the field of the inductive head.

Equations (7.26–7.29) may be utilized directly for any MR configuration. Examples for unshielded single and dual heads are given in (Smith & Wachenschwanz, 1987; Smith, *et al.*, 1992). In these expressions, the potential or field above the head in the vicinity of the recorded medium, is determined by establishing an equivalent current $J_s^0(y)$ in (7.22) and then solving for the fields including the effects of all permeable regions. Permeable regions always include the MR element, but permeable shields may also be included. The expressions were motivated by assuming that the bias magnetization does not vary across the track width (with z). In that case the effect of finite track width can be directly determined using (7.26–7.29) with suitable expressions for the fields off-track or at the track edges. For realistic finite track heads the bias magnetization will vary at the edges of the MR element. In that case, if the current is presumed uniform in the cross-track direction, then (7.22) applies with the local current equal to the bias magnetization:

$$J_s^0(y,z) = M_y^0(y,z) \qquad (7.30)$$

A visualization of (7.30) is not as simple as in Fig. 7.4(b); however, the fields are calculated in terms of the equivalent surface current (7.30) in the presence of *all* permeable media. If the current is not simply unidirectional, then the analysis is more involved (Smith, 1993).

It is important to note that the replay expressions derived in this section are best visualized in a 2D approximation. The replay voltage of a wide-track MR head with the media above the track can be directly determined, as discussed in the next section, utilizing simplified approximations for the surface potential. Off-track response is more complicated, since a biased, saturated MR element possesses a tensor intrinsic permeability that is large only in directions orthogonal to the bias magnetization. A simplified extension of reciprocity is given at the end of this chapter in the section on off-track response.

Application to shielded heads

In this section these expressions are evaluated for the example of shielded MR heads (Fig. 7.1(c)). Shielded heads are of practical interest and provide simple analytic evaluation of the expressions derived in the previous section. In addition, the response of a shielded head is closely approximated by that of an inductive head so direct numerical comparisons of relative replay voltage may be made. For shielded heads, as in the utilization of reciprocity for inductive heads in Chapter 5, it is convenient to define a normalized field h_{MR} in terms of the line integral of the field at the element surface across either gap from element to shield:

$$h_{MR}(r') = \frac{H(r')}{\int_{t/2}^{g+t/2} dx' h_x^s(x')} \qquad (7.31)$$

The efficiency of a shielded head may be defined similarly as for an inductive head:

$$E_{MR} = \frac{\int_{t/2}^{g+t/2} dx' h_x^s(x')}{\int_{0}^{D} dy' J_s(y')} \qquad (7.32)$$

or, utilizing (7.22):

$$E_{MR} = \frac{\int_{t/2}^{g+t/2} dx' \, \boldsymbol{h}_x^s(x')}{\int_0^D dy' \, M_y^0(y')} \qquad (7.33)$$

The above definition of the efficiency is most reasonable for shielded structures. If, for simplicity, the denominator in (7.33) is written in terms of the average:

$$\langle M_y^0 \rangle = \frac{1}{D} \int_0^D dy' \, M_y^0(y') \qquad (7.34)$$

then substitution of (7.31), (7.33), and (7.34) into (7.26) yields:

$$V_{MR}(x) = \frac{2\Delta\rho J E_{MR} \langle \sin\theta_0 \rangle}{t M_s} \int_{Vol} \boldsymbol{h}_{MR}(\boldsymbol{r}' + x\hat{\boldsymbol{x}}) \cdot \boldsymbol{M}^{rec}(\boldsymbol{r}') d^3 r' \qquad (7.35)$$

where $\sin\theta = M_y/M_s$ is the component of equilibrium or bias magnetization along the vertical direction in the element. An alternative form in terms of the magnetic charge ρ^{rec} (2.3) in the recording medium and the scalar potential of the MR equivalent field (7.27) is:

$$V_{MR}(x) = \frac{2\Delta\rho J E_{MR} \langle \sin\theta_0 \rangle}{t M_s} \int_{Vol} \Phi_{MR}(\boldsymbol{r}' + x\hat{\boldsymbol{x}}) \rho^{rec}(\boldsymbol{r}') d^3 r' \qquad (7.36)$$

In (7.36) Φ_{MR} is the scalar potential of the equivalent MR head field in the medium, normalized to a unit potential drop from the element to either shield.

In order to evaluate the MR voltage quantitatively, as well as to compare it with that of an inductive head, estimates of the efficiency E_{MR} and average bias magnetization orientation $\langle \sin\theta_0 \rangle$ must be given. The bias magnetization orientation for a typical (wide-track) film is shown in Fig. 7.2(a). The resulting $\langle \cos^2\theta_0 \rangle$ is shown in Fig. 7.2(b) for $H = 0$. Even though the magnetization is reasonably uniform over the depth of the element, (7.11) will be utilized to estimate $\langle \sin\theta_0 \rangle$. If the bias field is chosen to give a magnetization direction that is at 45° at the film center, then:

$$\langle \sin \theta_0 \rangle = \frac{1}{\sqrt{2}D} \int_{-D/2}^{D/2} dy' \sqrt{1 - (2y'/D)^2} = \frac{\pi}{4\sqrt{2}} \approx 0.55$$

$$\langle \cos^2 \theta_0 \rangle = \frac{1}{\sqrt{2}D} \int_{-D/2}^{D/2} dy' (1 + (2y'/D)^2) = \frac{2}{3} \approx 0.66$$

(7.37)

in good agreement with Figs. 7.2(a), (b). The average of the normalized vertical magnetization $\langle \sin \theta_0 \rangle$ is only slightly less than the average that would occur if the film magnetization were uniformly directed at 45° to the cross-track direction over the film depth. Thus, (7.37) is a reasonable approximation for a variety of configurations including shielded heads of finite track width and shows that the averages are relatively insensitive to specific configuration, as long as the bias magnetization is approximately at 45° over most of the film depth.

The head efficiency may be determined from an 'effective' record process as in (7.26) or directly by determining the replay flux in the head (Thompson, 1975). In Fig. 7.6(a) the direct playback flux from the medium is sketched. If the permeability of the MR element is not sufficiently high, the flux will not be conducted uniformly down the element. Fringing to the nearby high permeability shields will occur: the flux in the bottom of the MR element can be significantly less than that entering at the top. Correspondingly, Fig. 7.6(b) shows the 'effective' playback field for reciprocity. For a given current distribution, insufficiently high permeability will cause non-vanishing fields in the element and fringing across to the shields along the element will occur. In this case the potential drop across either gap at the head surface will be much less than the total current, and the efficiency will be small.

The efficiency is estimated by utilizing (2.11) in conjunction with an 'effective' field analysis. In Fig. 7.7 an integration path is shown that passes through the element over the top of one of the gaps (I) through the shield and returns back to the element across the gap at the bottom of the element (II). Application of (2.11) in conjunction with (7.32), (similar to the analysis at the beginning of Chapter 3 for an inductive head), yields:

$$E_{\mathrm{MR}} = 1 - \frac{\int_{\mathrm{MR}} dy' H_y^{\mathrm{MR}}(y')}{\int_0^D dy' J_\mathrm{s}(y')} - \frac{\int_{\mathrm{II}} dx' H_x(x')}{\int_0^D dy' J_\mathrm{s}(y')}$$

(7.38)

In (7.38) the permeability of the shields is assumed to be infinite so that no potential drop occurs. Transmission line analysis may be utilized to

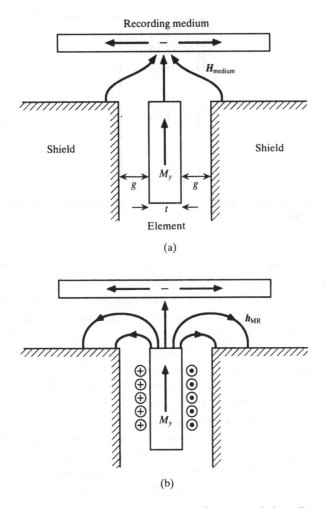

Fig. 7.6. Illustration of (a) field into element from recorded medium and (b) field produced by fictitious current distribution of Fig. 7.4(b).

show that the second term is small if the characteristic length for flux propagation is less than the depth of the element (Thompson, 1975):

$$D < l_{MR} = \sqrt{\mu_r g t / 2} \qquad (7.39)$$

or in scaled form:

$$\frac{D_{max}}{t} \approx \sqrt{\mu_r g / 2t} \qquad (7.40)$$

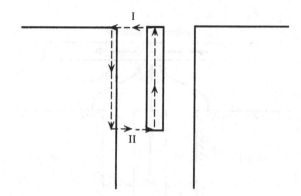

Fig. 7.7. Illustration of path of integration for efficiency definition for a shielded element.

For $\mu_r \sim 1000$ and $g/t \sim 6$ this yields $D_{max}/t \sim 55$. For an element of thickness $t \sim 200\text{Å}$ the element depth should be less than $D_{max} \sim 1\mu\text{m}$.

Even if the potential drop in the MR element is negligible, the third term in (7.38) will remain finite and will depend on the reluctance of the region above the MR head relative to that below the element. If the shield extended only as deep as the element, then the reluctance would be equal, the potential drop to the shield above and below the element would be identical and the efficiency would be 50%. If the shield closed across the bottom of the element, virtually contacting the element, then the potential drop at the bottom would vanish and the efficiency would be 100%. If the element thickness is much less then the shield-to-element spacing, Green's function analysis (Lindholm, 1977) may be utilized to show that the line integrals at the top and bottom are about equal. Thus, with an element depth that is sufficiently smaller than the characteristic length (7.35) the efficiency is approximately $E_{MR} \sim 0.5$.

With a head of reasonable efficiency, which entails an MR element of high permeability, neither the efficiency E_{MR} nor the average equilibrium magnetization $\langle \sin\theta_0 \rangle$ depends on the head depth. Thus, the MR replay voltage (e.g. (7.37)) does not depend, to first order, on the element depth D.

Evaluation of the playback voltage

An explicit playback expression utilizing (7.36) is developed for a shielded head and compared to that of an inductive head (Potter, 1974). First a qualitative discussion is given. In Fig. 7.6(a), (b) a sketch of a recorded transition of longitudinal magnetization passing the MR head is shown. In Fig. 7.6(a) the fields from the transition are sketched at the element. Since the MR voltage is proportional to the (vertical) signal field or flux in the element, it is clear that the voltage has the symmetric 'bell' shape form like that due to playback with an inductive head. The voltage is maximum when the transition is over the center of the element and decreases as the transition center is moved to either side. In playback with an inductive head the voltage also maximizes when the transition is over the gap center, however, at that position, the flux in the inductive head vanishes. The voltage is maximum because the temporal derivative of the flux (spatial derivative times the speed v) is maximum at the center location. The sign of the voltage is positive for the example in Fig. 7.6(a) of a positive-going transition (4.6) and a element biased so that $M_y^0(y')$ is positive.

In Fig. 7.6(b), the reciprocity view of MR playback is sketched. A positive bias magnetization yields surface currents $J_s(y')$ from (2.15) that flow counterclockwise around the element as viewed from the top. This current distribution yields an MR external flux distribution that flows from the element into the space above and returns to the shields as shown. Correlation of this field with a positive-going recorded magnetization transition yields a positive playback voltage.

The essence of the MR equivalent field as sketched in Fig. 7.6(b) is that the field is approximately that of two conventional inductive 'ring' heads of opposite deep-gap field displaced by a distance $g + t$. Thus, compared with reciprocity for an inductive head, reciprocity for an MR head corresponds to an approximate spatial differentiation of the flux into an inductive head. Thus, the MR voltage corresponds to spatial differentiation of the inductive head voltage. The playback voltage of an inductive head corresponds to spatial differentiation of the flux in the inductive head, but times the relative speed. Therefore, the MR replay voltage is similar in form (and almost identical for small gap shielded MR heads) to that of an inductive head, but without the velocity factor. This generalization holds even for unshielded elements, because the MR equivalent field is asymmetric with respect to that of an inductive head. The MR field always has the form of a spatial derivative of an inductive 'ring' head. An MR head spatially differentiates while an inductive head temporally differentiates.

Even though there is no time derivative in the playback voltage for an MR head, the effective spatial differentiation does not permit a DC response. As can be seen in Fig. 7.6(b), a uniformly magnetized medium (without poles) will not give an MR playback voltage. It is to be noted that the field from an MR head utilized in the reciprocity calculation is similar in form to that of the pole inductive heads suggested for perpendicular recording (e.g. Mallinson & Bertram, 1984).

Equation (7.36) may be utilized to give a simple form for the playback voltage from a thin film medium recorded with a longitudinal magnetization pattern of arctangent shape at location x_0 in the medium. With these assumptions, neglecting track edge effects, and utilizing superscripts to distinguish the magnetization for the recording medium and the MR element, (7.36) can be written (similar to (7.29)):

$$V_{MR}(x) = \frac{4\Delta\rho J W E_{MR} \langle \sin \theta_0 \rangle M_r^{rec} \delta\phi_{MR}(x + x_0, d + a)}{t M_s^{MR}} \qquad (7.41)$$

In (7.41) the normalized MR field potential ϕ_{MR} is evaluated at spacing $d + a$ for an arctangent transition with parameter a, as shown in Chapter 5. Equation (2.62) can be utilized to obtain the potential at any spacing in terms of the surface potential.

In Fig. 7.8(a),(b) the surface potential (a) and the corresponding surface longitudinal field component (b) are plotted from a numerical solution for a shielded head with infinite permeability for both the MR element and the shields. The normalized potential varies from zero on the shields up to unity at the element. The potential variation is approximately linear across each gap, but shows non-linear curvature similar to that of an inductive head (Fig. 3.10(a)). The surface longitudinal field vanishes above the permeable surfaces and shows the 'cusp' behavior across each gap of the surface field of an inductive head (Fig. 3.10(b)). Note that the field near the element corners approaches infinity more rapidly than that at the shield corners. In the vicinity of an isolated corner the potential varies as $\phi \propto r^{\pi/(2\pi-\beta)}$ where β is the interior angle on the corner and r is distance from the corner. Thus, the field approaching the MR element corner varies as $(x - t/2)^{-\frac{1}{2}}$ and at the shield corners as $(x - t/2)^{-\frac{1}{3}}$.

It is convenient to write the surface potential in terms of the surface field that exists only above the gaps:

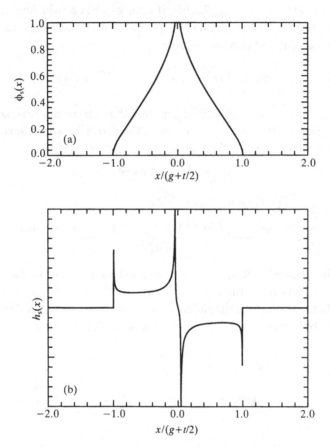

Fig. 7.8. (a) Surface scalar potential variation from numerical analysis for a shielded element; the potential varies from zero on the shields to unity at the element. (Courtesy of N. Smith.) (b) Corresponding surface longitudinal field component.

$$\phi^{s}_{MR}(x) = \int_{-\infty}^{x} dx'(h^{S-L}_{x}(x' + (g_{L} + t)/2) - h^{S-R}_{x}(x' - (g_{R} + t)/2)$$

(7.42)

In (7.42) h^{S-L}_{x}, H^{S-R}_{x} denote, respectively, *positive* surface field functions on the left and right hand side of the elements, and the center of each gap corresponds to the zero argument of each function. g_{L}, g_{R} allows for possible off-center placement of the element between the shields. If the element is centered so that $g_{L} = g_{R}$, then symmetry occurs:

$H_x^{S-L}(x) = H_x^{S-R}(-x)$ (Fig. 7.8(b)). If each gap has a field function that can be approximated by the surface gap field of an inductive head, then (7.42) for a centered element becomes:

$$\phi_{MR}^s(x) = \int_{-\infty}^x dx'(h_x^{Ind}(x' + (g+t)/2) - h_x^{Ind}(x' - (g+t)/2) \quad (7.43)$$

h_x^{Ind} is the inductive head surface gap field that can be approximated by (3.44) to include the cusps or by (3.16) in the Karlqvist approximation. If the gap is small, then (7.43) is approximately (Potter, 1974):

$$\phi_{MR}^s(x) \approx (g+t)h_x^{Ind}(x) \quad (7.44)$$

Substitution of (7.44) into (7.41) yields:

$$V_{MR}(x) \approx \frac{4\Delta\rho JWE_{MR}\langle\sin\theta_0\rangle M_r^{rec}\delta(g+t)h_x^{Ind}(x+x_0, d+a)}{tM_s^{MR}} \quad (7.45)$$

Thus, the shielded MR head voltage, to good approximation, has exactly the same functional form as that for an inductive head (5.49). It is to be noted that (7.27) can be simplified to give a form similar to (7.45), which includes both magnetization components. In the spirit of (7.43) and (7.44):

$$h_{MR}(x, y, z) \approx h^{Ind}(x' + (g+t)/2, y', z')$$

$$- h^{Ind}(x' - (g+t)/2, y', z') \approx (g+t)\frac{\partial h^{Ind}(x, y, z)}{\partial x}$$
$$\quad (7.46)$$

Substitution into (7.35) and integration by parts yields:

$$V_{MR}(x) \approx \frac{2\Delta\rho JE_{MR}\langle\sin\theta_0\rangle(g+t)}{tM_s}\int_{Vol} h^{Ind}(r' + x\hat{x}) \cdot \frac{\partial M^{rec}(r')}{\partial x'}d^3r'$$
$$\quad (7.47)$$

which has the same form as the playback voltage (5.34) for an inductive head. Note that the above comparisons with an inductive head refer to a long-pole 'ring' head without finite length undershoots. Because of the even symmetry of the MR head potential, an analysis of undershoots is different from that of a thin film head. Since the reciprocity potential vanishes on both shields, the undershoots are likely to be negligibly small.

Similar to analysis of the isolated pulse discussed in Chapter 5, measurement of the Fourier transform is useful for a determination of the replay constants. Evaluation of (7.26) or (7.28) in general, or the approximate forms of (7.45) and (7.47) at long wavelengths yields:

$$V_{MR}(k \to 0) \approx \frac{4\Delta\rho JWE_{MR}\langle \sin\theta_0 \rangle M_r^{rec}\delta(g+t)}{tM_s^{MR}} \qquad (7.48)$$

Depending on which constants are known (7.48) may be utilized to determine the unknowns. Most likely the product $E_{MR}\langle \sin\theta_0 \rangle$ can be estimated.

The peak MR voltage for a thin medium recorded with a longitudinal magnetization transition from (7.45) is:

$$V_{MR} \approx \frac{9\Delta\rho JWM_r^{rec}\delta(g+t)}{8\sqrt{2}tgM_s^{MR}}\tan^{-1}\frac{g}{2(d+a)} \qquad (7.49)$$

where the Karlqvist field approximation (3.16) has been utilized along with (7.37) and $E_{MR} \sim 0.5$. Evaluation of (7.49) for reasonable values of head and medium parameters: $d+a \sim g/2$, $\delta/t \sim 2$, $g/t \sim 5$, $M_r^{rec} \approx M_s^{MR}/2$, $W = 10\mu m$, $J = 5 \times 10^6$ A/cm^2, and $\Delta\rho = 4 \times 10^{-7}$ ohm-cm yields voltages on the order of mvolts.

Care must be taken that the MR element remains linear. A simple estimate may be made comparing (7.2) with (7.49) or with general forms (7.35), (7.36). Assuming that the average angular deviation due to a peak signal is limited by $\pm 15°$ from the ambient $45°$, an approximate criterion for linearity, utilizing (7.2) and (7.49) is:

$$\frac{M_r^{rec}\delta(g+t)}{tgM_s^{MR}}\tan^{-1}\frac{g}{2(d+a)} < 0.5 \qquad (7.50a)$$

For $g/t \sim 5$ and $d+a = g/2$, the medium remanent flux upper limit necessary to maintain approximate linearity and to avoid saturation is related to the MR saturation flux by:

$$M_r^{rec}\delta \approx 0.4M_s^{MR}t \qquad (7.50b)$$

Encoding schemes may be designed where isolated transitions do not occur, so that (7.50) need not be satisfied (Schneider, 1985). Note that the condition for linearity does not involve the product $\Delta\rho JW$.

A quantitative comparison of the MR replay voltage to that of an inductive head may be made by taking the ratio of (7.49) to (5.51):

$$\frac{V_{MR}}{V_{Ind}} \approx \frac{2\Delta\rho J}{NvB_s^{MR}}\frac{E_{MR}\langle \sin\theta_0 \rangle}{E_{Ind}}\frac{g+t}{t} \qquad (7.51)$$

Note that (7.51) does not depend on the trackwidth W. Assuming $E_{MR} = 0.5E_{Ind}$, with $J = 10^7$ A/cm^2, $\Delta\rho = 4 \times 10^{-7}$ ohm-cm, $g/t = 5$, $B_s^{MR} = 1$T, and $\langle\sin\theta_0\rangle = \pi/4\sqrt{2}$, yields:

$$\frac{V_{MR}}{V_{Ind}} \approx 1333/Nv(m/s)^{-1} \approx 53\,300/Nv(in/s)^{-1} \qquad (7.52)$$

Thus, for an inductive head with $N = 267$, $v = 5$ m/s, the voltages are comparable.

Fourier transforms and accurate pulse shapes

In the simplified approximation made in (7.44), the Fourier transform of the replay voltage due to recorded magnetization patterns with a shielded, single element MR head is identical to that of an inductive head. However, such an approximation is not accurate at short wavelengths. The Fourier transform of (7.43), assuming symmetric head field functions for each gap and a centered element, is:

$$\phi_{MR}^s(k) = g\,\frac{\sin(kg/2)}{kg/2}\,h_x^S(k) \qquad (7.53)$$

where h_x^S denotes the surface field above either gap. In (7.53) the thickness of the element has been neglected ($t \ll g$). If the Fourier transform assuming a linear potential drop (Karlqvist approximation, (3.17)) is utilized, (7.53) becomes:

$$\phi_{MR}^s(k) = g\left(\frac{\sin(kg/2)}{kg/2}\right)^2 \qquad (7.54)$$

The squared factor yields spectral minima rather than 'nulls' that correspond to where a change in the sign of the Fourier transform occurs. The first minimum occurs at $\lambda \sim g$ where g is the shield-to-element distance, which emphasizes the correspondence between the gap length of an inductive head and half the shield-to-shield spacing. In Fig. 7.9 spectra corresponding to square wave recording $k\phi_{MR}^s(k)$ are plotted. The dashed curve corresponds to (7.54). The solid curve is a Fourier transform of the surface field in Fig. 7.7(b). Note that the effect of an accurate surface field is to fill in the gap minimum and reduce the maximum mid-band response somewhat. Measured responses of MR heads show very shallow gap minima (Schwarz & Decker, 1979) that can be ascribed to off-center elements or, as shown here, the effect of an asymmetric gap field (Problems 7.5, 7.6).

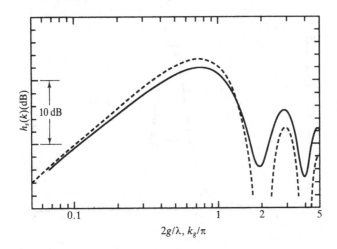

Fig. 7.9. Fourier transform of the surface field of Fig. 7.7(b) versus normalized wavelength $g/\lambda = kg/2\pi$. The dashed curve assumes a linear potential variation across the gaps.

The pulse shape is also slightly different to that given by (7.45). Utilizing (7.43) with the Karlqvist approximation (3.16) for the inductive head field and thin longitudinal media (5.5) yields (Potter, 1974; Smith, 1991; Yuan & Bertram, 1993):

$$V(x) \propto (g+x)\tan^{-1}\frac{g+x}{y} + (g-x)\tan^{-1}\frac{g-x}{y} - 2x\tan^{-1}\frac{x}{y}$$
$$- \frac{y}{2}\ln\frac{(1 + ((g+x)/y)^2)(1 + ((g-x)/y)^2)}{(1 + (x/y)^2)^2} \tag{7.55}$$

where y denotes a generalized spacing $y = d + a$ (or $\sim d + a + \delta/2$ including the thickness). In Fig. 7.10 (7.55) is plotted along with the Karlqvist field (3.16) for $y_{\text{eff}}/g = 0.25$. The more accurate pulse height is about 12% larger, and the PW_{50} is larger as well. An accurate fitting to (7.55) is:

$$\frac{PW_{50}}{g} \approx \sqrt{1 + \left(\frac{2y}{g}\right)^2 + \frac{(y/g)^{1.3}}{0.13 + (y/g)^{1.3}}} \tag{7.56}$$

The difference of (7.50) with the simplified approximation $PW_{50}/g \approx \sqrt{(1 + (2y/g)^2)}$ (5.53) is $\sim 15\%$ for a typical value of

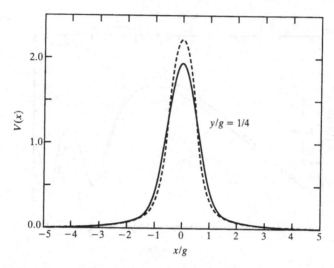

Fig. 7.10. MR response utilizing the spatial difference of the shielded MR element. The dashed curve approximates the difference by a derivative.

$y_{eff}/g = 0.5$. An analysis by (Smith, 1991) gives:

$$PW_{50} \approx \sqrt{2g^2 + 4(d+a)^2} \qquad (7.57)$$

which is reasonable for moderate effective spacings. Equation (7.57) shows that the 'effective gap' for the pulse width is $\sqrt{2}g$ or the geometric mean of the shield-to-shield spacing and the shield-to-element spacing.

Dual MR heads

Utilization of two MR elements in a differential mode, either with or without shields is attractive. In Fig. 7.11(a),(b) two proposed configurations are illustrated. For clarity the element spacing $(\Delta \sim t)$ is exaggerated and possible shields are not shown. Figure 7.11(a) corresponds to a 'dual' MR head (Smith, *et al.*, 1992a). Current is applied to both elements in the same direction. The fields generated by the current magnetize the elements asymmetrically as shown. The magnetostatic fields produced by the vertical component of each element cause each element to be a bias layer (SAL) to the other. Playback flux rotates the magnetization of one element toward the current direction, increasing the resistance, and the other away from the current flow, decreasing the resistance. A voltage therefore results only

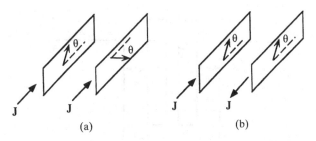

Fig. 7.11. Configurations for double element differential MR heads: (a) a dual head, (b) a gradiometer head.

from fields produced by media that are spatially asymmetric with respect to the structure. For example a recorded transition of longitudinal magnetization whose center is positioned exactly between the elements produces no replay voltage. The dual head is spatially sensitive in an unshielded format.

In Fig. 7.11(b) the configuration for a 'gradiometer' MR head is shown (Gill, *et al.*, 1989; Indeck, *et al.*, 1988). In the gradiometer head the current applied to the elements is in opposite directions, and the resulting current fields bias the element magnetizations in a common direction. A replay flux in the elements therefore rotates both magnetizations in the same direction, yielding a common change in resistance. However, because the current is in opposite directions, the net voltage across the two elements cancels for uniform or symmetric replay fields.

The application of reciprocity to these differential structures is straightforward. For example (7.36) can be utilized, written here in an explicit 2D form:

$$V_{\mathrm{MR}}(x) = \frac{2\Delta\rho JWE_{\mathrm{MR}}\langle\sin\theta_0\rangle}{tM_{\mathrm{s}}} \int_{-\infty}^{\infty} \mathrm{d}x' \int_{d}^{d+\delta} \mathrm{d}y'\phi_{\mathrm{MR}}(x+x',y')\rho^{\mathrm{rec}}(x',y')$$

$$(7.58)$$

E_{MR} is now the efficiency of the dual structure defined as in (7.32). For a shielded structure the numerator in (7.32) is the surface field integral from one element across the gap to the closest shield (Fig. 7.12(a)), and the denominator is the average effective magnetic equivalent current in either element. The average vertical component of the bias magnetization $\langle\sin\theta_0\rangle$ is evaluated in either element, and the absolute value is utilized. Both E_{MR} and $\langle\sin\theta_0\rangle$ will depend on which dual structure is analyzed.

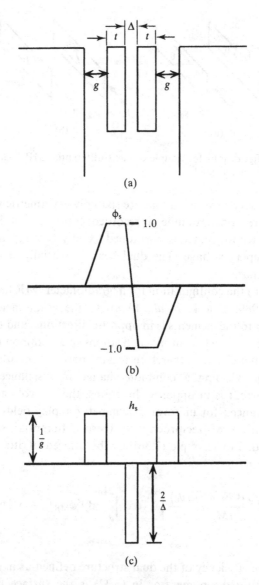

Fig. 7.12. (a) Configuration for a double element MR head, (b) the surface potential for a differential configuration (assuming a linear potential variation for simplicity) and (c) the corresponding surface longitudinal field.

First, the surface potential and playback pulse shapes are discussed. For both types of differential heads the surface potential has the form illustrated in Fig. 7.12(b). The potential varies from zero at the left shield up to unity at the nearest element. Between the elements the potential varies from $+1$ to -1 and then from -1 to zero across the second gap to the other shield. In Fig. 7.12(b) linear variations are shown for simplicity. For a thin medium recorded with an arctangent of longitudinal magnetization the replay voltage is (as in (7.41)):

$$V_{\text{MR}}^{\text{Diff}}(x) = \frac{4\Delta\rho J W E_{\text{MR}}^{\text{Diff}} \langle \sin \theta_0^{\text{Diff}} \rangle M_{\text{r}}^{\text{rec}} \delta}{t M_{\text{s}}^{\text{MR}}} \, \phi_{\text{MR}}^{\text{Diff}}(x + x_0, d + a) \qquad (7.59)$$

where explicit notation for a differential head is given. Utilizing a linear approximation for the potential variation (Fig. 7.12(b)) and (2.62), the replay pulse has the asymmetric form shown in Fig. 7.13(a). The Fourier transform of the isolated pulse for a differential head is explored in Problem 7.8.

For a thick medium recorded with a perpendicular magnetization or arctangent shape centered at x_0, (7.58) becomes:

$$V_{\text{MR}}^{\text{Diff}}(x) = \frac{4\Delta\rho J W E_{\text{MR}}^{\text{Diff}} \langle \sin \theta_0^{\text{Diff}} \rangle M_{\text{r}}^{\text{rec}}}{t M_{\text{s}}^{\text{MR}}} \int_{-\infty}^{x_0} dx' (\phi_{\text{MR}}^{\text{Diff}}(x + x', d + a)$$
$$- \phi_{\text{MR}}(x + x', d + a + \delta)) \qquad (7.60)$$

A simple way of visualizing the form of (7.60) is to realize that a sharp transition located at x_0 in a medium can decomposed into two media: one with a uniform magnetization of $-2M_{\text{r}}$ extending from $-\infty$ to x_0 and zero magnetization from x_0 to ∞ and a second medium uniformly magnetized at M_{r} over its entire length. The second medium does not yield a replay voltage. Perpendicular media are usually utilized with a keeper above the medium. For a high permeability keeper the potential at the keeper vanishes. In that case (7.60) becomes:

$$V_{\text{MR}}^{\text{Diff}}(x) = \frac{4\Delta\rho J W E_{\text{MR}}^{\text{Diff}} \langle \sin \theta_0^{\text{Diff}} \rangle M_{\text{r}}^{\text{rec}}}{t M_{\text{s}}^{\text{MR}}} \int_{-\infty}^{x_0} dx' \phi_{\text{MR}}^{\text{Diff}}(x + x', d + a) \qquad (7.61)$$

For keepered perpendicular media the voltage varies as an integral of the potential, evaluated at a distance $d + a$ from the head surface. Note that the medium thickness does not enter directly. With a keepered structure the potential is changed and, of course, (2.62) may not be utilized to

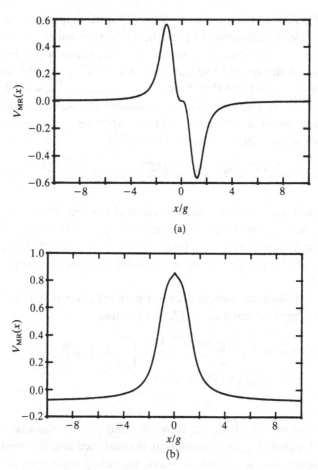

Fig. 7.13. Output voltage pulse shape for (a) a longitudinal magnetization and (b) a perpendicular magnetization. The pulse shapes are idealized since an arctangent transition shape is utilized for both transitions.

determine the potential (Chapter 5). However, if the medium thickness is on the order of the gap or greater $(\delta \geq g)$, the potential near the surface is only slightly changed. Utilizing the surface potential in Fig. 7.12(b) as an approximation for $d + a \ll 2g$, the pulse shape for perpendicular recording is determined from (7.61) and is plotted in Fig. 7.13(b). In contrast to that for a longitudinal pattern, with a differential head a symmetric pulse shape occurs for a (symmetric) transition of perpendicular magnetization. It will be shown in Chapter 8 that the characteristic magnetization transition for perpendicular recording does not possess the perfect asymmetry of an arctangent transition. Thus, in general, the replay voltage for perpendicular recording with a differential head is never perfectly symmetric.

The efficiency for keepered differential structures may be analyzed by transmission line techniques as in the case of single element heads. The average equilibrium vertical magnetization $\langle \sin \theta_0^{\text{Diff}} \rangle$ depends on the magnetization profile. For a 'gradiometer' head (Fig. 7.11(b)) the parallel equilibrium magnetizations imply that the result for a single element (7.39) applies to first order. For the 'dual' head the antiparallel bias magnetizations yield a fairly uniform bias magnetization pattern (Smith, *et al.*, 1992a).

Off-track response – finite track widths

Micromagnetic analysis of true finite track-width elements yields magnetizations that vary at the track edges as well as at the top and bottom of the element (Yuan & Bertram, 1993). Generally the magnetization is pinned by 'exchange tabs' to be along the cross-track direction at the track edges. In that case the sensitivity is less than that given by 2D analysis shown in Fig. 7.2(b), especially if the track width is narrow. In addition, MR heads generally have asymmetric off-track responses (Yeh, 1982). The MR magnetization configurations corresponding to a finite track width unshielded SAL are utilized to calculate an off-track response. A thin recording medium with a longitudinal magnetization transition extending over a finite track width W is assumed. A direct calculation of the playback response was performed. The fields from the recording medium were expressed analytically, assuming a ramp magnetization profile (Problem 4.2). In Fig. 7.14(a) the average resistance change $\langle \cos^2 \theta \rangle$ is plotted versus off-track displacement Δz. The response is asymmetric with respect to off-track displacement. Not only does the maximum response not occur for no displacement ($\Delta z = 0$), but the response decreases at different rates depending on the direction of displacement. In fact, a voltage minimum occurs for displacements towards negative Δz.

The reason for this asymmetric behavior is illustrated in Fig. 7.14(b). A sketch of the magnetic field from the magnetized medium in the (y, z) plane of the element is shown. The field flows away from the transition center asymmetrically. Also shown is the MR element, biased nominally at 45° to the z direction, in off-track positions on either side of the finite track width recording. For the element on the right (corresponding to $\Delta z > 0$) the signal field is generally orthogonal to the MR bias

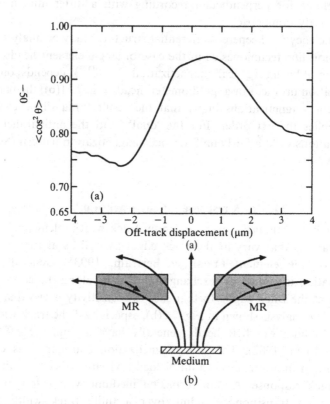

Fig. 7.14. (a) MR response versus off-track position (Δz) for the unshielded element with SAL shown in Fig. 7.3. (b) Illustration of fields in the (y,z) MR element plane from a recorded longitudinal medium. The biased MR element is shown in two positions to illustrate off-track asymmetry.

magnetization. In this case the torque on the magnetization is large, and large rotations occur giving large resistance changes. The response is not as great as that for perfect on-track playback ($\Delta z = 0$) since the field magnitude has decreased. For the element shown on the left ($\Delta z < 0$), the signal field is generally parallel to the MR magnetization, yielding small rotations and correspondingly small resistance changes. A specific off-track position yields a minimum in the response. As the element is moved off track toward $\Delta z < 0$, the average signal field angle rotates from along the y direction to along the z direction at large displacements. At a specific location the average field is approximately parallel to the MR magnetization minimizing the response. The overall response decreases since the field magnitude decreases with off-track position.

Spatially averaging yields a minimum rather than a vanishing of the response. On track ($\Delta z = 0$) the response is not the largest due to non-uniform bias magnetization (Fig. 7.2). Configurations exist that yield symmetric off-track response (Shelledy & Nix, 1992). An example is the 'dual' head illustrated in Fig. 7.11(a).

The application of reciprocity to determine off-track voltages is complicated (Smith, 1993). Here a schematic interpretation is given, assuming that the current is uniform along the cross-track (z) direction. In order to apply expressions such as (7.26), the relation between the signal flux and MR element magnetization incremental variation (7.18) must be examined. In an MR film the bias magnetization is approximately uniform and directed at 45° to the cross-track direction. The susceptibility of the magnetization to external fields depends on the direction of the net internal field to the bias direction. If the field is along the bias magnetization then no torque will occur and consequently no small signal MR magnetization rotation will occur. If the net signal flux is perpendicular to the magnetization, then maximum small signal magnetization rotation will occur. An anisotropy of the susceptibility can be analyzed by a more general form of (7.3). The analysis leading to (7.5) is only for an intermediate angle, although it is clear from (7.5) that for fields larger than that to bring θ to 90°, no further change in the magnetization can occur. If the response of the magnetization is approximated by a susceptibility tensor $\tilde{\chi}$ and permeability tensor $\tilde{\mu} = \mu_0(\tilde{\chi} + \tilde{1})$, then (7.18) can be written generally as:

$$M^{\text{sig}} = \tilde{\chi}^{-1} \cdot \tilde{\mu} \cdot B^{\text{sig}} \tag{7.62}$$

independent of the demagnetizing tensor. If the susceptibility in one principal direction is very large, even though it may vanish in the other, then (7.62) yields (7.18). In that case the general expressions given by (7.25–7.28) still hold. However, the evaluation of the reciprocity field is more complicated. A current source as illustrated in Fig. 7.4(b) still holds, however the field that results is affected by the anisotropic susceptibility. For example, if the bias magnetization is at an angle θ to the cross-track direction, then an approximate susceptibility tensor is given by:

$$\tilde{\chi} = \chi_\perp \begin{pmatrix} \cos^2\theta_0 & \cos\theta_0 \sin\theta_0 \\ \cos\theta_0 \sin\theta_0 & \sin^2\theta_0 \end{pmatrix} \tag{7.63}$$

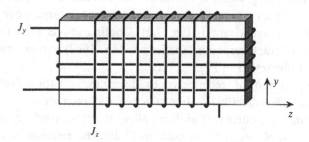

Fig. 7.15. Illustration of double current distribution that approximates the anisotropic MR susceptibility.

where $\chi_\perp$ is the susceptibility perpendicular to the equilibrium magnetization. This tensor has equal entries for $\theta_0 = 45°$. Note that (7.5) does not yield a simple tensor and (7.63) is only an approximation.

Application of current distribution J_y and corresponding current field in the y direction yields, via (7.63), element magnetization in both the vertical, y, and cross-track, z, directions. A simple way to view this anisotropic effect is to consider source currents J_y and J_z and a diagonal susceptibility. In that case the extension of Fig. 7.4(b) is shown in Fig. 7.15 where currents J_y and J_z are illustrated for a single MR element. The fields from J_z added to those from J_y yield the asymmetric response shown in Fig. 7.14(a). The dual head as illustrated in Fig. 7.11(a) will not exhibit, by superposition, off-track asymmetry.

MR resistance noise and SNR

Noise voltage due to the resistance of the MR element is given by:

$$NV = \sqrt{4kTR\Delta f} \qquad (7.64)$$

where R is the element resistance, k is the Boltzmann constant, T is the absolute temperature and Δf is the band width. Utilizing (7.2) the power signal to noise ratio may be written as:

$$SNR_{power} = \frac{J^2 \rho D W t}{64 k T \Delta f} \left(\frac{\Delta \rho}{\rho}\right)^2 \qquad (7.65)$$

In (7.65) it is assumed that the maximum change in the angular term of (7.2) is 0.25. This SNR definition is peak signal to broad band noise. However, independent of definition, the SNR will always vary as the ratio of power in the element to 'thermal power' $kT\Delta f$ times the square of the relative MR resistance change $\Delta\rho/\rho$. Increasing $\Delta\rho/\rho$ improves the ultimate MR signal to noise ratio and thus, this ratio provides an ultimate MR figure of merit. Utilizing typical values given in this chapter, the SNR is enormously large: the signal to noise ratio is currently limited by medium noise (Jeffers & Wachenschwanz, 1987). Nevertheless, assuming the MR element can be driven to the extreme of linearity, (7.2) entails that performance improvement occurs from developing media with increased $\Delta\rho$.

Problems

Problem 7.1 Consider a thin film ($t \ll D$ in Fig. 7.1), but with ($W \gg D$) to form a 2D structure. Assume that magnetization is uniform across the thickness and the long direction, but that it varies with depth. Let the variation of the depth component be $M(y) = M$, $-s < y < s$ and let it vanish linearly at the surfaces: $M(y) = M(D/2 - |y|)/(D/2 - s)$ for $s < |y| < D/2$. The origin $y = 0$ is at the film center and $s < D/2$. Use the 2D form for the magnetostatic field given in Chapter 2 to solve for the field at all y in the film. Sketch the demagnetization field for various representative choices of s.

Problem 7.2 Utilize (2.14) to average the magnetostatic field over the depth of a uniformly magnetized film:

$$\langle H_D(y) \rangle = -\langle N \rangle M_y = -\frac{tM_y}{\pi D}\left(1 + \ln\frac{2D}{t}\right)$$

or

$$\langle H_D(y) \rangle \approx -\langle N \rangle M_y \approx -\frac{tM_y}{\pi D}\ln\frac{D - \Delta}{\Delta}$$

if the average is taken to within Δ of the surfaces ($\Delta^2 \gg (t/2)^2$).

Problem 7.3 The magnetization in an element due to a constant bias field varies with depth (Smith, *et al.*, 1993):

$$\frac{M_y^0(y)}{M_s} = \frac{H_{bias}D}{t}\sqrt{1 - (2y/D)^2}$$

In contrast to the rather uniform magnetization due to a permanent magnet or saturated SAL bias, a constant bias field yields a magnetization that increases rapidly with depth into the medium. Show that the parabolic form of the variation of resistance with field holds until the element first saturates at the film center. What is the average angle at this condition? For increases in field beyond this critical point, the magnetization will gradually saturate toward the element surfaces with a configuration as in Problem 7.1 with successively increasing s. Develop a relation between external field and position s by examining the energy in the unsaturated region. Plot the final total resistance versus field.

Problem 7.4 Show that for a finite permeability the right hand side in (7.18) is multiplied by the factor $(\mu_r - 1)/\mu_r$ where μ_r is the relative permeability. Derive (7.62).

Problem 7.5 Derive an expression for the voltage and Fourier transform of the isolated pulse for a shielded MR head with an off-center element. Assume a linear gap potential and plot both the voltage and its transform.

Problem 7.6 Give a qualitative discussion of the effect on the Fourier transform minimum at $\lambda \sim g$ for a centered element of a surface gap field that has: (a) symmetric cusps and (b) asymmetric cusps.

Problem 7.7 Use the transmission analysis in Thompson (1975) to derive an expression for the 'effective' record efficiency of a shielded MR head.

Problem 7.8 Derive an expression for the Fourier transform of the isolated pulse for a differential head utilizing the simple linear potential variation illustrated in Fig. 7.12(b). Note that the isolated pulse transform has vanishing response for a differential head at long wavelengths ($k \to 0$). At what wavelengths are the response minima?

Problem 7.9 Sketch how the surface potential and surface field (Fig. 7.12(b), (c)) change as a high permeability keeper is brought close to the head surface.

8

Record process:
Part 1 – Transition models

Introduction

In this chapter, models will be discussed that give insight into the recording process. In contrast to the playback process, the recording of magnetization patterns is a non-linear phenomenon. Thus, except for special cases, analysis must be by computer simulation. It is the long-range magnetostatic fields that cause computer simulations to be iterative and extremely time consuming. There are two philosophical approaches to computations of the record process. One is to neglect the fine details and develop reasonably approximate models that capture the main physical features of the process. Such simplified models allow for analytic solutions of magnetization patterns or solutions that require a minimum of computer simulation. This simplified approach is useful for developing guidelines in media development (e.g. the effect of coercivity or remanence changes) or in head–media interface geometry development (effect of head–medium spacing, record gap, medium thickness). Parameters can be easily changed in simplified models.

The other approach is to develop full numerical micromagnetic models. These are required in order to compute detailed, or second order effects, of the recording process, such as noise, edge-track writing, and non-linear amplitude loss and thus, for example, give fundamental input to error-rate calculations. Simulations can give detailed information about pulse asymmetry in tape recording (Bertram, *et al.*, 1992) or the fluctuations of the transition center across the track that leads to position jitter or transition noise in thin film media (Zhu, 1992). For example, in Fig. 8.1 the vector magnetization distribution from micromagnetic simulation of a recorded transition in thin film media is shown. The medium is presumed to be polycrystalline, as in Fig. 1.2, with

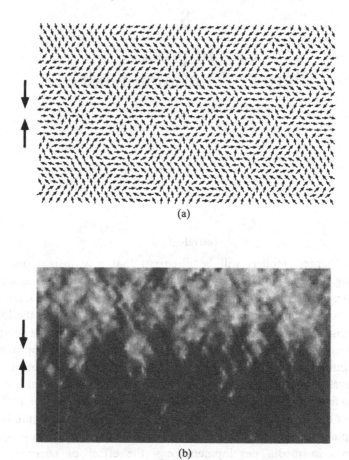

(a)

(b)

Fig. 8.1. (a) Grain magnetizations in a polycrystalline thin film from micromagnetic simulation of a recorded transition. (b) Gray scale plot of the component of magnetization along the recording direction. The vectors at the right of the figures denote the average position of the transition so that the cross-track direction is from left to right. For the simulation the grain crystalline anisotropy axes were randomly oriented in 3D and no intergranular exchange coupling was assumed. Taken from Bertram & Zhu (1992).

individual grain anisotropy axes randomly oriented. In Fig. 8.1(a) the magnetization vector orientation is shown corresponding to each grain (simulated by a hexagonal lattice). On either side of the transition the magnetization is in the remanent state ($\pm M_r$): the magnetizations are approximately parallel to the track direction except for a small ripple pattern. In the transition region the equilibrium magnetizations of each grain rotate away from the recording direction to form localized vortex patterns. The magnetization configuration of the transition results from a reduction of magnetostatic energy against the random crystalline anisotropy. The magnetization averaged across the track width plotted versus track position will show a general antisymmetric shape that can be approximated, for example, by the functional forms in Table 4.1. Note that at the transition center where the average component of magnetization along the track direction vanishes, the grain magnetizations are correlated. The magnetization directions, however, are equally distributed in the film plane (Bertram, *et al.*, 1993). Even though the anisotropy axes are randomly oriented, magnetostatic fields from typical Co based media force the grain magnetizations to lie in the film plane.

In Fig. 8.1(b) a gray scale picture of Fig. 8.1(a) is shown in which the component of magnetization along the track direction varying between $\pm M_r$ is represented by a variation from black to white. The approximate 'zigzag' wall form separating regions of the two different remanent magnetization directions is readily seen. Simplified modeling assuming zigzag walls can give reasonable analytic models of transition shape as well as noise (Minnaja & Nobile, 1972; Freiser, 1979; Muller & Murdock, 1987; Middleton & Miles, 1991; Semenov, *et al.*, 1991).

In this chapter simplified, macroscopic models of the recording process are discussed. The essential simplification is to assume a (possible vector) hysteresis loop and then calculate the effect of the net field including the external field and the macroscopic magnetostatic field (2.16). Here only 2D models are discussed, so that magnetization variations are restricted to the (x, y) plane. First an overview of the basic writing process is given, followed by transition modeling for thin longitudinal media. Comments about problems in simplified modeling of thick tape media are included. The chapter also deals with modeling of perpendicular media and demagnetization limits of longitudinal transition shapes.

The magnetic recording process

Here, a qualitative discussion of the writing process in longitudinal recording is given, including the formation of 'hard' and 'easy' transitions. The essence of the recording process is to produce reversals of magnetization according to the input record-head current. The writing process yields alternately 'hard' and 'easy' transitions. In Fig. 8.2, a positive record-head field is sketched that is oppositely directed to an assumed initial 'DC' reversed magnetization (at remanent level-M_r). In a region ($H_h > H_c$) near the gap, the head field exceeds the coercivity, and

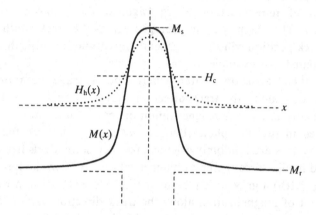

Fig. 8.2. Writing of a 'hard' transition against uniformly saturated media. This case forms transitions at either side of the gap the instant the head current is applied. The magnetization is solid and the head field is dashed.

the magnetization is reversed in that region (and saturated to M_s). In this case, as illustrated in the figure, two transitions are written, one on either side of the gap. As the medium moves in the presence of the constant head field (fixed current), the 'upstream' or right transition in the figure moves with the medium while the left transition remains fixed to the gap edge continuously formed by new media moving into the gap region and becoming saturated. The right transition is the recorded transition that will eventually be part of a data pattern. This written transition is termed a 'hard' transition and is formed simultaneously with a transition at the opposite gap edge.

In Fig. 8.3 the magnetization pattern is shown after a time interval when the medium has moved with respect to the configuration formed at the instant of field reversal and transition formation. It is assumed that the recording current is held constant during this motion. The recorded or right hand transition moves with the medium while the left hand transition remains fixed to the gap. Note that the recorded transition after medium motion relaxes somewhat away from the presence of the head fields, so that the remanent transition magnetization varies from $-M_r$ to $+M_r$. In Fig. 8.4 the instant of writing a second transition is shown. The (negative) head field that writes this transition is now in the same direction as the 'DC' magnetized medium and, to good approximation, only the recorded or right hand transition is written. The head field saturates the gap region in the same direction as the

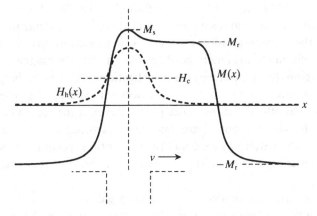

Fig. 8.3. Magnetization configuration as time passes with head current on. The 'upstream' transition is fixed to the medium and the left edge transition is fixed to the gap.

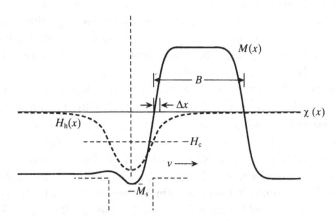

Fig. 8.4. Writing of an 'easy' transition in the direction of incoming magnetization when the head current is reversed. Left edge transition does not occur or is minimal. The writing transition spacing between the first and second transition in this 'dibit' is $B = v/T$ where T is the time interval between head-current reversal.

incoming magnetization, and no appreciable left edge transition is written. This second transition written in the medium is termed an 'easy' transition. Subsequent transitions alternate between 'hard' and 'easy'.

The terms 'hard' and 'easy' arise from the effect of demagnetization fields on the recording process. As will be discussed in this chapter, the 'hard' transition requires more head field to saturate the magnetization in the gap region due to increased demagnetization fields (Bloomberg, *et al.*, 1979). In addition, non-linear bit shift occurs in the writing of a 'hard' transition that affects the overwrite process, even if the record current is sufficient to saturate both hard and easy transitions, as discussed in Chapter 9. The pattern of hard and easy transitions is complicated if the medium entering the head region contains a previously written data pattern.

Utilizing simplified models that give averaged properties, the shape of a track width averaged recorded transition depends on the head field during recording, the magnetostatic fields associated with the transitions, and the medium *M-H* (vector) hysteresis loop. Generally the magnetization at any position along the medium (x), or at any depth into the medium (y) is given by:

$$M(x,y) = F_{\text{loop}}(H_{\text{h}}(x,y) + H_{\text{d}}(x,y)) \qquad (8.1)$$

In (8.1) F_{loop} is a generalized vector hysteretic function that relates the total vector fields to the resultant vector magnetization at each point in the medium. A vector *M-H* loop representation for F_{loop} that includes remanent as well as reversible properties has been developed using the Preisach formulation (e.g., Bhattacharyya, *et al.*, 1989) or particle assemblies (Beardsley, 1986). Since F_{loop} is, in general, a non-linear function, as well as hysteretic and field direction dependent, numerical analysis must be utilized to solve (8.1). Each of N discretization cells is represented by (a possibly spatial varying) F_{loop}, and the set of N coupled equations is solved iteratively. The problem is numerically intensive due to computation of the magnetostatic fields that depend on the magnetization pattern (2.8, 2.22). Computation time varies as N^2 for direct field evaluation, but only as $N \ln N$ when Fast Fourier transform techniques are utilized (Mansuripur, 1989; Yuan & Bertram, 1992a). Equation (8.1) can be extended to analyze track edge recording (z). Because this formulation utilizes average magnetization properties at each point in the medium, medium noise does not result. Thus, (8.1) is convenient to study overwrite (Bhattacharyya, *et al.*, 1991) and non-linearities arising from loop hysteresis (Simmons & Davidson, 1992).

For thin film recording media where the magnetization does not vary with depth into the medium and a single component of magnetization is of interest, (8.1) simplifies to:

$$M(x) = F_{\text{loop}}(\bar{H}_{\text{h}}(x) + \bar{H}_{\text{d}}(x)) \qquad (8.2)$$

where $M(x)$ is the magnetization component of interest and $\bar{H}$ represents the field component along the magnetization direction at each position x averaged through the depth y of the medium. The M–H loop appropriate to F_{loop} corresponds to the conventional unidirectional form illustrated in Fig. 1.3. In general, numerical analysis simulates the writing of a transition sequence, including reversible and irreversible magnetization changes that occur during medium motion.

The writing of a single transition against a reversed background remanent magnetization (Fig. 8.2) involves utilization of only half of the major or outer loop. For media magnetized initially negatively, the loop portion illustrated in Fig. 8.4 (solid) applies. Application of a spatially varying, but generally positive field, changes the magnetization along this curve. As the recording process proceeds by the increase of the head field to its final value, demagnetization fields form corresponding to the changing magnetization pattern. Even if the head field has regions of negative value (as in perpendicular recording with a ring head) or the magnetostatic field is opposite to and exceeds the head field, the development of the magnetization will be along the monotonic curve in Fig. 8.4. Changes along minor loops will not occur. Thus, in an iterative solution to (8.2) in which the magnetization at each position x might oscillate due to the iteration procedure, a decrease of the magnetization will follow the major loop and not a minor loop. Decreases in magnetization from a given stage in the iteration is a mathematical, not a physical phenomenon. The writing of a single transition, in essence, involves a monotonic change of magnetization away from an initial remanent state. Simple approximations to the major loop are shown in Fig. 8.4 of a linear slope (dashed) and a perfectly square loop (dotted).

Examples of the iterative solution of (8.1) are given in (Potter & Schmulian, 1971; Barany, 1989; Speliotis & Chi, 1978) for thin longitudinal media and (Beusekamp & Fluitman, 1986; Zhu & Bertram, 1986) for perpendicular recording. If a perfectly square medium is assumed, Fig. 8.4 (dotted), then the magnetization transition shape that develops is one in which the magnetostatic field, at least in the central portion of the transition, just balances the head field:

$$H_c = H_h(x) + H_d(x) \qquad (8.3)$$

This form may be inverted analytically to solve for the magnetization pattern in many cases. In this chapter analytical analysis for perpendicular recording is discussed; however, attempts at an analytic solution for longitudinal recording have been made (van Herk & Wesseling, 1974). For longitudinal recording the assumption of (8.3) is not valid over the regions of near saturation where the demagnetizing field vanishes. With the assumption of a square loop with no reversible susceptibility, the effects of medium motion and subsequent field reversal are readily included. As the medium moves away from the gap region (as in Fig. 8.3), the head fields decrease. The magnetization remains fixed as the net field decreases, until at some stage the magnetostatic field reaches a value of $-H_c$. At this point (8.2) must be solved, but the iteration is more complicated than simple utilization of a monotonic transfer curve, as illustrated in Fig. 8.5.

Equation (8.3) can also be written in a derivative form:

$$\frac{\mathrm{d}M(x)}{\mathrm{d}x} = \frac{\mathrm{d}M}{\mathrm{d}H_{\mathrm{loop}}} \left(\frac{\mathrm{d}\bar{H}_h(x)}{\mathrm{d}x} + \frac{\mathrm{d}\bar{H}_d(x)}{\mathrm{d}x} \right) \qquad (8.4)$$

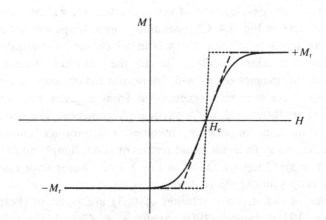

Fig. 8.5   Sketch of major loop from Fig. 1.3, which is the transfer curve for the writing of a transition from a medium initially saturated at $-M_r$. The dashed curve is a linear slope approximation where the magnetization varies from $-M_r$ to $+M_r$ with no distinction between M_r and M_s. The dotted curve represents a square-loop material.

Equation (8.4) can be solved iteratively in terms of the transition derivative. dM/dH_{loop} is simply the field derivative of the F_{loop} or the M–H loop derivative and is a function of the total field at each position along the medium. Corresponding to (8.3), if the magnetization changes are confined to the region of the loop where $H = H_c$, then the transition shape may be determined, assuming a square loop, by solving:

$$\frac{d\bar{H}_d(x)}{dx} = -\frac{d\bar{H}_h(x)}{dx} \tag{8.5}$$

Note that in a macroscopic formulation, the magnetization at each point in the medium is a function of the net field via the M–H loop. The net field in magnetic recording is due to the sum of the applied recording field and the magnetostatic field. If the spatial change of the fields is on the order of the film grain size or the spatial scale of whatever microscopic phenomena that determines the bulk M–H loop, then micromagnetic analysis must be utilized in the determination of recorded magnetization patterns.

Models of longitudinal recording

In the first part of this section a simplified analytic model for the longitudinal recording in thin media will be discussed in detail (Williams & Comstock, 1971; Maller & Middleton, 1973). The model gives great insight into the longitudinal recording process and provides a simple analytic formula for the transition length. The magnetization is assumed to be longitudinal, which is a good approximation for high-moment thin film media and the differential form (8.4) will be utilized with only longitudinal field components. The simplification arises from the *a priori* assumption of the transition shape. In the simplest form of the model the magnetization is assumed to vary between $\pm M_r$ in a functional form where the position x is scaled by the transition parameter 'a' (4.17). The transition parameter is taken to be an unknown to be determined. Here the transition shape wil be left arbitrary, but typical shapes are listed in Table 4.1. For an arctangent transition the form:

$$M(x) = \frac{2M_r}{\pi} \tan^{-1} \frac{x - x_0}{a} \tag{8.6}$$

is utilized where x_0 denotes the location of the center of the transition. The demagnetization fields associated with (4.6) or any antisymmetric transition have the form shown in Fig. 4.6(a).

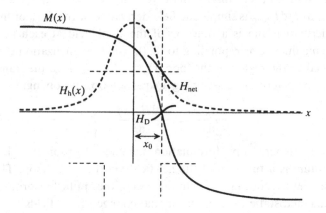

Fig. 8.6. Magnetization transition, head field and demagnetizing field near the transition center ($x = x_0$) for a simplified analysis of longitudinal recording.

A simple view of a recorded transition without regard to 'hard' or 'easy' effects is shown in Fig. 8.6. After the head field is applied and before medium motion is considered, a transition is written with center location at x_0. Assuming a perfectly asymmetric transition the magnetostatic field vanishes at the transition center so that the center location is given, to first order, by the position where the (depth-averaged) head-field magnitude equals the medium coercivity:

$$\bar{H}_x^h(x_0) = H_c \tag{8.7}$$

The magnetostatic fields at the transition center are sketched in Fig. 8.6, and it is seen that the slope of the net field at the transition center is reduced. If a transition shape is specified with one unknown parameter, then (8.4) can be utilized only at one location to solve for the unknown. It is assumed in this model that the significant region for specifying 'a' is at the transition center. At high densities, when transitions overlap, the magnetization slope at the transition center provides the major contribution to the replay voltage (6.35); at low densities, the transition region is important for an accurate determination of the peak voltage (5.36).

A general transition shape is assumed (Table 4.1) with center slope given by:

$$\frac{dM}{dx} = \frac{2M_r}{\pi a} \tag{8.8}$$

The *M–H* loop derivative at the transition center where the net field equals the medium coercivity may be expressed as:

$$\frac{dM}{dH_{\text{loop}}} = \frac{M_r}{H_c(1 - S^*)} \tag{1.2}$$

The gradient of the magnetostatic field at the transition center is given for thin media ($\delta \ll a$) by (4.19):

$$\frac{dH_x^d}{dx} = -\frac{M_r\delta}{\pi a^2}\int_0^\infty \frac{ds}{s}\frac{d^2 f(s)}{ds^2} = -\frac{M_r\delta I}{\pi a^2} \tag{8.9}$$

where I denotes the integral. From Table 4.1 I depends on the transition shape, but is close to the value of $I = 1$ for an arctangent transition shape.

The head-field gradient may be determined utilizing the Karlqvist approximation for the longitudinal field (3.16). Inverting (8.7) at spacing y (assuming thin media) yields:

$$x_0 = \pm(g/2)\sqrt{1 - (2y/g)^2 + (4y/g\tan^{-1}(\pi H_c/H_0))}$$
$$x^2 + y^2 > (g/2)^2$$

$$\tag{8.10}$$

$$x_0 = \pm(g/2)\sqrt{1 - (2y/g)^2 - (4y/g\tan^{-1}\pi(1 - H_c/H_0))}$$
$$x^2 + y^2 < (g/2)^2$$

(The transition between the two regimes in (8.10) occurs when $H_0 = 2H_c$.) Taking the derivative of the field (3.16) with evaluation at $x = x_0$ given by (8.10) yields:

$$\frac{dH_x^h}{dx} = -\frac{QH_c}{y} \tag{8.11}$$

where Q is given by:

$$Q = \frac{2x_0 H_0}{\pi g H_c}\sin^2(\pi H_c/H_0) \tag{8.12}$$

Note that Q depends on the deep-gap field relative to the coercivity (H_0/H_c) and the relative spacing ($2y/g$) through (8.10). Q is plotted in Fig. 8.7 versus H_0/H_c for spacing to gap ratios of $y/g = 0.125, 0.25, 0.5$. Although the head-field gradient reaches a maximum versus location x at fixed deep-gap field (Fig. 8.6), the gradient also has a maximum versus

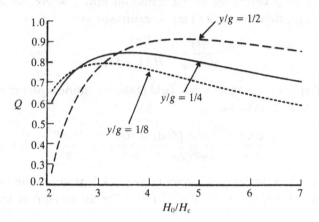

Fig. 8.7. Plot of head-field gradient factor Q versus normalized deep-gap field H_0/H_c, for spacings $y/g = 0.125, 0.25, 0.5$. Using (8.10) the location for the maximum in Q can be determined and is $x_0/g = 0.643$ for $y/g = 0.25$.

deep-gap head field at a fixed field $H_h = H_c$. An optimum value of $Q \sim 0.84$ occurs at $H_0/H_c \sim 3.75$ for a typical value of $y/g = 0.25$ with recording location $x_0 \sim 0.64g$ a distance $0.14g$ past the gap corner over the core surface.

Substitution of (1.2), (8.8), (8.9), and (8.11) into (8.4) yields:

$$a = \frac{(1 - S^*)(d + \delta/2)}{\pi Q} + \sqrt{\left(\frac{(1 - S^*)(d + \delta/2)}{\pi Q}\right)^2 + \frac{M_r\delta(d + \delta/2)I}{\pi Q H_c}}$$

$$(8.13)$$

(Note that for a positive slope transition as in (8.8), the deep-gap field is negative (Fig. 8.6), yielding a positive head-field gradient.) The spacing for evaluation of the head-field gradient is taken from the head surface to the center of the medium: $y \sim d + \delta/2$ (d is the non-magnetic head–medium net spacing that includes medium overcoat or possible pole-tip recession). The transition parameter given by (8.13) applies to any transition shape with center slope given by (8.8) and demagnetization field gradient characterized by I (8.9). The transition length is reduced, in general, by a reduction in head–medium spacing d, an increase in the loop squareness S^*, an increase in the medium coercivity H_c, and a decrease in medium flux content $M_r\delta$. The transition length varies with the deep-gap field or record-head current via Q (8.8): an increase in head-field gradient decreases the transition parameter. The expression given in (8.13) does not take micromagnetic effects into account, such as grain orientation, intergranular exchange interaction, and local magnetostatic effects (except indirectly via S^*, M_r and H_c). However, (8.13) successfully approximates transition lengths computed by micromagnetic analysis (Bertram & Zhu, 1992).

A slightly modified form for the transition parameter occurs if medium motion is accounted for that takes the transition into a region away from the head field (Williams & Comstock, 1971). This effect is small for well oriented media. For typical high-moment thin longitudinal films the last term in (8.13) dominates so that:

$$a \approx \sqrt{\frac{M_r \delta (d + \delta/2) I}{\pi Q H_c}} \qquad (8.14)$$

Equation (8.14) also arises directly by solving (8.5) using (8.9) and (8.11), assuming a square loop ($S^* = 1$) so that (8.3) holds. In this approximation, in the vicinity of the transition center, the transition magnetostatic field plus the head field yield a constant net field equal to the coercivity. The demagnetizing field gradient just balances the head-field gradient to yield a zero net field slope (Fig. 8.6). Equation (8.14) is plotted in Fig. 8.8 (solid) in scaled form a/δ versus d/δ. A typical ratio of $M_r/H_c = 8$ was used with $Q = 0.8$ and $I = 1$. Note that for typical ratios of spacing to medium thickness $d/\delta \sim 3$, the transition parameter is

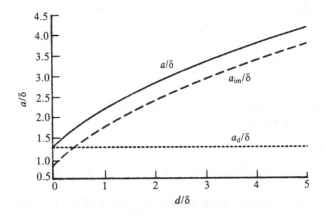

Fig. 8.8. Normalized transition parameter a/δ versus normalized spacing d/δ assuming square-loop ($S^* = 1$) media. a_{im} denotes the inclusion of head imaging. a_d is a demagnetization limited result. $M_r/M_c = 8$ was utilized.

approximately equal to the head–medium spacing ($a \sim d$). An expression for the transition parameter assuming a square loop and an arctangent transition shape for medium thickness that may not be small is considered in Problems 8.1 and 8.2. The result, given by the indirect expression in Problem 8.2, is more accurate than (8.14) for exceptionally small spacings ($d < \delta$).

The effect of imaging on the transition parameter can be included. Either (8.4) or (8.5) can be resolved by modifying the demagnetizing field gradient to include head imaging. Since the recording of a transition for optimal head current (maximum Q in Fig. 8.7) occurs at a center location over the gap core, perfect imaging will be assumed in the analysis presented here. An arctangent transition is assumed, and center film values for field derivatives will be utilized. A square loop material is also assumed so that (8.5) is solved with a modified magnetostatic field gradient:

$$\frac{\partial H_{\mathrm{d}}^{\mathrm{net}}}{\partial x} = \frac{\partial H_{\mathrm{d}}(0,0)}{\partial x} - \frac{\partial H_{\mathrm{d}}(0, 2d + \delta)}{\partial x} \tag{8.15}$$

The net demagnetizing field gradient at the transition center in the center of the film is given by the direct term due to the medium poles plus the image term, which is a transition of opposite sign spaced a distance $2d$ from the main transition to account for (perfect) imaging from a high-permeability head (Chapter 2). Utilization of (4.14) yields the demagnetization field gradient at any vertical spacing with respect to the transition center (4.15). (In Chapter 4, y is referred to the medium center.) Utilizing (4.15) with $y = 0$ and $y = 2d + \delta$, the net field gradient at the transition center evaluated at the medium center is:

$$\frac{\partial H_{\mathrm{d}}}{\partial x} = -\frac{M_{\mathrm{r}}\delta}{\pi}\left(\frac{1}{a(a + \delta/2)} - \frac{1}{(a + 2d + \delta)^2 - (\delta/2)^2}\right) \tag{8.16}$$

Utilizing the expression for the head field gradient (8.11) and (8.16) in (8.5) yields the quartic form:

$$\frac{1}{a^2} = \left(\frac{1}{a_{\mathrm{im}}(a_{\mathrm{im}} + \delta/2)} - \frac{1}{(a_{\mathrm{im}} + 2d + \delta)^2 - (\delta/2)^2}\right) \tag{8.17}$$

In (8.17) a_{im} is the transition length with imaging and a is the non-imaged solution given by (8.14). In Fig. 8.8 a_{im}/δ is plotted versus d/δ (dashed) for a typical demagnetization ratio $H_{\mathrm{r}}/H_{\mathrm{c}} = 8$ (or $B_{\mathrm{r}}/H_{\mathrm{c}}$ in cgs). Note that the effect of imaging is to reduce the transition parameter. For nominal spacings $(d/\delta > 1)$ the percentage change is moderate. As discussed in Chapter 2, the field from a pole source depends on the spatial extent of the poles relative to the distance from the source to the observation point. Thus, as shown in Problem 8.3, if $a/(2d + \delta)$ is small, the effect of imaging is small, and conversely, if the spacing d is small compared to the charge spread represented by a, the effect of imaging is large. As shown in Fig. 8.8, a moderate change due to imaging occurs except for very low spacings. Slightly more accurate approximations may be obtained by using field gradients averaged over the medium thickness rather than center values (Problem 8.4).

Voltage–current relations

The transition parameters determined above may be utilized in the replay voltage expressions developed in Chapters 5 and 6. However, here the explicit relation of voltage to input current is discussed, since a measurement of the 'V–I' curve is one of the basic system characterizations (Fig. 8.9(a)). The voltage of an isolated pulse as well as that of the spectrum are analyzed for thin longitudinal media recorded with an arctangent transition. The discussion is for recording and playback with a long pole inductive ring head, but the extention to, for example, MR playback heads is obvious. From (5.51) and (6.39) the isolated pulse voltage and the spectrum are proportional, respectively, to:

$$V^{\text{peak}} = C_1 2M_r \delta \tan^{-1} \frac{g}{2(d + a)} \tag{8.18}$$

and

$$V(k) = C_2 2M_r \delta e^{-ka} \tag{8.19}$$

where C_1 and C_2 are constants irrelevant to this discussion that can be obtained from (5.51) and, for example, (6.39) respectively. The factor of $2M_r$ denotes that the transition is from $-M_r$ to $+M_r$ or, equivalently, a magnetization change of $2M_r$.

In the analysis for the transition parameter it was assumed that the write current was sufficient to yield a full transition of magnetization change $2M_r$. The write current produces fields in the medium, and it is the maximum field along the gap center $(x = 0)$ that produces the magnetization change (Fig. 8.2) via the appropriate portion of the M–H loop (Fig. 8.5). For low currents there is a range where the field is insufficient to cause any appreciable change in magnetization and the

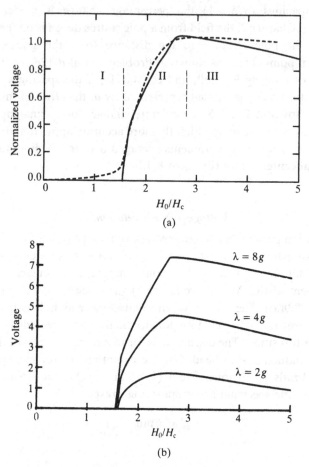

Fig. 8.9.   Voltage versus record current ('V–I' curves) for (a) an isolated pulse and (b) a general Fourier analysis of any pattern (isolated pulse, squarewave, etc.) at various wavelengths. The axes are plotted versus normalized deep-gap field (H_0/H_c) since $H_0 = NIE/g$. The dashed curve in (a) is a sketch of measured data. All other curves represent a simplified model.

voltage is virtually zero. When the current exceeds a value that produces maximum record fields slightly less than the coercivity, then magnetization changes begin to occur with transitions written with small total changes in magnetization $\Delta M \ll 2M_r$. As larger record currents are utilized, the total magnetization change increases as well as the replay voltage. For example, when the current amplitude is set so that the maximum record field equals the coercivity (or precisely the remanent coercivity), the magnetization change over the transition is about M_r. At this current the replay voltage via (8.18) would be about half that of maximum except that the transition parameter (and shape) is not given by that for full current recording. In addition, the maximum head field exceeds the coercivity, because demagnetization fields are present.

It is important to know the field required to saturate the medium (to yield a change of $2M_r$), since the corresponding record current is the minimum that would yield reasonable overwrite, as discussed in Chapter 9. Let H_{sat} denote that field on the M–H loop (Fig. 1.3) that brings the magnetization from $-M_r$ along the major loop to $+M_r$ remanent value. In Fig. 1.3 this field 'closes the loop' so that a subsequent reduction in field leaves the magnetization in the saturation remanent state. Using the Karlqvist field approximation (3.16) and magnetostatic fields for a thin film (4.13), the net field at the gap centerline is:

$$H \approx \frac{2H_0}{\pi} \tan^{-1} \frac{g}{2d} - \frac{4M_r\delta}{\pi g}\left(\frac{(4d/g)^2}{1+(4d/g)^2}\right) \quad (8.20)$$

The second term assumes that the transition is written close to the gap edge, a distance $g/2$ from the field evaluation point. Perfect imaging is included following Problem 8.3 for the field derivative where the transition shape does not enter, because it is assumed that the field evaluation point ($g/2$) is large compared to the transition parameter (a). The second factor in the magnetostatic field term is the effect of perfect imaging for a thin film. A factor of two is included in the demagnetizing field term to account for the worst case of saturation of a hard transition. The results in (Bloomberg, *et al.*, 1979) are accurately characterized by (8.20). Inverting (8.20) gives the deep-gap field for saturation:

$$\frac{H_0}{H_c} = \frac{\dfrac{H_{sat}}{H_c} + \dfrac{4M_r\delta}{\pi H_c g}\left(\dfrac{(4d/g)^2}{1+(4d/g)^2}\right)}{\dfrac{2}{\pi} \tan^{-1} \dfrac{g}{2d}} \quad (8.21)$$

Utilizing $H_{sat}/H_c \sim 1.5$, $d = g/4$, $g/\delta = 6$, $M_r/H_c = 6$ yields $H_0/H_c \sim 3$, which is close to that for optimum recording (Fig. 8.7).

A simplified model for the 'V–I' curve is derived here to give insight into the 'V–I' curve. The linear approximation to the major loop in Fig. 8.5 is assumed where S^* (1.2) characterizes the slope. In that case recording begins when the gap center field reaches $H_L = S^*H_c$. For lower currents (region I in Fig. 8.9(a)) the replay voltage is zero. In this simple model the medium saturates at a field $H_U = (2 - S^*)H_c$. For region II in Fig. 8.9(a) where the maximum field is less than that to fully saturate the magnetization, (8.18) is utilized to determine the voltage, but with $2M_r$ replaced by:

$$\Delta M = \frac{M_r}{(1 - S^*)} \left(\frac{H}{H_c} - S^* \right) \tag{8.22}$$

and a is determined from (8.13) by replacing M_r by $\Delta M/2$ (with Q determined from (8.12)). The variation of ΔM with deep-gap field for region II, where $\Delta M < 2M_r$, is obtained by utilizing (8.20) for the maximum medium field H, where in (8.20) M_r is replaced by ΔM. For a simple approximation, the change in transition location with record current is neglected, perfect imaging is assumed and a fixed location at $x_0 = g/2$ is assumed. In this case (8.22) becomes:

$$\Delta M = \frac{\dfrac{M_r}{(1 - S^*)} \left(\dfrac{2H_0}{\pi H_c} \tan^{-1} \dfrac{g}{2d} - S^* \right)}{1 + \dfrac{2M_r \delta}{(1 - S^*)\pi g} \left(\dfrac{(4d/g)^2}{1 + (4d/g)^2} \right)} \tag{8.23}$$

For region III where the medium is saturated ($\Delta M = 2M_r$), (8.18) gives the voltage with a determined from (8.14) and Q from (8.12).

The solid curve in Fig. 8.9(a) is the result of the simple model with $S^* = 0.8$, $M_r/H_c = 6$, $g/d = 4$ and $g/\delta = 6$. In region III where the medium is saturated, the slight decrease in voltage for $H_0/H_c \sim 3.75$ corresponds to the increase in the transition parameter caused by a decrease in Q (Fig. 8.7). This decrease is seldom seen on a 'V–I' curve due to head saturation. For a thin film head with $M_s \sim 1$T, depending on the design, the deep gap field does not exceed $H_0^{max} \sim 7000$ Oe, so that for $H_c \sim 1500$ Oe the curve stops changing (decreasing) at about $H_0/H_c \sim 5$.

In Fig. 8.9(b) the simple model is utilized to plot the spectra components after Fourier analysis. Equation (8.19) is utilized, and curves for $\lambda = 2g$, $4g$, $8g$ are shown. For the saturation region (III) the voltage decreases more rapidly at shorter wavelengths. Even though the change in Q, and therefore the transition parameter a, with current is slight, the exponential factor in (8.19) yields strong changes at small k or short wavelengths. Note that the spectra discussed in Chapters 5 and 6 are for fixed a or fixed current in Fig. 8.9(b). System evaluation is generally at that current that maximizes the band edge (typically $\lambda = 2g$) output or at slightly higher currents, as discussed in Chapter 9, that yield reasonable overwrite ratios.

Degmagnetization limits

It is possible that the demagnetization fields associated with a recorded transition might be large enough to cause irreversible demagnetization when the recorded transition moves into free space away from the recording head fields and the influence of the high-permeability head. For an arctangent transition the peak field, location, and magnetization are given by (4.12) and are illustrated in Fig. 4.6. If the demagnetization field exceeds the major loop field at the corresponding level of magnetization, the medium will demagnetize. For reasonably square-loop media this field is approximately less than, but close to, the coercivity. Assuming a square loop for an estimate, the value of the transition parameter where the peak field just equals the coercivity is given from (4.12) by:

$$a_{\rm d}/\delta = \frac{\sqrt{1 + \cotan(\pi H_{\rm c}/2M_{\rm r})} - 1}{2} \approx \frac{M_{\rm r}}{2\pi H_{\rm c}} \qquad (8.24)$$

In (8.24) the final approximate term is for thin media ($\delta \ll a$). This limit does not depend on the head field gradient or head–medium spacing. It will depend slightly on S^* if (8.24) is rederived utilizing the constant slope loop illustrated in Fig. 8.5. Various forms for this limit have been derived (Potter, 1970; Middleton, 1966; Lindholm, 1973). Note that, as discussed in Chapter 4, the maximum demagnetizing field is $H_{\rm c} = M_{\rm r}$ corresponding to a perfectly sharp transition ($a = 0$). The thin film form of (8.24) is plotted in Fig. 8.8 for $M_{\rm r}/H_{\rm c} = 8$. At typical values of $M_{\rm r}/H_{\rm c}$ the demagnetization limit is reached only at very small head–medium spacings.

A comparison of (8.14) and (8.24) indicates that an approximate criterion necessary for irreversible demagnetization effects to occur is:

$$\frac{M_{\rm r}}{H_{\rm c}} > \frac{2\pi(1 + 2d/\delta)}{Q} \qquad (8.25)$$

The development of high density recording systems is toward near contact ($d \sim \delta \sim 200\text{Å}$) with increased coercivity with $M_{\rm r}/H_{\rm c} \leq 5$. Thus, irreversible demagnetization effects should not occur. It is important to design a head–medium configuration so that the peak medium fields are less than the medium coercivity. If the fields are approximately equal to the coercivity, then extraneous fields and thermal demagnetization can demagnetize a recorded transition. One example is demagnetization of a recorded transition or dibit as it passes the reversed field at the edge of an energized thin film head (Che & Bertram, 1993b).

Thick media tape recording

Modeling of the recording process in thick particulate media that is nominally longitudinally oriented (Fig. 1.1) is complicated because of the vectorial nature of the magnetic fields and magnetizations and because of variation of the magnetization configuration with depth into the medium. In longitudinal tape recording the vertical demagnetization fields are not sufficiently large to keep the magnetization in the tape plane during the recording process. The approximation of longitudinal magnetization and fields as utilized in thin film modeling is not accurate. Numerical simulation of the recording process in thick media with moderate demagnetization ($M_r/H_c \sim 2$) has illuminated many aspects of the process (Bertram & Beardsley, 1988; Bertram, *et al.*, 1992). A discussion of essential elements for any simplified modeling learned from numerical simulation will be discussed here. Focus will be on a simple vectorial model that provides an explanation to the 'V–Γ' curve that is distinctly different from that of thin films.

It is characteristic of longitudinal tape recording that the isolated pulse is asymmetric: the initial rise of the voltage versus time is more rapid than the subsequent decay (Fig. 8.10). Numerical simulation reveals that the asymmetry is not due to a small rotation of the magnetization toward the perpendicular direction over the transition (Fig. 5.7(b)). The asymmetry can be shown to be caused by the shift in the center of the recorded transition with depth into the medium even in the longitudinal component (Problem 5.10) as illustrated in Fig. 8.11(a). In addition,

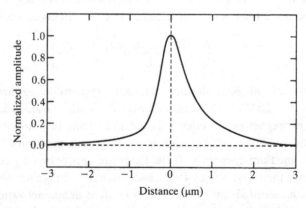

Fig. 8.10. Isolated pulse voltage measured on thick longitudinal tape. Taken from Armstrong *et al.* (1991), © IEEE.

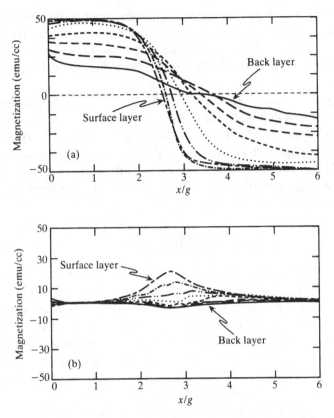

Fig. 8.11. Numerical simulation of an isolated transition in thick longitudinal particulate media. (a) Longitudinal component versus distance for eight lamina seperated by $g/8$. (b) Vertical component. Taken from Bertram *et al.* (1992), © IEEE.

the transition parameter for the longitudinal component increases with depth into the medium: the medium particles are not exchange coupled in the vertical direction, as in thin films, to prohibit a depth variation of magnetization. The thin film expression (8.13) gives insight into this variation. The transition parameter increases as the medium is made thicker. Alternatively, if (8.13) is thought to represent different lamina in the medium ($\delta \rightarrow dy$), then the transition parameter is seen to increase with depth into the medium. If a solely longitudinal model were appropriate for tape recording, then solution of (8.4) as an integral equation in (8.7), as well as in the more general form of (8.9) for an arctangent given by (4.14), yields a depth variation of the transition parameter $a(y)$. Such estimates have been made assuming a linear increase of *a* with depth into the medium (Middleton & Wisely, 1976). In addition, it is clear that the transition parameter $a(y)$, and thus the replay voltage via (8.18) and (8.19), increases monotonically with demagnetization ratio M_r/H_c (Koester & Pfefferkorn, 1980; Bertram & Niedermeyer, 1978).

A major difference between tape recording on thick media and thin film recording is that the depth of recording does not equal the magnetic layer thickness. As discussed in (Bertram & Beardsley, 1988) and indicated in Fig. 8.11(a), the depth of recording is set approximately where the value of the maximum field equals the coercivity. Utilizing the Karlqvist approximation (3.16), this depth of recording is given approximately by:

$$\delta = \frac{g/2}{\tan(\pi H_c/2H_0)} - d \tag{8.26}$$

The depth of recording increases monotonically with deep-gap field, which affects the 'V–I' curve, as discussed below. As seen in Fig. 8.11(a), the effect of a finite slope to the major loop (Fig. 1.3) yields a gradual decrease in the remanent recorded magnetization in the deep layers, so that (8.26) corresponds to where the change in magnetization over a transition is M_r rather than $2M_r$. As in the discussion of saturation requirements for thin film media, demagnetization fields affect the depth of recording as well as the variation of remanent magnetization with depth in the back layers. This is an important factor in the analysis of overwrite in thick tape media.

It is clear from numerical simulation that a successful simplified model of the tape recording will require four ingredients: (1) utilization of the total head-field gradients as well as both components of the magneto-static field, (2) inclusion of the effect of the high-permeability head as illustrated in Problem 8.4, (3) incorporation of the effect of phase shift of the recorded transition with depth, which is dominated by the magnetostatic fields, and (4) inclusion of the vertical component of magnetization, even if small.

Tape recording in longitudinally oriented particulate media shows evidence of vertical magnetization (Lemke, 1982, 1979). Although the vertical component does not appreciably affect the optimized isolated pulse voltage (Fig. 8.10), there is recorded in longitudinal media a vertical component in the surface layers at the transition center (Fig. 8.11(b)). The vertical component has the character of a Hilbert transform of the longitudinal component (Fig. 2.14(b)); however, the spectral behavior is different. The vertical magnetization as shown in Fig. 8.11(b) is confined to the transition center and does not exhibit the long tails of a Hilbert transform. For example, if the vertical component is approximated by the derivative of the longitudinal component, then the voltage spectrum (8.19), neglecting depth effects, is modified by:

$$V(k) \propto \delta k e^{-ka}(M_x^r + CkM_y^r) \qquad (8.27)$$

where C is an appropriate scaling constant with the dimensions of length. In agreement with experiments on isotropic media (Lemke, 1982), a spatially confined vertical component adds to the spectrum of the longitudinal component only at medium to short wavelengths and does not affect the long wavelength voltage.

The voltage–current curves for longitudinal tape recording exhibit a more pronounced variation than those for thin film media (compare Fig. 8.9(b) with Fig. 8.12). In Fig. 8.12 'V–I' curves are shown for longitudinal tape for a wide range of M_r/H_c ratios. Even though the wavelength is short ($\lambda \sim g$) compared to that in Fig. 8.9(b), the voltage decreases rapidly with current past the peak and exhibits a null. This

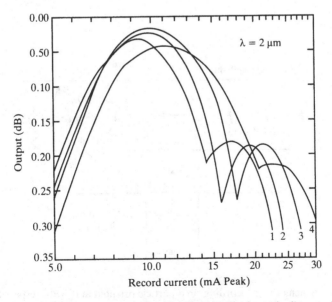

Fig. 8.12. Voltage versus current in thick longitudinal particulate media with varying M_r/H_c. Taken from Bertram & Neidermeyer (1978), © IEEE.

behavior is approximately independent, to first order, of demagnetization ratio M_r/H_c; and, in fact, a series of nulls can be measured (Iwasaki & Takemura, 1975; Bertram & Niedermeyer, 1978).

The presence of vertical components, as illustrated in Problem 6.6, can be utilized to explain a pronounced '$V–\Gamma$' curve with current nulls. In Fig. 8.13 a particle is shown oriented at angle θ_p with respect to the recording direction. When a transition is being recorded with a positive field, as shown, the vector head field is generally oriented toward the head at the recording location. Thus, the head field makes a large angle with respect to the recording direction $\theta_H + \theta_p$. The switching field of the particle is generally a monotonic increasing function of the net field angle to the particle easy axis (Knowles, 1978; Schabes, 1991). Thus, for the subset of particles in the tape oriented at $+\theta_p$, a transition will be written near the head gap where the field is large. For particles oriented at $-\theta_p$ the transition will be written further away from the gap since the head field is along or close to the particle axis (Bertram, 1984a,b). With a general distribution of particle orientations and intrinsic switching fields, as well as the inclusion of magnetostatic fields, the writing process is more complex, but this essential feature describes well the vector magnetization distribution (the medium coercivity as well as average switching field as used in (8.26) is an ensemble average) (Bertram & Beardsley, 1988).

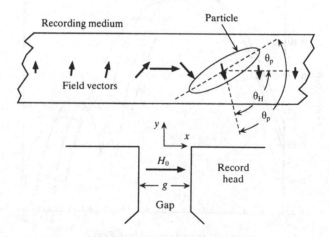

Fig. 8.13. Schematic of recording on a particle oriented at θ_p with respect to the recording direction. The head field vectors are shown making an angle θ_H with respect to the x axis.

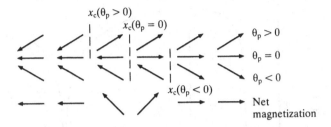

Fig. 8.14. Schematic of recording transitions on assemblies of particles at three orientations in the (x,y) plane. The net effect is shown that yields a magnetization that rotates in direction through the transition. Taken from Bertram (1984b), © IEEE.

In Fig. 8.14 a schematic of the writing of a transition for three orientations $\theta_p > 0$, $\theta_p = 0$, $\theta_p < 0$ is shown. The essential feature is that the transition center for each orientation group is sequentially at $x_c(\theta_p > 0) < x_c(\theta_p = 0) < x_c(\theta_p < 0)$. The net vectorial sum yields a net vector magnetization as shown, where the magnetization 'rotates' through the transition center in agreement with Fig. 8.11(a), (b). The effect of these vector phase shifts on the spectra is discussed here in terms of a two-particle picture with orientations solely at $\pm\theta_p$. In this case the spectra for thick media, neglecting depth variations, (6.35) is:

$$V(k) \propto (1 - e^{-k\delta})e^{-ka(\delta)} \cos(\theta_p - k\Delta x(\delta)/2) \qquad (8.28)$$

Δx is the recording location difference of the two populations and is written as a function of recording depth (8.26) to denote a monotonic increasing variation with record current. Equation (8.28) is plotted versus record current, or equivalently H_0/H_c, in Fig. 8.15, assuming for simplicity $\Delta x = (\delta + d)/2$, $a = \delta$, $\theta_p = 30°$, $d/g = 0.2$, $\lambda/g = 2$. The voltage vanishes for low currents (I_{th}) below which even the surface layers are not switched. Above I_{th} the depth of recording continuously increases, and correspondingly the voltage increases due to the first term in (8.28), which varies as $k\delta$ for small thicknesses. Increasing current causes the first term to level for $\lambda < 2\pi\delta$. However, at these higher currents the latter two terms cause a decrease in voltage, and thus an optimum current occurs (I_{opt}). The transition loss term yields only a gradual decrease, while the last term that results from the vector phase shift dominates. As the current is increased, Δx increases causing the voltage to decrease rapidly between $\lambda\theta_p/\pi < \Delta x < \lambda(\theta_p/\pi + 1/2)$. The

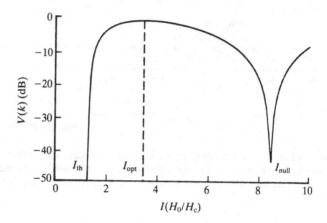

Fig. 8.15. Simple model for '$V - I$' curve for longitudinal tape for the voltage spectra at short wavelengths.

first null (I_{null}) occurs at $\Delta x = \lambda(\theta_{\text{p}}/\pi + 1/2)$. In agreement with data (Iwasaki & Takemura, 1975), (8.28) predicts that the nulls occur at lower currents for shorter wavelengths, because of the monotonic increase of Δx with record current.

The 'V–I' curve measured at very long wavelengths in thick tape can be utilized for estimates of the spacing d and the turns-efficiency product NE (Armstrong, *et al.*, 1991; Beardsley, *et al.*, 1992). The simplified analysis above has been utilized to study the effect of head saturation (Wachenschwanz & Bertram, 1991).

Models of perpendicular recording

The recording process in media with vertically oriented magnetization is more complicated than that in nominally longitudinal media. In perpendicular recording the transition is formed both during 'standstill' recording at the instant when the head current is changed and during the subsequent movement of the transition along the head away from the gap or pole region. At the instant the head field is changed, the leading edge of the transition, as well as most of the transition center region, is written. Subsequent motion of the medium results in demagnetization of the trailing edge of the transition, which had experienced field levels that caused the magnetization to exceed demagnetization limits. In this section a simplified model of the complete recording process for thin media will be presented first. Following that, the effect of medium thickness will be examined by a simple model of the 'standstill' recording process that utilizes Fourier analysis. Results of numerical simulation of subsequent medium motion will be presented. These simplified models exhibit the essence of the process and do not depend on whether a probe-keeper or ring head is utilized for recording; different head structures affect primarily the gradient of the recording field, yielding, of course, transitions with different gradients. Approximate models that assume arctangent magnetization forms yield general parameter trends, but miss the essential asymmetry of the isolated pulse (Middleton & Wright, 1983). As in models for longitudinal thin films, the magnetization is assumed to be invariant with depth into the medium. Media considered for perpendicular recording often include a high permeability keeper layer below the recording layer. The effect of this keeper layer on the playback process is discussed (Problem 8.13), however, the effect of a keeper on the record process is neglected since the layer is likely to be saturated by the record head.

The demagnetization field for a depth invariant vertical magnetization is given in 2D from (2.22) evaluated by the medium center by:

$$H_y^d(x) = -\frac{\delta}{2\pi} \int_{-\infty}^{\infty} dx' \frac{M(x')}{(x-x')^2 + (\delta/2)^2} \tag{8.29}$$

The field averaged through the medium thickness is:

$$\bar{H}_y^d(x) = -\frac{1}{2\pi\delta} \int_{-\infty}^{\infty} dx' M(x') \ln \frac{(x-x')^2 + (\delta/2)^2}{(x-x')^2} \tag{8.30}$$

If the spatial variation or scale of the magnetization is large compared to the medium thickness, all forms yield a simple form for the demagnetizing field.

$$H_y^d(x, y) = -M(x) \tag{8.31}$$

The demagnetizing field for thin media is everywhere equal to the local magnetization, which is the result for a thin film uniformly magnetized in the perpendicular direction ($N = 1$ in (2.16)). Perpendicular media often have square intrinsic $M-H$ loops (Fig. 8.5 dotted) as studied by micromagnetic analysis (Zhu & Bertram, 1989a). A square loop, as plotted versus external applied field, will be transformed into the sheared loop shown in Fig. 8.16(a), since the internal field is simply $H_{\text{int}} = H_{\text{ext}} - M$. The loop slope is $dM/dH = -1$ and the remanent magnetization is $M_r = H_c$. For a completely sheared loop it is assumed

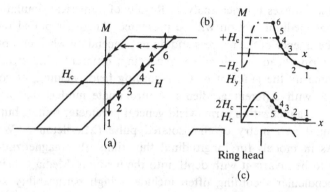

Fig. 8.16.   Plot of fields and magnetization for thin film model of perpendicular recording. (a) Sheared loop of magnetization versus external field. (b) Magnetization of a transition when field is switched (dashed) and after medium motion (solid). (c) Vertical-recording head field. Taken from Bertram (1986a), © IEEE.

that $M_r > H_c$, a condition that will also be assumed for the modeling discussed in this section.

Thin medium model

First a model of this process is presented that utilizes a simple form of the demagnetizing field (Middleton & Wright, 1983; Wright & Middleton, 1985). It is assumed that the medium is thin with respect to the spatial variation of the magnetization through the transition so that (8.31) applies. Thus, the recording process may be completely described, in this approximation, by simply evaluating the effect of an applied field on the sheared loop at each point in the medium (Fig. 8.16). In Fig. 8.16(c) a typical vertical field profile is shown plotted for an inductive head (the following applies equally well utilizing a probe-head field). It is assumed that a positive head field has been applied to a negatively saturated medium $M(x) = -M_r = -H_c$. The magnetization that results from the application of the head field (negative deep-gap field in this example) before medium motion is considered first. The magnetization is simply a mapping of the field profile onto the sheared M–H loop. At position 1 the field is slightly positive, resulting in a magnetization slightly increased from $-M_r$. As points 1–6 are examined sequentially, the magnetization increases monotonically, as shown in Fig. 8.16(b). Position 3 corresponds to a head field equal to the coercivity so that $M(x_3) = 0$. At position 5 the head field is $2H_c$ so that $M(x_5) = H_c$. At position 6 the field exceeds $2H_c$ so that $M(x_6)$ exceeds H_c. The magnetization at positions further to the left need not be examined since a change occurs with medium motion.

The medium moves in the direction of positive x. Thus, for a fixed recording current, as the transition moves away from the position where it was initially written, the field history is simply the field experienced from Fig. 8.16(c) by moving from a given recording location along positive x toward $x \to \infty$. For example, consider the medium point x_1 relative to the head when the current was switched. As the medium subsequently moves, that point experiences a decreasing field continuously from the recording field H_1. The magnetization that results follows from the change of magnetization along the minor loop in Fig. 8.16(a). This minor loop is simply a straight line for a square-loop material so that the magnetization moves from point '1' along a horizontal line to the same magnetization level but with the field zero. All magnetization will move along a horizontal minor loop to the $H = 0$ axis. For all magnetizations at positions greater than $x_5 (H_h \leq 2H_c)$ the remanent magnetizations after medium motion are unchanged from that in the 'standstill' process. For all positions where the head field exceeded $2H_c$ and the medium magnetization exceeded $M_r = H_c$, medium motion reduces the magnetization to the saturation remanent state. An example is shown in Fig. 8.16(a) for position 6 where medium motion and subsequent field decrease cause the magnetization to move along the horizontal minor loop until the opposite side of the major loop is reached. At that point a decrease in field reduces the magnetization along the major loop to the saturation remanent magnetization $M_r = H_c$.

There are other medium locations far to the left of the transition center, shown dashed in Fig. 8.16(b), where switching of the field gave a small change in magnetization or drove the magnetization further into negative saturation. Subsequent medium motion in the presence of a constant head field current would leave the magnetization in the remanent saturation state of $+ H_c$. This final state can be seen by following the field history in Fig. 8.16(c). All these medium locations will move first into large positive field regions greater than $2H_c$, bringing the magnetization to a level greater than $+ H_c$. Subsequent motion causes the field experienced to reduce to zero, which, by the above arguments, leaves the medium in the saturated remanent state of $+ H_c$.

The final magnetization transition after initial switching and subsequent motion away from the field region is shown (solid) in Fig. 8.16(b). On the leading edge of the transition the magnetization varies as the head field offset by the initial saturation level $-H_c$. For all trailing edge points where the recording field exceeded $2H_c$, the remanent magnetization was reduced to a level equal to the coercivity. Thus the transition is very asymmetric but still varies between the asymptotic demagnetization limits of $+H_c$ and $-H_c$. The field required to saturate the transition is $2H_c$. The transition is given by:

$$
\begin{aligned}
M(x) &= H_c & x &< x_{2H_c} \\
&= H_y^h(x) - H_c & x &> x_{2H_c}
\end{aligned}
\tag{8.32}
$$

where x_{2H_c} is the first location from positive x where the head field equals twice the coercivity. There are several interesting points to make about the magnetization pattern that will hold, at least to first approximation, even for thick media. First, for media where typically $H_c < M_r$, the final magnetization pattern does not depend on M_r and scales with H_c. The transition shape depends on the head fields and demagnetization fields via the M–H loop. Secondly, adjusting the record current to a value so that the maximum positive field experienced by the medium is $2H_c$ yields the transition with the sharpest gradient. For an increased head field the region of changing magnetization occurs further away from the head center. Thus, the field decreases more slowly with a smaller gradient and the transition is broader (Problem 8.10).

The effect of an asymmetric transition on the voltage pulse shape is illustrated in Fig. 8.17(a)–(d). In Fig. 8.17(a) the vertical magnetization pattern from Fig. 8.16(b) is plotted along with a simple, sharp, symmetric transition (dashed) for comparison. The spatial derivative $\partial M / \partial x$ is plotted in Fig. 8.17(b). The replay voltage is given by a correlation of the magnetization derivative with the replay head. As illustrated in Fig. 8.17(c) for an inductive ring head, at each time step the voltage arises from integrating the overlap regions of the product of the head field and the magnetization derivative. The respective replay voltage shapes are indicated in Fig. 8.17(d). For a symmetric, sharp transition the replay voltage shape follows the replay head field evaluated at the medium spacing. (If the symmetric transition were broad, then the approximate voltage shape would be the head field evaluated at the effective distance $d + a$.) For the asymmetric transition pattern the resulting voltage exhibits an initial undershoot that is less than the subsequent positive peak. As the magnetization pattern moves to the right with respect to the head as indicated by the arrow in Fig. 8.17(c), the correlation increases in the negative direction as the negative magnetization derivative begins to

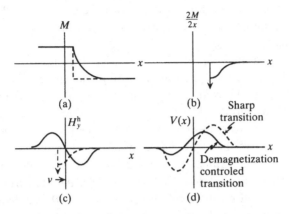

Fig. 8.17. Determination of the isolated pulse voltage in perpendicular recording. (a) Magnetization transition shape (dashed is sharp transition for comparison), (b) magnetization transition derivative (shape transition yields an impulse indicated by an arrow), (c) illustration of correlation of magnetization derivative with the replay head via reciprocity, and (d) illustration of final pulse shape with initial minimum less than subsequent maximum. Taken from Bertram (1986a), © IEEE.

overlap the positive region of head field. The voltage minimum occurs approximately when the magnetization derivative overlaps the positive head field region. However, the voltage minimum is limited by the tail of the magnetization derivative that extends toward positive x where the head field is of opposite sign. Later on, when the transition has moved under the negative head field region, the overlap integral and the corresponding voltage maximum are larger, since the extended tail of the magnetization derivative does not sample a region of head field of opposite sign. Note that the isolated pulse shape due to an imperfectly asymmetric magnetization transition with complete vertical orientation is similar to replay from a perfectly asymmetric transition with magnetization not quite vertically oriented (Fig. 5.7(c)).

Thick medium modeling

It was shown in Chapter 4 that perpendicular magnetization patterns have smaller demagnetization fields when the medium is made thicker (Fig. 4.6(b)). Here we discuss, with simple models (Wielinga, *et al.*, 1983; Lopez, 1984; Wright & Middelton, 1985), how the magnetization transition created for thin media is narrowed by utilization of thick media. Results of numerical simulation will be discussed following the simplified techniques. As in the discussion for thin media, it will be presumed that $M_r > H_c$.

The simplified models utilize a Fourier transform technique to determine the written magnetization during the instant that the head current is changed. Therefore, for discussion purposes, it is convenient to consider a head whose field does not cause any abrupt changes in the magnetization. A probe head is considered since the perpendicular field component always has the same polarity everywhere along the longitudinal, x, axis. Thus, given a medium that is initially negatively saturated (at remanent state $-H_c$ for $M_r > H_c$), the field from a probe head when switched positive will, in combination with the demagnetization fields, bring the medium at each point up the M–H loop from the initial saturation state. This will be a linear process if a square intrinsic loop is assumed, as in Fig. 8.5 (dotted), and if the head field magnitude is nowhere great enough to saturate the medium. Nevertheless, as discussed earlier, this assumption is convenient only for simplicity of the model, and the conclusions actually apply well to all heads utilized in perpendicular recording. It is further assumed that the microstructure of the medium forces the magnetization not to vary with depth into the medium (which may occur in columnar CoCr perpendicular films). This assumption is probably not critical for recording with keepered-probe heads (Perlov & Bertram, 1987).

The thin medium approximation that yields the simple form of (8.31) and hence, the sheared loop of Fig. 8.16(a) does not hold for thick media. However, for a square M–H loop where the magnetization is confined entirely along the vertical outer section of the loop at the instant of field reversal, the condition (8.3) applies requiring that the total field must equal the medium coercivity:

$$H_c = H_h(x, y) + H_d(x, y) \qquad (8.33)$$

Equation (8.33) applies everywhere in the medium whether or not the magnetization is assumed to vary with depth. For a medium with no magnetization variation with depth, a first approximation might be to choose the medium center for the evaluation of (8.33). From micromagnetic energy considerations, if the columns switch by uniform rotation, then a depth average of (8.32) should be utilized. If nucleation of reversal occurs near the surface where the largest fields are experienced, then perhaps a nucleation depth near the surface should be chosen. Here the medium center will be utilized as the evaluation point of (8.33) with the depth averaged field approximation discussed in Problem 8.14.

Because (8.33) is a linear expression, it may be Fourier transformed by applying (2.42) as an operator to both sides. This yields:

$$H_c\delta(k) = H_h^s(k)e^{-k(d+y)} + H_d(k, y) \tag{8.34}$$

In (8.34) the y axis origin is taken at the medium surface, separated at distance d from the head surface. A keeper is not included here but is considered in Problem 8.14. The vertical component of the demagnetization field for a vertically magnetized medium uniformly magnetized with depth may be determined from (4.23):

$$H_d(k, y) = -\frac{M(k)}{2}(e^{-ky} + e^{-k(\delta-y)}) \tag{8.35}$$

Substituting (8.35) into (8.34) yields:

$$H_c\delta(k) = H_h^s(k)e^{-|k|(d+\delta/2)} - M(k)e^{-|k|\delta/2} \tag{8.36}$$

so that:

$$M(k) = H_h^s(k)e^{-|k|d} - H_c\delta(k) \tag{8.37}$$

The first term in (8.37) is the head field at the surface of the medium closest to the head. The second is the dc offset due to the initial state of negative saturation at level $-H_c$. Equation (8.37) can be immediately transformed to yield:

$$M_y(x) = H_y^h(x, d) - H_c \tag{8.38}$$

This expression is remarkable in its simplicity with regard to the magnetization written during the first stage of 'standstill' recording when the head current is changed. It states that: *the written magnetization is given by the head field at the medium surface nearest the head offset by the initial saturation level.* Equation (8.38) holds at all depths into the medium, since the magnetization was assumed not to vary with depth.

A sketch of (8.38) is shown in Fig. 8.18(a). For a probe head a double transition is written at the instant the current is changed. The head field is evaluated at the medium center, but the magnetization pattern is evaluated using the surface head field. Thus, the effect of the reduction of demagnetization fields by increasing the medium thickness is to

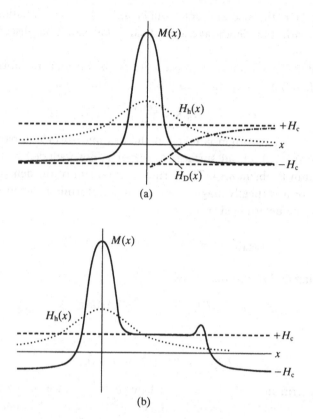

Fig. 8.18. (a) Effect of thick film demagnetization on the formation of a perpendicular transition. The long dashed curve is the vertical head field versus position along the motion direction evaluated at the film center. The solid curve is the magnetization that follows the vertical field evaluated at the film surface offset by a level shift equal to the medium coercivity. The dotted curve is the demagnetizing field plotted only in the region of the recorded transition. (b) Final written transition after medium motion. On the left side of the transition demagnetization to yield a level of H_c occurs except for a region near the transition center.

'sharpen' the magnetization transition so that it always follows the head surface field gradient. Even though the medium thickness is not explicitly part of (8.38), a thicker medium yields reduced demagnetization fields near the transition center (8.35 with $y = \delta/2$). It can be shown that (8.38) also results if the criterion of volume averaged fields is utilized (Problem 8.15).

If it is believed that magnetization reversal nucleates at the surface closest to the head where the field is largest then a modified form of (8.37) results:

$$M(k) = \frac{2H_{\mathrm{h}}^{\mathrm{s}}(k)\mathrm{e}^{-|k|d}}{1 + \mathrm{e}^{-|k|\delta}} - H_{\mathrm{c}}\delta(k) \qquad (8.39)$$

Back transformation is difficult; however, it is clear that the extra factor multiplying the head field enhances high-frequency components. Thus, the effect of evaluating (8.33) at the medium surface is to yield a transition that is sharper than the surface head field.

As the medium moves along the head after the instant of field reversal, to first order the magnetization will reduce everywhere on the head side of the transition where the magnetization exceeds the coercivity, a result similar to the result for medium motion for thin perpendicular media. The final transition is illustrated in Fig. 8.18(b). The reduction of magnetization is simple for regions far to the left of the transition center where the demagnetizing field follows (8.31); the magnetization reduces to a level equal to $+H_{\mathrm{c}}$. However, near the transition the fields are reduced for thick media from the result given by (8.31) so that the magnetization need not be limited to $+H_{\mathrm{c}}$. This phenomenon results in an overshoot of the magnetization above a saturation value of $+H_{\mathrm{c}}$ as illustrated in Fig. 8.18. Step by step demagnetization during motion is illustrated in Beusekamp & Fluitman (1986).

The overshoot that occurs during demagnetization may also be understood by regarding the dotted curve in Fig. 8.18(a). This curve is the demagnetization field after the instantaneous standstill recording process is complete; it is determined simply from (8.33) evaluated at the medium center. In Fig. 8.18(a), moving along the transition from the leading edge side (to smaller x), the point where the magnetization field exceeds the coercivity ($H_{\mathrm{D}} < -H_{\mathrm{c}}$) occurs is where the magnetization exceeds the coercivity limit ($M > H_{\mathrm{c}}$). Thus, approximately, demagnetization during medium motion will cause an overshoot approximately on the head field side of the transition in the region where the magnetization can exceed $+H_{\mathrm{c}}$.

Numerical simulation of perpendicular recording utilizing a ring head has been performed (Zhu & Bertram, 1986). Considerable overshoot occurs since the case of virtual contact recording is considered. The surface head field is shown, which is close to but somewhat sharper in spatial variation than the magnetization pattern in the transition region. During standstill recording with a ring head, the negative head field

($x < 0$) saturates the magnetization near the left gap edge, yielding an extremely sharp secondary transition above the gap. This sharp transition reduces the demagnetization fields in the primary transition and yields a sharper transition than that predicted by (8.39).

In a pattern similar to the results developed for thin media, a sharpest transition will occur when the maximum field at the medium center (or the depth averaged head field) is approximately $2H_c$. In that case the magnetization transition will vary from $-H_c$ to $+H_c$ and the shape of the transition (apart from an overshoot) will follow (8.32) with the field evaluated at the medium surface ($y = d$). This is generally true whether a probe or ring head is utilized for recording. Thus, neglecting overshoot, a simple expression for the transition for perpendicular recording in media of any thickness with a ring head using (3.16) is:

$$M_y(x) = + H_c \quad x < 0$$

$$H_c \left(2 \left(\ln \frac{(x+g)^2 + d^2}{x^2 + d^2} \middle/ \ln \frac{g^2 + d^2}{d^2} \right) - 1 \right) \quad x > 0 \qquad (8.40)$$

In (8.40) it is assumed that the head–medium spacing is small so that the peak of the head field occurs at $x \sim g/2$. The arbitrary location of the transition has been placed so that the saturation point occurs at $x = 0$. The form of (8.40) emphasizes the complete scaling of the transition magnetization with the coercivity (H_c) as long as $M_r > H_c$. Since the form of (8.40) in essence yields the same asymmetry as that for thin media (8.32) the voltage pulse shape with a ring head will be similar to that illustrated in Fig. 8.17(d). Numerical simulation of micromagnetic processes in perpendicular media yield transitions in agreement with the simplified ideas presented here (Bertram & Zhu, 1992).

Problems

Problem 8.1 Utilizing (4.19), show that the thickness averaged demagnetizing field gradient for an arctangent transition with parameter a is given by:

$$\frac{\mathrm{d}\bar{H}_x^d}{\mathrm{d}x} = -\frac{2M_r}{\pi a} \left(1 - \frac{a}{\delta} \ln \frac{a + \delta}{a} \right)$$

Problem 8.2 Show that an implicit expression for the transition parameter not assuming thin media and utilizing depth averaged fields is given by:

$$\frac{a/\delta}{\left(1 - \frac{a}{\delta}\ln\left(1 + \frac{\delta}{a}\right)\right)} = \frac{2M_r}{\pi Q H_c \ln\left(1 + \frac{\delta}{d}\right)}$$

Assume that magnetostatic fields dominate the effect of loop shape, the transition center is not a function of depth, the transition shape is an arctangent, and that the range of y/g for the average of the head field gradient is such that $Q \sim$ constant. Note that the scaled transition parameter a/δ is a function of M_r/H_c and the scaled spacing d/δ. Show that $M_r/H_c \sim 6$, $d/\delta \sim 1/4$, $Q \sim 0.85$ yields $a/\delta \sim 0.9$ so that the assumption $\delta \ll a$ is not true for these parameters.

Problem 8.3 Using (2.22) show that the demagnetizing field gradient at the transition center ($x = x_0$) for a general transition (Table 4.1), but at any vertical separation y from the transition center, is given by:

$$\frac{dH_x^d}{dx} = -\frac{M_r \delta}{\pi a^2} \int_0^\infty \frac{s\,ds}{s^2 + (y/a)^2} \frac{d^2 f(s)}{ds^2}$$

Utilizing (4.19) show that the net demagnetizing field gradient including imaging is given by:

$$\frac{dH_x^d}{dx} = -\frac{M_r \delta}{\pi a^2} \int_0^\infty \frac{(y/a)^2 ds}{s(s^2 + (y/a)^2)} \frac{d^2 f(s)}{ds^2}$$

Thus, the ratio y/a sets the extent of demagnetization field reduction by head imaging. For magnetic recording $y = 2d + \delta$, so that the effect of the high-permeability head is important only for $2d + \delta < a$.

Problem 8.4 Repeat Problem 8.2 including perfect imaging to obtain:

$$\frac{a/\delta}{\left(1 - \frac{a}{\delta}\ln\left(\frac{(a + \delta)(a + 2d)^{1/2}(a + 2d + 2\delta)^{1/2}}{a(a + 2d + \delta)}\right)\right)} = \frac{2M_r}{\pi Q H_c \ln\left(1 + \frac{\delta}{d}\right)}$$

Show that for the parameters in Problem 8.2, a/δ with imaging is ~ 0.7. This result is close to that obtained using medium center values (8.17).

Problem 8.5 Develop a formulation for the 'V–I' curve that includes details discussed in the derivation of the simple model. Assume an arctangent transition and a linear M–H loop, but include accurate magnetostatic fields and imaging for a finite gap (Chapter 2) and the change in recording location with record current. Reformulate the head-field gradient factor Q. Discuss the effect of transition shift (Chapter 9) for a hard transition at low currents.

Problem 8.6 Repeat the simple derivation for the short wavelength 'V–I' curve in Fig. 8.9(b), but plot V_{rms} versus record current, including head saturation as discussed in Problem 3.9.

Problem 8.7 Utilizing (8.19) and (8.14) find the $M_r\delta$ product that maximizes the short wavelength output ($\lambda = 2g$).

Problem 8.8 Utilize the simple model as expressed by (8.27) to derive an approximate vertical contribution to the isolated pulse (Fig. 8.10) and fit to the results of Fig. 8.11(b).

Problem 8.9 Derive (8.28) and compare with the results of Problem 6.6. Give a physical picture of why the phase term in (8.28) yields a maximum voltage at a finite phase shift ($\Delta x = \lambda\theta_p/\pi$). Assuming a vanishingly small gap (far-field head approximation) show that $\Delta x = \Delta\theta(\delta + d)/2$.

Problem 8.10 Use (8.32) and the Karlqvist head field (3.16) to plot magnetization transitions for various deep-gap fields. Derive an expression for the transition slope at $M(x) = 0$ (approximate transition center) and plot versus deep-gap field.

Problem 8.11 Utilize (2.8) to generalize (8.31) to consider a film of any thickness where the vertical magnetization component may vary with depth, but the variation in the film plane occurs over much larger scales than the film thickness. Show that with this assumption, the demagnetizing field in a film due to the vertical magnetization component is equal to the negative of the local magnetization:

$$H_y^d(x, y, z) = -M(x, y, z)$$

Problem 8.12 Repeat the arguments leading to Fig. 8.17(d) for a pole replay head that has a vertical field that is symmetric (i.e. similar to the longitudinal field of an inductive ring head). Qualitatively sketch the replay pulse and indicate the asymmetric regions.

Problem 8.13 Consider a thick medium uniformly recorded with depth with a vertical magnetization pattern of any variation along the recording direction, with an infinitely permeable keeper against the bottom surface of the medium away from the head. Show that the replay voltage with an inductive head is simply a convolution of the normalized magnetization derivative (4.4) with head-field potential at the medium surface closest to the head (Problem 5.12). Thus, the pulse varies monotonically from a positive value to a negative value through the transition, much like the transition shape itself.

Problem 8.14 Show that (8.40) also results using depth-averaged head fields.

Problem 8.15 In this problem a simplified quasi-analytic formula for recording and playback with a probe-keeper head will be developed. Consider a probe-keeper head as sketched in Fig. 3.20. Show that if the keeper has an infinite permeability, the replay voltage is proportional to:

$$V(x) \propto \int_{-\infty}^{\infty} dx' \phi(x' + x, d) \frac{\partial M_y(x')}{\partial x'} \propto \int_{-\infty}^{\infty} dx' h_x(x' + x, d) M_y(x')$$

where ϕ is the potential at the medium surface ($y = d$), h_x is the longitudinal component of the reciprocity field, and M_y is the (presumed depth independent) perpendicular magnetization transition. The head field may be simply approximated by viewing Fig. 3.20 from the side and imagining the image of the probe at a distance $d + \delta$ into the keeper. With this side view the 'head' looks like a shallow gap (gap length T), long pole ring head. The 'gap' length is $2(d + \delta)$ and the gap centerline is along the keeper surface. Approximate the field outside the 'gap' by the Karlqvist approximation to show that:

$$h_x^{P-K}(x, y) \approx \frac{1}{2\pi(d + \delta)} \ln \frac{((2(d + \delta) - y)^2 + (x + T/2)^2)}{(y)^2 + (x + T/2)^2} \qquad |x| > T/2$$

and show that an approximate 'deep-gap' field is given by $h_x \sim 1/(d + \delta)$. The (x, y) coordinate origin is taken at the center of the probe surface.

For the magnetization assume that (8.32) holds where:

$$H_y^{\text{P-K}}(x,y) \approx \frac{H_0}{\pi}\left(\tan^{-1}\frac{y}{x+T/2} + \tan^{-1}\frac{2(d+\delta)-y}{x+T/2}\right) \quad |x| > T/2$$

The deep-gap field is chosen to give a maximum field at the medium center of $2H_c$ or $2H_c \sim H_y^{\text{P-K}}(T/2, d+\delta/2)$. With these assumptions do a simple numerical integration of the voltage and compare with (Perlov & Bertram, 1987).

9

Record process:
Part 2 – Non-linearities and overwrite

Introduction

This chapter is devoted to a discussion of non-linearities and overwrite in digital magnetic recording. The magnetic recording process is inherently non-linear as discussed in Chapter 8. The term 'non-linearity' in magnetic recording technology refers to phenomena that cause linear superposition to be invalid. These non-linearities arise from interbit magnetostatic interactions that occur during the write process. Two essential non-linear effects occur: non-linear bit shift and high-density, non-linear amplitude loss. In this chapter the example of dibit recording is discussed to illustrate these non-linearities. In high-density digital disk and tape systems new information is written over previous data. Separate erase heads to ensure complete erase are not utilized. The overwrite process is a form of erasure, which at sufficiently high record currents is dominated by non-linear bit shift effects of the 'hard' and 'easy' transitions. A simple model of overwrite is presented that agrees well with experiment. Although numerical models may be utilized to determine these non-linearities, simplified analytic models are presented here, except for the case of non-linear amplitude loss.

Non-linear bit shift

Non-linear bit shift occurs during the writing process. The magnetostatic fields from previously written transitions cause the location of a transition currently being written to be shifted away from that determined solely by the recording head field. These shifts depend on the data pattern as well as on the location of each transition in the sequence. The term 'non-linear bit shift' is utilized here to refer to shifts

245

that occur due to previously written transitions. Shifts caused by 'hard' transitions, or overwrite phenomena, also affect the net bit shift, as clarified in Problem 9.6. In this section non-linear bit shift of a dibit pattern in longitudinal recording in thin film media is discussed. All fields are evaluated along the medium center line.

Fig. 8.4 shows a sketch of the writing of a second transition in the presence of a first transition separated by bit spacing B. x_0 denotes the recording location of an isolated transition, which is set by the deep-gap field H_0, the record geometry and the medium parameters (8.10). The writing of the second transition of a dibit occurs in the additional field from the first transition. Assuming a perfectly asymmetric transition as exemplified in Table 4.1, the transition center location of the second transition is given by:

$$H_c = H_x^h(x) + H_{\text{pert}} \qquad (9.1)$$

H_{pert} is any extraneous field that exists at the nominal writing location at the time the record current is changed to write a transition. The self-demagnetizing field of the transition is not included, because it vanishes at the center of a perfectly asymmetric transition. Expansion of (9.1) to first order about the unperturbed transition location x_0 yields a shift $\Delta x \equiv x = x_0$ of:

$$\Delta x = -\frac{H_{\text{pert}}}{\dfrac{\partial H_x^h}{\partial x}(x_0)} \qquad (9.2)$$

Equation (9.2) is obtained assuming the perturbing field is small compared to the head field. For non-linear bit shift in dibit recording H_{pert} is the magnetostatic field due to the first transition evaluated at the nominal writing location. In that case (9.2) becomes:

$$\Delta x = -\frac{H_x^l(-B)}{\dfrac{\partial H_x^h}{\partial x}(x_0)} \qquad (9.3)$$

For the case illustrated in Fig. 8.4, both the head-field gradient $\partial H_x^h/\partial x$ and the interaction field H_x^l are positive. Thus the bit shift is positive and the second transition of the dibit is written closer to the first transition than the writing separation B. The identical result occurs for the reversed pattern. It is generally true in longitudinal recording that non-linear bit shift 'pulls' the transitions together, since the dibit pattern represents the case of greatest interaction field.

The interaction field given by (4.11) for an arctangent transition may be written in the form:

$$H_x^1(-B) = -\frac{2M_r}{\pi} \tan^{-1} \frac{\delta B}{2(B^2 + a(a + \delta/2))} \tag{9.4}$$

Generally, the bit separation is greater than the transition parameter $(B > 2a)$, and with $\delta < a$ a good approximation for thin films is:

$$H_x^1(-B) \approx -\frac{M_r \delta}{\pi B} \tag{9.5}$$

In the writing of the second transition of a dibit, both transitions are over the high-permeability core of the record head (Fig. 8.4). Thus, imaging must be included. Perfect imaging is assumed and (4.8) is utilized for the image field with $y = 2d + \delta$. Assuming a transition length small compared to B, as in the approximation leading to (9.5), the net interaction field may be written as:

$$H_x^1(-B) \approx -\frac{M_r \delta}{\pi B} + \frac{M_r \delta B}{\pi(B^2 + (2d + \delta)^2)} \approx -\frac{M_r \delta(2d + \delta)^2}{\pi B^3} \tag{9.6}$$

The interaction fields are independent of specific shape, since they are evaluated at distances large compared to the transition parameter, as can be seen by direct evaluation of (2.22). Substitution of (9.6) into (9.3) and utilization of (8.11) for the head field gradient yields:

$$\Delta x = \frac{4M_r \delta(d + \delta/2)^3}{\pi Q H_c B^3} \tag{9.7}$$

The bit shift is often normalized by half the cell size $(B/2)$: for error rate calculations in peak detection, the probability of shifting a transition from the cell center to the edge of the adjacent cell is desired (Katz & Campbell, 1979). In that case the normalized, dimensionless fractional shift is:

$$\frac{2\Delta x}{B} = \frac{8M_r \delta(d + \delta/2)^3}{\pi Q H_c B^4} \tag{9.8}$$

Equation (9.8) applies to a dibit, but the dependence on head–medium parameters is general for all patterns. Note that, to first order, the non-linear bit shift is independent of the transition shape. Non-linear bit shift is reduced by an increase in medium coercivity or a decrease in medium flux content $(M_r \delta / H_c)$. Reducing the head–medium spacing, d, also results in a reduction of non-linear bit shift due to a simultaneous increase in the head-field gradient and reduction of the interbit interaction field by enhanced imaging. The bit shift increases rapidly with a decrease in the bit spacing. An increase of head current beyond that for nominal optimum recording reduces this shift through the reduction in Q (Fig. 8.7).

An alternative form for (9.8) for thin film media where (8.14) applies is:

$$\frac{2\Delta x}{B} = \frac{8a^2(d + \delta/2)^2}{IB^4} \tag{9.9}$$

An example of typical system design where $a \sim d \sim g/4$ ($\delta \ll d$) with $B \sim 2g$ yields (for $\delta \ll d$) $2\Delta x/B \sim 0.2\%$. However, for high density PRML channels where $B \sim PW_{50}/2$, (5.53) yields a shift as large as $2\Delta x/B \sim 12\%$. Operation at such densities will require narrower transitions and lower flying heights compared to the playback gap than the estimate used in this example. In Fig. 9.1 percentage bit shift using (9.9) is plotted versus density $(D = 1/B)$. It is assumed for simplicity that $a = d + \delta/2$, $I = 1$; curves for different effective spacings $d + \delta/2 = 12.5$nm, 25nm, 50nm, and 100nm are shown. Bit shift is plotted only over the range of $a/B < 0.5$ where (9.9) is accurate. The fourth power dependence on density yields the rapid increase at high densities.

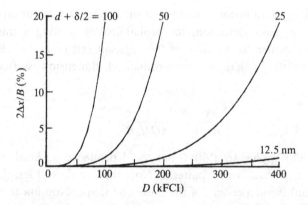

Fig. 9.1. Percentage non-linear bit shift for a dibit pattern versus recording density or bit separation. Equation (9.9) is utilized with $a = \delta + d/2$.

Non-linear amplitude loss

In thin film media the writing of patterns at high densities yields non-linear amplitude loss as well. This phenomenon occurs in addition to a voltage decrease of the dibit pattern produced by the non-linear bit shift. A technique to measure non-linearities is by the recording of 'pseudo-random' sequences (Palmer, *et al.*, 1987). The dibit pair in this sequence produces an 'echo' dibit whose amplitude and precise location depend on the extent and type of total non-linearity (Che & Bertram, 1993). It is difficult to distinguish between a dibit echo caused by non-linear bit shift and that caused by amplitude loss since the location shift is only one half a bit cell (Fig. 9.2). Simulation shows that simple bit shift predicted by (9.8) is too small to predict the observed 'effective' non-linear bit shift as estimated by this technique. At high densities 'percolation' occurs, by which regions across the track become erased and the amplitude is reduced by an 'effective' track width narrowing (Arnoldussen & Tong, 1986; Zhu, 1992; Che & Bertram, 1993).

High density erasure or 'percolation' is shown in Fig. 9.3(a), (b) as determined by numerical simulation. The gray scale corresponds to the two different orientations of magnetization in the dibit pattern. Figure 9.3(a) corresponds to no intergranular exchange coupling between adjacent grains in polycrystalline thin films (Fig. 1.2), and Fig. 9.3(b) corresponds to grains that have slight exchange coupling. In both figures various regions across the track exhibit a uniform magnetization where

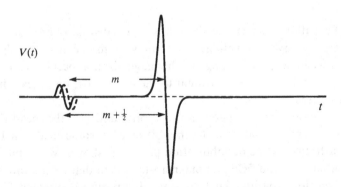

Fig. 9.2. Illustration of dibit echo in recording of a pseudo-random sequence. The echo dibit occurs at $m + 1/2$ bit cells displaced from the main echo for non-linear bit shift where m is an integer. For non-linear amplitude loss the echo occurs displaced by m cells (dashed).

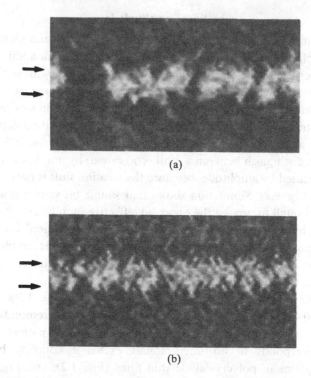

Fig. 9.3.   Planar view of numerical micromagnetic simulation of dibit recording in thin film media. The reversed magnetization in the dibit center is bounded by the 'zigzag' boundaries of the two transitions. Percolation regions are shown where the dibit pattern is erased. (a) and (b) refer respectively to exchange and non-exchange coupled grains in polycrystalline films. Taken from Zhu (1992), © IEEE.

the magnetization outside the dibit has percolated through the dibit region. Upon playback, voltage arises only from the non-percolated regions in which a spatial change of the magnetization occurs. Thus the net voltage amplitude is proportional to the area across the track where the dibit pattern remains.

To first order, this phenomenon is produced by the head-field gradients in combination with the interbit magnetostatic fields. In Fig. 9.4 an illustration of the destabilization process is shown. When bit 2 is being written, the head field and magnetostatic field that extend into the region of previously written bit 1 are small, but positively oriented. They are largest at the 'zigzag' tips that are closest to the bit being written. The net field acts in a direction to reverse the magnetization on the side closest to the new transition. When reversal occurs, the magnetization reverses in

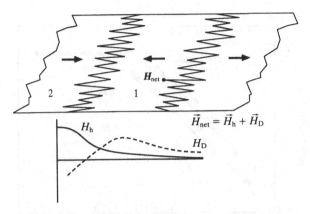

Fig. 9.4. Illustration of destabilizing fields in the writing of the second dibit. The positive fields due to the second transition and the reversed head field are shown at the closest zigzag tip of the first transition.

percolation chains reaching across to the transition being written, partially erasing the dibit and yielding uniform magnetization in strips at various regions across the track (Fig. 9.3(a), (b)). It has been shown that the magnetization configuration due to the local grain anisotropy dispersion yields low coercivity tips that percolate during the dibit writing process (Che & Bertram, 1993). Simultaneous destabilization of bit 2 also occurs, due to the interbit interaction field.

This erasure phenomenon increases with decreasing bit density since the interaction field increases as well as the reversed head field. Non-linear amplitude loss decreases with reduced head–medium spacing because the high-permeability record head reduces the magnetostatic field and reduces the spatial extent of the head field (increases the head-field gradient). A reduction in medium relative magnetostatic fields $(M_r\delta/H_c)$ reduces erasure. Experimental measurements show that the voltage amplitude in dibit recording compared to that of linear superposition decreases with decreasing scaled bit separation B/a for a fixed flying height (Lin, *et al.*, 1992); although, since a is difficult to determine as well as the precise spacing, the scaling may only be approximate. The measurements plotted in Fig. 9.5 were determined with a variety of media of varying $M_r\delta/H_c$ and measured at two different flying heights. Numerical micromagnetic analysis agrees well with these measurements (Che & Bertram, 1993). Non-linear amplitude loss may also be calculated from non-linear hysteresis phenomena (Simmons & Davidson, 1992).

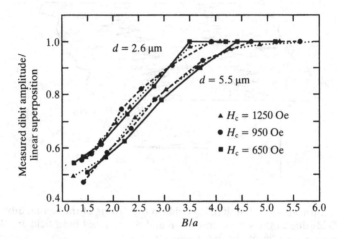

Fig. 9.5. Measured non-linear amplitude loss versus normalized bit separation B/a. The loss is the ratio of the measured dibit amplitude to that of linear superposition. Taken from Lin, *et al.* (1992), © IEEE.

Percolation appears to be independent of the exchange interaction to first order. Thus, in Figs. 9.4(a), (b), the amount of effective track width reduction is similar for both cases of high and low intergranular exchange coupling. Large exchange coupling produces large, but few, percolation regions, and low exchange coupling produces many narrow percolation 'fingers'; the net percolation as a percentage of the recorded track width is about the same for both cases. It is to be noted that in general large intergranular exchange produces large interaction clusters and 'slow zigzag' periodicity across the track giving rise to increased noise, as discussed in Chapter 12, also see Bertram, *et al.* (1992). Thus, experimental measurements of non-linear amplitude reduction in high- and low-noise media with the same macroscopic properties (M_r, H_c, δ, and thus, transition parameter a) yield approximately the same loss (Lin, *et al.*, 1992).

Overwrite

In digital magnetic recording the erasure of old information is accomplished by direct overwrite of new data patterns. Under certain conditions the writing fields are sufficient to reduce any residual original information to levels low enough not to cause errors while reading the new data. Overwrite is often specified by an 'f_1/f_2' ratio. f_1 represents an 'all-ones' or square wave pattern written at frequency f_1. This original information is overwritten by an all-ones pattern at f_2 frequency where $f_2 = 2f_1$. The overwrite ratio is then the level of playback of the residual f_1 signal after overwriting to the original f_1 signal level. With record currents somewhat larger than that necessary to optimize the signal, as well as with medium design optimization, overwrite ratios of -40 dB are readily achieved.

Residual overwrite signal is caused predominantly by three phenomena: residual recorded signal, timing shifts of f_2 at the f_1 rate, and residual edge track effects. If insufficient record current exists, then the f_1 magnetization is not completely erased and actual signals from residual recorded patterns occur. Even if the original signal is erased, the magnetostatic fields from the incoming f_1 pattern extend across the gap and modulate the zero crossings of the newly recorded f_2 pattern, yielding a frequency component at the f_1 rate. Further, incomplete erasure of edge-track magnetization can leave a residual f_1 pattern that is somewhat independent of current (Palmer, *et al.*, 1988).

In tape recording on thick particulate media, the first two phenomena occur simultaneously, yielding complicated interference phenomena (Wachenschwanz & Jeffers, 1985). In the surface layers of the tape the recording fields are sufficient to yield only phase modulation; in the back layers insufficient current yields residual magnetizations. In thin films sufficient current yields predominantly phase effects. It is to be emphasized that overwrite is a recording phenomenon in which magnetostatic fields play a large role. Incomplete erasure is exacerbated during the writing of 'hard' transitions due to the inter-transition interactions across the gap (Bloomberg *at al.*, 1979). Magnetostatic fields from the incoming pattern produce a non-linear write shift of the zero crossings of the new information, an effect similar to the non-linear bit shift discussed above. Iterative numerical analysis of overwrite for thin films has been performed (Bhattacharyya, *et al.*, 1989).

A simplified analytic analysis of the bit shift phenomena in overwrite in thin longitudinal film media will be presented here. The analysis depends on the relative shift of 'hard' and 'easy' transitions and agrees well with measurements (Lin, *et al.*, 1993). It is assumed that insufficient erasure does not occur and that edge track effects are negligible. This analytical analysis is an expansion of a previous numerical model (Fayling, *et al.*, 1984). Figures 9.6(a)–(d) show the writing of two complete cycles of new information of bit cell length B corresponding to four successive

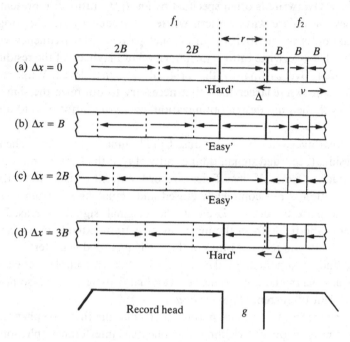

Fig. 9.6.   Schematic of magnetization pattern during the overwrite of an all-one's pattern at frequency f_1 by an all-one's pattern at f_2 at ($f_1 = f_2/2$). Configurations at four time shifts are shown where the medium has moved in distances equal to the f_2 bit cell length $B(B = 2v/f_2)$. Fig. 9.6(a): initial position, (b) shift by B, (c) shift by $2B$, (d) shift by $3B$.

transitions at f_2. The four half cycles at f_2 correspond to one complete cycle of the incoming information at f_1 with bit cell length $2B$. Figs. 9.6(a)–(d) show the writing process of two complete cycles of f_2 magnetization pattern. The bit cell length of f_1 is B. Two cycles of f_2 correspond to one cycle of f_1 magnetization. r is the size of the writing zone, which is defined as the region over the head gap in which the total field is greater than the medium coercivity ($2|x_0|$ in (8.10)). In Fig. 9.6(a), a hard transition is written since the head field is oriented in the opposite direction to the magnetization entering the writing zone. The magnetostatic field from the left-side transition (plus fields from the entire f_1 pattern) shifts the recorded transition toward the gap (denoted by Δ). In Fig. 9.6(b) the medium is shown moved to the right by a distance B and the head field reverses its direction to write the next transition in the f_2

pattern. The head field is now in the same direction as the magnetization entering the writing zone. An easy transition is then written and no hard transition shift occurs. However, the f_2 magnetization transition being written is still shifted by the magnetostatic field of the f_1 magnetization. The medium is then further moved to the right by a distance B (a shift of $2B$ from the pattern in (a)), as shown in Fig. 9.6(c). The head field again reverses, and the f_1 magnetization entering the writing zone reverses simultaneously. Consequently, an easy transition is written similar to (b) (with only the transition shift due to the magnetostatic field from the f_2 magnetization). Using the same analysis, Fig. 9.6(d) (with the magnetization pattern shifted a distance $3B$ from that in Fig. 9.6(a)) is similar to Fig. 9.6(a): both the hard transition shift and the shift due to the f_1 magnetization pattern occur.

In general the location of the f_2 transitions can be calculated by the bit shift formula (9.2) where the numerator is the fields from all the incoming f_1 transitions, including any extra 'hard' transition interaction field that might occur. Here the effects of the oscillating f_1 field sources are neglected. These fields are small at low flying heights since they form a decreasing sum alternating in sign and are clearly well imaged by the high-permeability head. Only the fields due to hard transitions will be included here, because they are largest due to their proximity to the source and because imaging is not complete at the gap edges.

In Figs. 9.6(a)–(d) two of the new transitions are 'hard' (a) and (d) and two are 'easy' (b) and (c). Both hard transition shifts are in a direction toward the gap, opposite to that intrinsic to the newly written data (as in the non-linear bit shift calculated in this chapter). If $r = 2|x_0|$ (8.10) denotes the separation of the two transitions, then the shift, utilizing (9.2), (8.11), and (2.22) including imaging, is given by:

$$\Delta = -\frac{M_r \delta(d + \delta/2)}{\pi Q H_c r} F(d, g, r) \qquad (9.10)$$

The function $F(d, g, r)$ is unity when imaging is not included ((9.5) with the evaluation distance B replaced by r) and equal to $(d + \delta/2)^2/r^2$ for perfect imaging (9.6). Since the magnetostatic fields arise from one transition near one gap corner and are evaluated across the gap near the opposite corner, finite gap imaging must be included. Utilizing the simplified approximation discussed in Chapter 2 and integrating (2.31):

$$F(d,g,r) = \frac{4(d+\delta/2)^2}{r^2} - \frac{(d+\delta/2)r}{\pi(4(d+\delta/2)^2 + r^2)}$$

$$\times \left(\ln\left(\frac{(r/2 - g/2)^2 + (d+\delta/2)^2}{(r/2 + g/2)^2 + (d+\delta/2)^2} \right) + \frac{r}{(d+\delta/2)} \left(\operatorname{atan}\left(\frac{r-g}{2(d+\delta/2)} \right) - \operatorname{atan}\left(\frac{r+g}{2(d+\delta/2)} \right) \right) \right)$$

$$(9.11)$$

For completeness, from (8.10):

$$r = g\sqrt{1 - ((2d+\delta)/g)^2 + 2(2d+\delta)/g \tan^{-1}(\pi H_c/H_0)}$$
$$r^2 + (2d+\delta)^2 > g^2$$

$$(9.12)$$

$$r = g\sqrt{1 - ((2d+\delta)/g)^2 + 2(2d+\delta)/g \tan^{-1}\pi(1 - H_c/H_0)}$$
$$r^2 + (2d+\delta)^2 < g^2$$

Increasing record current, via the deep-gap field H_0 in (9.12), yields a monotonic increase in the hard transition pair separation r and yields a corresponding monotonic decrease in the shift. Following Fig. 2.10 the variation of F with increasing record current causes the shift in (9.10) to vary between that for no-imaging to that for perfect imaging.

The overwrite ratio for the f_2/f_1 measurement can be calculated simply utilizing the Fourier analysis techniques for multiple transitions in Chapter 6. From Figs. 9.6(a), (d) the shift pattern over two cycles of f_2 is $(\Delta, 0, 0, \Delta)$. The relative phase of f_1 and f_2 is not important if only 'hard' and 'easy' relative shifts are considered. Let h(x) and e(x) stand for the playback voltages corresponding to hard and easy transitions, respectively. The spectrum of the playback waveform for the original f_1 signal is:

$$V_{f_1}(k) = \text{h}_{4B}(k)\exp(ik\Delta) - \text{e}_{4B}(k)\exp(ik2B) \qquad (9.13)$$

where k is the wave number, $\text{h}_{4B}(k)$ and $\text{e}_{4B}(k)$ are the Fourier transforms of a sequence h(x) and e(x) of opposite polarity and each with bit length $4B$. These two square waves are shifted by $2B$ with respect to each other to form the f_1 signal before overwrite. Δ is the hard transition shift (9.10), which yields a constant phase shift for the $\text{h}_{4B}(k)$ spectrum. Measured at the corresponding f_1 frequency, $k_{f_1} = 2\pi/4B$, the signal level of the fundamental is:

$$V_{f_1}(k_{f_1}) = \mathrm{h}_{4B}(k_{f_1})\exp(ik_{f_1}\Delta) + \mathrm{e}_{4B}(k_{f_1}) \tag{9.14}$$

After overwrite, the modulated f_2 playback signal can be analyzed in the same fashion. Following Figs. 9.6(a), (d):

$$V_{f_2}(k_{f_1}) = \mathrm{h}_{4B}(k)\exp(ik\Delta) - \mathrm{e}_{4B}(k)\exp(ikB)$$
$$+ \mathrm{e}_{4B}(k)\exp(ik2B) - \mathrm{h}_{4B}(k)\exp(ik(3B+\Delta)) \tag{9.15}$$

The remaining f_1 signal strength after overwrite, again measured at $k_{f_1} = 2\pi/4B$, is:

$$V_{f_2}(k_{f_1}) = (1+i)(\mathrm{h}_{4B}(k_{f_1})\exp(ik_{f_1}\Delta) - \mathrm{e}_{4B}(k_{f_1})) \tag{9.16}$$

The overwrite ratio measured by spectral analysis is therefore the ratio of (9.16) to (9.14):

$$\mathrm{OW}_{f_1/f_2} = 20\log_{10}\left|\frac{(1+i)(\mathrm{h}_{4B}(k_{f_1})\exp(ik_{f_1}\Delta) - \mathrm{e}_{4B}(k_{f_1}))}{\mathrm{h}_{4B}(k_{f_1})\exp(ik_{f_1}\Delta) + \mathrm{e}_{4B}(k_{f_1})}\right| \tag{9.17}$$

where in (9.17) the ratio is given in dB. In most magnetic recording systems, the shape difference between a hard and an easy transition is negligible. Thus (9.17) simplifies using $\mathrm{h}_{4B}(k) = \mathrm{e}_{4B}(k)$ and $k_{f_1} = \pi/2B$ to:

$$\mathrm{OW}_{f_1/f_2} = 20\log_{10}\left(\frac{\sqrt{2}\pi}{4B}\Delta\right) \tag{9.18}$$

Overwrite in terms of the hard transition shift has been calculated by slightly different means (Tang & Tsang, 1991). The overwrite ratio, including the expression for the shift (9.10) in terms of the magnetostatic interaction field, is given by:

$$\mathrm{OW}_{f_1/f_2} = 20\log_{10}\left(\frac{\sqrt{2}M_r\delta(d+\delta/2)}{4BQH_cr}F(d,g,r)\right) \tag{9.19}$$

where $F(d,g,r)$ is given in (9.11) and (9.12).

To first order the overwrite ratio is proportional to the 'hard' transition shift. It also improves for the same reasons the transition length is reduced: increasing coercivity, reduced medium flux content and lower flying height. Overwrite decreases with increasing record current, apart from the effect of medium saturation, primarily because the 'hard' transition pair separation 'r' increases with increasing record current.

Figure 9.7 shows a typical overwrite curve versus deep-gap field (or record current) relative to coercivity. As a reference a corresponding 'V–I' curve for the isolated pulse peak is shown dashed. The overwrite decreases to levels below -30dB for currents beyond that required to saturate the recorded transitions. In this high current regime, transition shift is expected to dominate. In Fig. 9.8 overwrite utilizing (9.19) is plotted versus deep-gap field. Measurements are also shown. Best agreement is found utilizing proper gap imaging (9.11). Figure 9.9 plots overwrite ratio versus medium demagnetization factor $M_r\delta/H_c$. The data were taken at a fixed head–medium spacing. Agreement is good except at low $M_r\delta/H_c$. The medium for this case was of very high coercivity, so that head saturation probably occurred during recording causing the actual deep-gap field to be less than that predicted by $H_0 = NIE/g$. In that case the actual value for the magnetostatic field evaluation distance r was less than that predicted by (9.12).

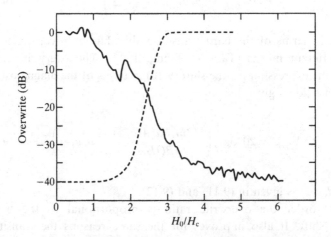

Fig. 9.7.   Measured overwrite ratio versus current (H_0/H_c). Dashed is '$V-I$' curve of isolated voltage peak. Taken from Lin, *et al.*, (1992) ©IEEE.

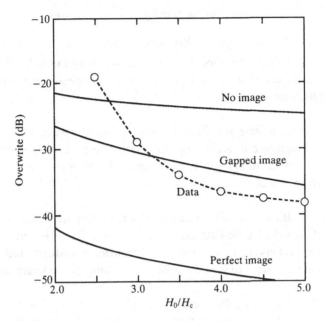

Fig. 9.8. Comparison of theoretical model and measurement corresponding to Fig. 9.7. Taken from Lin, *et al.*, (1992) ©IEEE.

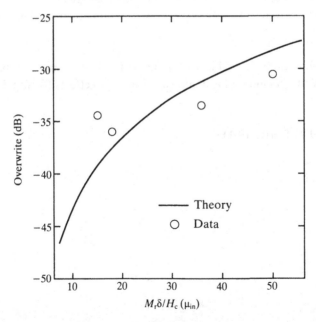

Fig. 9.9. Overwrite ratio versus $M_r\delta/H_c$ at fixed head–head spacing. A deep gap field of $4H_c$ was used. Taken from Lin, *et al.*, (1992) ©IEEE.

Problems

Problem 9.1 Using the model presented for longitudinal recording, show qualitatively that non-linear bit shift in perpendicular recording 'pushes bits apart'.

Problem 9.2 Calculate the fixed bit shift in square wave recording in thin film longitudinal media by calculating the magnetostatic field (including perfect imaging) from a semi-infinite train of previously recorded transitions.

Problem 9.3 Bit shift is often measured by recording the repeated NRZI pattern of '101101'. Calculate the linear and non-linear bit shift of the center dibit versus density. Hint: Use Fourier transform techniques discussed in Chapter 6 to find the high-density limit of the linear bit shift.

Problem 9.4 Following Problem 8.4 develop a formula for non-linear bit shift for thick longitudinal media, including perfect imaging.

Problem 9.5 Pick a data pattern and examine both linear and non-linear bit shift for longitudinal recording in thin film media. If a tri-bit is chosen, compare results with those from the numerical modeling of Chi (1980).

Problem 9.6 Consider dibit recording and calculate the net bit shift versus density, including hard and easy transition effects (Tsang & Tang, 1991).

Problem 9.7 Derive (9.11).

10

Medium noise mechanisms:
Part 1 – General concepts, modulation noise

Introduction

Noise in magnetic recording arises from three predominant sources: the playback amplifier, the playback head, and the recording medium. Amplifier noise depends on current or voltage noise sources. Head noise arises from the loss impedance of the head due to the complex part of the permeability (Figs. 3.2, 3.3). Since the head impedance is matched to the amplifier, inductive head noise results as Johnson noise with the loss impedance as the effective noise resistor (Davenport & Root, 1958). Playback head loss impedance and head noise limited system signal-to-noise ratios have been discussed in detail (Smaller, 1965). In Chapters 10, 11 and 12, analysis of the predominant medium noise mechanisms will be presented. The discussion will focus on calculations of the power spectral density. Measurements of noise spectra can be utilized readily to identify and analyze medium noise sources.

Medium noise arises from fluctuations in the medium magnetization. This noise can be separated into three somewhat distinct sources: amplitude modulation, particulate or granularity noise, and phase or transition noise. An illustration of modulation and transition noise is shown in Figs. 10.1(a), (b), respectively. In conventional amplitude modulation noise, the fluctuations are proportional to the recorded medium magnetization or flux levels. As the recording density is increased, the noise regions decrease relative to the bit length or transition separation and the noise decreases, as measurements in Fig. 10.2(a) show. Transition noise refers, in general, to fluctuations that are concentrated near the recorded transition centers (Fig. 10.1(b)). Transition noise increases with recording density since reduced bit separation reduces the proportion of low noise regions, as illustrated by

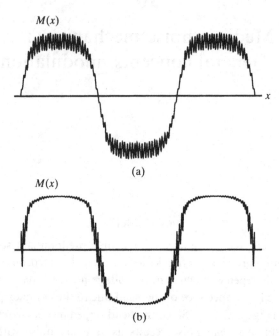

Fig. 10.1. Illustration of two distinct types of noise phenomena: (a) conventional modulation noise where the fluctuations vary in proportion to the recorded magnetization or medium flux level and (b) transition noise where the fluctuations are concentrated near the transition centers.

measurements shown in Fig. 10.2(b). Amplitude modulation effects dominate tape noise and arise from tape non-uniformities, such as thickness variations, particle concentration fluctuations and medium surface roughness. An important modulation source, important in FM video recording ('flicker noise'), is surface asperity noise (Daniel, 1964; Coutellier & Bertram, 1987). In this case non-magnetic asperities cause a fluctuation in the head–medium spacing. This effect is seen predominantly at long wavelengths and occurs only for partial penetration recording, since the asperities produce an effective fluctuation in medium depth of recording.

The fundamental noise source is the granularity of all recording media (Figs. (1.1), (1.2)). As each single-domain grain passes the playback head, a voltage di-pulse is produced. The random dispersion of grain orientations, sizes and locations yields a background noise to recorded signals (Thurlings, 1980, 1983; Daniel, 1972). Particulate noise is seen primarily in tape systems. Tape systems are generally equalized, boosting

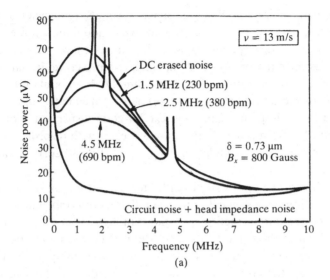

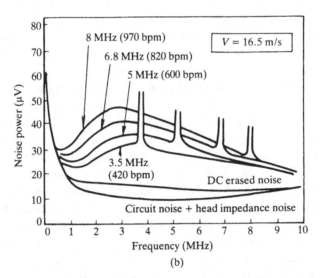

Fig. 10.2. Power spectral measurements at a variety of recording densities: (a) tape and (b) thin film sputtered oxide media. The signal is shown as narrow-banded, large amplitude voltage 'spikes', and the noise is the broad band variation. In (a) the total noise is seen to decrease with increasing recording density corresponding to Fig. 10.1(a). In (b) the total noise is seen to increase with increasing recording density corresponding to Fig. 10.1(b). Taken from Nakanishi *et al.*, (1982).

high frequency noise as well as the signal. At high frequencies or short wavelengths, particulate noise dominates over modulation noise and is heard as 'hiss' in analog audio systems. In magnetic tape, magnetization correlations alter the purely random, single grain picture, in particular, the large-scale correlations that cause amplitude modulation noise. However, in the ac erased state, uncorrelated particulate noise can be measured for high quality tapes that exhibit good dispersion (Coutellier & Bertram, 1987). In general, tape noise decreases with recording density due to a dominance of modulation phenomena (Fig. 10.2(a), Nakanishi, *et al.*, 1982).

Thin film media possess virtually no amplitude modulation or particulate noise. These media are generally smooth and uniform with strong intergranular magnetic correlations. In digital recording 'at the bit cell edges' where the magnetization is saturated between transitions, the media are extremely quiet. The grains are so densely packed that interparticle spatial correlations cancel the individual particulate noise. At the transitions noise will occur (Zhu & Bertram, 1988b; Arnoldussen, 1992). Transition noise can produce transition voltage position and amplitude fluctuations (Bertram & Zhu, 1992). This noise increases as the magnetization decreases from the saturated state and is maximum in the demagnetized state at the transition center, as illustrated in Fig. 10.1(b). In thin film media additional transition noise can occur due to medium coercivity fluctuations, which cause recorded transition position jitter via the *M–H* loop. Such transition jitter can also occur due to substrate roughness. Thin film media exhibit noise that increases with increasing density, which implies that transition centered noise dominates (Fig. 10.2(b)) (Belk, *et al.*, 1985). In addition, intergranular magnetic interactions, predominantly exchange interactions, will cause large-scale fluctuations in cross-track magnetization at the transition leading to large fluctuations. Interbit interaction fluctuations will produce a super-linear noise increase at very small bit spacings (Zhu, 1992). Noise in well-oriented, high-moment perpendicular media differs from longitudinal thin films in that the noise will occur in the non-saturated regions at the bit cell edges ($M_r = H_c < M_s$) as well as at the transition centers; this noise decreases with increasing recording density (Belk, *et al.*, 1985).

In this chapter the formalism for noise analysis will be presented briefly. Following that discussion amplitude modulation noise spectra will be calculated. Modulation noise is presented first, since it provides an easy visualization of the effect of noise sources and correlation phenomena on the power spectrum. The presentation in this chapter is similar to a previous study of tape noise resulting from density fluctuations and surface roughness (Tarumi & Noro, 1982).

Noise formalism

The expressions summarized in this section will be presented in terms of the spatial Fourier transform variables x and k instead of the usual variables time t and frequency f. Many discussions of these expressions are readily available (Lathi, 1968; Davenport & Root, 1958). All noise processes in the presence of recorded magnetization patterns are non-stationary. The formalism presented here is in terms of stationary processes. The focus in this chapter will be on average spectral properties of modulation noise. Non-stationary phenomena, as exhibited by transition noise in thin film media, are explicitly addressed in Chapter 12.

A function $s(x)$ is considered where x is the head–medium motion direction and $s(x)$ represents the playback voltage or flux. $s(x)$ could also be the magnetization in the medium, but then would be a function of x, y, z. $s(x)$ contains all the recorded information including signal and noise. Because $s(x)$ often contains limitless sinusoids, it is customary to define a new function:

$$s_L(x) = \left\{ \begin{matrix} s(x), -L/2 < x < L/2 \\ 0, \text{ otherwise} \end{matrix} \right\} \tag{10.1}$$

With this definition the power spectral density converges in the limit $L \to \infty$ and may be written as:

$$PSD_s(k) \equiv \lim_{L \to \infty} \frac{\langle |s_L(k)|^2 \rangle}{L} \tag{10.2}$$

where $s_L(k)$ is the Fourier transform of $s_L(x)$, and $\langle \rangle$ denotes an ensemble average over the possible random variables that comprise $s_L(x)$. $PSD_s(k)$ represents the power in a bandwidth $\Delta k/2\pi$ with the subscript denoting to which function the power refers. Since measured noise spectra sum positive and negative k, and refer to a frequency bandwidth Δf, the measured noise power $P_s(f)$ in Δf is:

$$P_s(k \text{ or } f) = PSD_s(k)\Delta k/\pi = 2PSD_s(f)\Delta f/v \tag{10.3}$$

The total power is given by:

$$TP_s = \frac{1}{2\pi} \int_{-\infty}^{\infty} PSD_s(k)dk \tag{10.4}$$

The autocorrelation function is defined as:

$$R_s(\eta) \equiv \lim_{L \to \infty} \frac{1}{L} \int_{-L/2}^{L/2} s(x)s^*(x+\eta)\mathrm{d}x \qquad (10.5)$$

The Wiener–Kinchin theorem gives the power spectral density as the Fourier transform of the autocorrelation function:

$$PSD_s(k) = \int_{-\infty}^{\infty} R_s(\eta)e^{-ik\eta}\mathrm{d}\eta \qquad (10.6)$$

The total power, therefore, is given by the useful form:

$$TP_s = R_s(0) \qquad (10.7)$$

Since $s(x)$ represents random processes, a statistical autocorrelation function may be defined:

$$R_s^{st}(\eta) \equiv \int_{-\infty}^{\infty} \mathrm{d}s(x) \int_{-\infty}^{\infty} \mathrm{d}s(x+\eta)s(x)s^*(x+\eta)P(s(x),s(x+\eta)) \quad (10.8)$$

$P(s(x), s(x+\eta))$ is the joint probability distribution for $s(x)$ and $s(x+\eta)$. By definition stationary processes depend only on the relative position of the two random functions so that $R_s^{st}(\eta)$ is not a function of x. Using the bracket notation $\langle\ \rangle$, (10.8) can be written compactly as: $R_s^{st}(\eta) = \langle s(x)s^*(x+\eta)\rangle$. (10.8) represents an ensemble average at given locations.

A random process is considered ergodic if:

$$R_s^{st}(\eta) = R_s(\eta) \qquad (10.9)$$

Physical noise processes that occur in magnetic recording are ergodic, except for the non-stationary transition noise predominant in thin films.

Tape density fluctuations

In this section the effect of fluctuations in tape density or particle packing is considered. The focus will be on the effect of packing cluster sizes on the noise power spectrum. Cross-track variations will not be considered in the recorded average magnetization, but cross-track variations in magnetization fluctuations will be included. Longitudinal magnetization will be assumed. The playback voltage given by (5.36) is required in order to include cross-track fluctuations:

$$V(x) = NEv\mu_0 \iiint_{medium} d^3r' h_x(\mathbf{r}' + x) \frac{dM_x(\mathbf{r}')}{dx'} \tag{10.10}$$

The statistical autocorrelation (10.8) is formed as follows:

$$R_s^{st}(\eta) = (NEv\mu_0)^2 \iiint_{medium} d^3r' \iiint_{medium} d^3r'' h_x(x' + x, y'')$$

$$h_x(x'' + x + \eta, y'') \cdot \frac{d^2}{dx'dx''} \langle M_x(\mathbf{r}')M_x^*(\mathbf{r}'') \rangle \tag{10.11}$$

In (10.11) the expectation averages are taken only over the magnetization, which is presumed to contain the fluctuations. The head field is taken to be invariant in the cross-track direction. The spatial derivatives may be placed outside the statistical average.

The magnetization can be written as:

$$M(x, y) = M_r m(x, y) p(x, y, z) \tag{10.12}$$

In (10.12) the subscript x has been deleted since this entire section will refer to the case of a longitudinal recorded magnetization. M_r is the longitudinal remanence of the medium that includes the average packing fraction. $m(x, y)$ is the normalized recorded signal level, which varies from -1 to $+1$. $p(x, y, z)$ is the normalized packing fraction of the particles, which is a random variable due to presumed fluctuations in the particle density. Although the recorded average magnetization does not vary across the track width, the packing density can vary in all three directions. $p(x, y, z)$ is taken to be a Gaussian distributed random variable with mean and rms values given by:

$$\langle p \rangle = 1$$
$$\langle p^2 \rangle = 1 + \sigma_p^2 \tag{10.13}$$

The average packing fraction has to be set to unity in (10.13) and enters as a linear factor in the tape remanence M_r. The second-order joint statistics of $p(x, y, z)$ are written as:

$$\langle p(\mathbf{r}')p(\mathbf{r}'') \rangle = 1 + \sigma_p^2 R_p(\mathbf{r}' - \mathbf{r}'') \tag{10.14}$$

Equation (10.14) is the expectation that location $\mathbf{r}'$ will have a packing density $p(\mathbf{r}')$ while location $\mathbf{r}''$ simultaneously has packing density

$p(r'')$. $R_p(r' - r'')$ is the correlation function which satisfies:

$$R_p(0) = 1$$
$$R_p(\infty) = 0 \tag{10.15}$$

Clustering of particles generally corresponds to $R_p(r' - r'')$ being a monotonically decreasing function of distance.

Substitution of (10.12), (10.14) into (10.11) yields:

$$R_s^{st}(\eta) = (NEv\mu_0 M_r)^2 \int\limits_{-W/2}^{W/2} dz' \int\limits_{-W/2}^{W/2} dz'' \int_{-\infty}^{\infty} dx' \int_{-\infty}^{\infty} dx'' \int_d^{d+\delta} dy' \int_d^{d+\delta} dy''$$

$$h(x' + x, y')h(x'' + x + \eta, y'') \frac{\partial^2}{\partial x' \partial x''} m(x', y')m(x'', y'')$$

$$(1 + \sigma_p^2 R_p(r' - r''))$$

$$\tag{10.16}$$

The power spectral density is obtained by Fourier transforming the autocorrelation function. A useful relation is:

$$\mathscr{F}_\eta \int\limits_{-\infty}^{\infty} dx' \int\limits_{-\infty}^{\infty} dx'' m(x')m(x'')h(x' + x, y')h(x'' + x + \eta, y'')R_p(r' - r'')$$

$$= \frac{1}{2\pi} |h_s(k)|^2 e^{-|k|(y' + y'')} \int\limits_{-\infty}^{\infty} dk' |m(k')|^2 R_p(k - k', y' - y'', z' - z'')$$

$$\tag{10.17}$$

In (10.17) a depth variation of the magnetization has been neglected, but the depth dependence of the head field is included by the two spacing loss terms. $\mathscr{F}_\eta$ denotes the Fourier transform operation. Utilization of (10.17) in the Fourier transform of (10.16) yields:

$$PSD(k) = \frac{(NEv\mu_0 k M_r)^2}{2\pi} |h_s(k)|^2 \int\limits_{-W/2}^{W/2} dz' \int\limits_{-W/2}^{W/2} dz'' \int_d^{d+\delta} dy' \int_d^{d+\delta} dy'' e^{-|k|(y' + y'')}$$

$$\cdot \int\limits_{-\infty}^{\infty} dk' |m(k')|^2 (2\pi\delta(k - k') + \sigma_p^2 R_p(k - k', y' - y'', z' - z''))$$

$$\tag{10.18}$$

Equation (10.18) contains two terms: the first is that due to the signal or average magnetization; the second is due to the fluctuations of the magnetization from the mean. The derivatives in (10.16) simply yield a k^2 factor. It is important to note that *all terms* in the power are proportional to the transform of the surface head field $|h_s(k)|^2$. Thus, head factors such as the gap loss or head length undulations, as seen in thin film heads, apply *both* to the signal and the noise.

The signal term is easily evaluated and yields:

$$PSD_{\text{sig}}(k) = (NEvW\mu_0 M_\text{r})^2 |h_s(k)|^2 e^{-2|k|d}(1 - e^{-|k|\delta})^2 |m(k)|^2 \qquad (10.19)$$

For square wave recording as discussed in Chapter 6, the fundamental component of the magnetization for an alternating series of transitions at separation $B = \pi/k_0$ is:

$$|m(k)|^2 = \frac{2\pi}{4}\left(\delta(k - k_0) + \delta(k + k_0)\right)\frac{16|F(k)|^2}{\pi^2} \qquad (10.20)$$

where $F(k)$ is given by the Fourier transform of (4.4). The measured signal power in band $\Delta k/2\pi$ sums negative and positive components. Thus, (10.19), (10.20) yields a signal pulse at wavenumber k_0 of amplitude:

$$P_{\text{sig}}(k_0) = \frac{8}{\pi^2}(NEvW\mu_0 M_\text{r})^2 |h_s(k_0)|^2 e^{-2k_0 d}(1 - e^{-k_0 \delta})^2 |F(k_0)|^2 \qquad (10.21)$$

This expression corresponds identically to that derived in (6.38) for magnetization with no depth variation.

In order to evaluate the noise term of (10.18) the form of the correlation coefficient must be known. For discussion purposes a simple form is chosen:

$$R_p(r) = \frac{1}{1 + (x/l)^2} \cdot \frac{1}{1 + (y/b)^2} \cdot \frac{1}{1 + (z/s)^2} \qquad (10.22)$$

This form is separable in the three coordinates with functional forms that are easily integrated and Fourier transformed. Equation (10.15) is satisifed and l, b, s are, respectively, independent correlation lengths in the x, y, z directions. The particle cluster volume defined by the spatial integral of the normalized correlation function is $\pi^3 lbs$. Equation (10.22) is readily Fourier transformed to yield:

$$R_p(k, y, z) = \pi l e^{-|k|l}\frac{1}{1 + (y/b)^2} \cdot \frac{1}{1 + (z/s)^2} \qquad (10.23)$$

Thus, the noise power term of (10.18) may be written as:

$$PSD_{\text{noise}}(k) = \frac{(NEv\mu_0 k M_r \sigma_p)^2}{2\pi} |h_s(k)|^2 \int\limits_{-W/2}^{W/2} dz' \int\limits_{-W/2}^{W/2} dz''$$

$$\cdot \frac{1}{1+((z'-z'')/s)^2} \int\limits_d^{d+\delta} dy' \int\limits_d^{d+\delta} dy'' e^{-|k|(y'+y'')} \quad (10.24)$$

$$\cdot \frac{1}{1+((y'-y'')/b)^2} \int\limits_{-\infty}^{\infty} dk' |m(k')|^2 \pi l e^{-|k-k'|l}$$

Utilizing (10.20) for the signal yields:

$$PSD_{\text{noise}}(k) = \frac{4l(NEv\mu_0 M_r \sigma_p)^2}{\pi} |h_s(k)|^2 |F(k_0)|^2 (e^{-|k-k_0|l} + e^{-|k+k_0|l}) e^{-2|k|d}$$

$$\cdot \int\limits_{-W/2}^{W/2} dz' \int\limits_{-W/2}^{W/2} dz'' \frac{1}{1+((z'-z'')/s)^2}$$

$$\cdot \int\limits_0^{\delta} dy' \int\limits_0^{\delta} dy'' k^2 e^{-|k|(y'+y'')} \frac{1}{1+((y'-y'')/b)^2}$$

$$(10.25)$$

First the dependence of the noise power spectrum on the transverse correlation length, s, is discussed. Two distinct cases are considered: (1) the correlation length is long with respect to the track width, $s \gg W$, and (2) the correlation length is short with respect to the track width, $s \ll W$. In the first case of long correlation length, the transverse correlation coefficient is unity across the track so that the z', z'' integrations yield:

$$\int\limits_{-W/2}^{W/2} dz' \int\limits_{-W/2}^{W/2} dz'' \frac{1}{1+((z'-z'')/s)^2} = W^2, \quad s \gg W \quad (10.26)$$

For transverse correlation lengths long with respect to the track width, the noise power varies as the square of the track width, or the noise voltage varies linearly with the track width. This dependence is identical to that of the signal (10.21), which is also an example of complete cross-track correlation.

If the cross-track correlation length is short with respect to the track width, then the first z'' integration is essentially over the entire span of the correlation (the limits can be set at $\pm$infinity). The second integral yields a single factor of W so that:

$$\int\limits_{-W/2}^{W/2} dz' \int\limits_{-W/2}^{W/2} dz'' \frac{1}{1 + ((z' - z'')/s)^2} = \pi s W, \quad s \ll W \qquad (10.27)$$

For transverse correlation lengths short with respect to the track width, the noise power varies linearly as the track width, and the noise voltage varies as the square root of the track width. With short correlation lengths, the noise power can be viewed as the sum of uncorrelated noise sources summed across the track. Within each correlation distance (10.26) applies, but the effective width is πs. The number of such units across the track is $W/\pi s$. Thus the total noise power varies as $(\pi s)^2 \times (W/\pi s) \propto Ws$.

Noise voltages that vary as the square root of the track width occur for virtually all recording media and for all types of noise mechanisms. Measurements of noise power versus decreasing track width should permit a determination of the transverse correlation length, but, to date the regime of (10.26) has not yet been seen (Jeffers & Wachenshwanz, 1987). It is to be noted that (10.26) and (10.27) apply, in general, to noise in magnetic recording and would arise if the spatial autocorrelation (10.16) were evaluated directly instead of the Fourier transform.

The dependence of the noise power on the vertical correlation length follows considerations similar to those of the transverse correlation length, but as seen in (10.25), the exponential thickness weighting factors must be considered. For the case of the correlation length long with respect to the medium thickness (or precisely the recording depth), $b \gg \delta$, the correlation factor is unity and the y', y'' integrations in (10.25) yield:

$$\int_0^\delta dy' \int_0^\delta dy'' k^2 e^{-|k|(y'+y'')} \frac{1}{1 + ((y' - y'')/b)^2} = (1 - e^{-k\delta})^2, \quad b \gg \delta$$

$$(10.28)$$

Long correlation lengths with respect to the thickness yield the same thickness spectral dependence as the signal (6.38). At long wavelengths the noise voltage as well as the signal voltage vary linearly as the medium thickness. For short correlation lengths it is necessary to specify not only that the correlation length is small with respect to the depth of recording, $b \ll \delta$, but that the correlation length is short with respect to the wavelengths measured $bk \ll 1$. In that case the Lorentzian correlation coefficient can be thought of as an impulse function of amplitude πb so that:

$$\int_0^\delta dy' \int_0^\delta dy'' k^2 e^{-|k|(y'+y'')} \frac{1}{1 + ((y' - y'')/b)^2} \tag{10.29}$$
$$= \pi b k (1 - e^{-2k\delta})/2 \qquad b \ll \delta, \quad kb \ll 1$$

For very short correlation lengths the characteristic factor $(1 - e^{-2k\delta})$ occurs. It can be thought of as a squared power spacing loss, weighting independent units of thickness πb. At long wavelengths, this power varies linearly as the thickness, with the same summation argument applying as in the case of correlation lengths small with respect to the track width. It is clear from a comparison of (10.28) and (10.29) that different correlation lengths in the thickness direction lead to different spectral content in the noise power. As discussed in Chapter 12, the magnetic layer thickness of thin film media generally corresponds to the thickness of the single-domain grains. Thus, the magnetization is always correlated through the recording thickness and (10.28) applies.

The correlation length in the longitudinal or medium-motion direction leads to side-bands around the carrier or signal frequencies at $\pm k_0$. The characteristic factor $(e^{-|k-k_0|l} + e^{-|k+k_0|l})$ is plotted in Fig. 10.3. The noise power peaks at the signal frequencies $\pm k_0$ and decays exponentially on both sides of the signal. The exponential decay is due to the assumption of a Lorentzian correlation function: a Gaussian correlation would

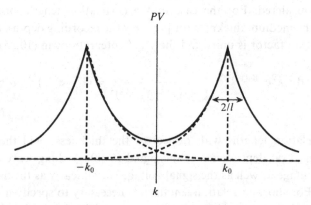

Fig. 10.3.   Spectral plot of amplitude modulation noise side-bands. The negative frequency term yields a folded side-band near the origin.

transform to Gaussian side-bands. These decays extend infinitely far so that the decay from $+k_0$ extends into the $-k$ region and conversely for the $-k_0$ side-band. These decays sum to form what appears to be a 'folded side-band' at very small k values. A spectral measurement that sums $\pm k$ will exhibit this folded side-band. Although the magnetization expressed by (10.20) is a sine wave at wavenumber k_0, square wave recording is represented simply by summing all the harmonics mk_0 where k_0 represents the fundamental. In that case (10.25) applies to each harmonic separately, with suitable weighting, and side-bands with appropriate overlap will occur at each $\pm mk_0$.

The side-bands extend in frequency, or wavenumber k, from either side of the carrier a distance approximately $1/l$, as indicated in Fig. 10.3. Long correlation lengths will yield side-bands that are compressed about the carrier. Short correlation lengths give side-bands that extend far from the carrier so that the spectral content of the correlation is almost flat or 'white'.

The complete noise power for positive k evaluated in a band Δk is given for the case of correlation lengths short with respect to both the track width, $s \ll W$, and the recording depth, $b \ll \delta$:

$$
\begin{aligned}
P_{\text{noise}}(k > 0) = {} & 2\Delta k l b s \sigma_p^2 W (N E v \mu_0 M_r)^2 |F(k_0)|^2 \\
& \cdot |h_s(k)|^2 e^{-2kd} k(1 - e^{-2k\delta})(e^{-|k-k_0|l} + e^{-|k+k_0|l})
\end{aligned}
\tag{10.30}
$$

The first line contains frequency independent terms, whereas the second contains the spectral-shaping terms. The noise power spectrum is proportional to the band width Δk in which it is measured. The level is also proportional to the correlation volume lbs as well as the fluctuation variance σ_p^2. The noise power varies linearly with the track width W in contrast to the signal power (10.21), but varies quadratically as does the signal with the factors $N E v \mu_0 M_r$. The noise power level varies as the magnitude squared of the normalized transition transform evaluated at the carrier frequency: $|F(k_0)|^2$ as does the signal (10.21). Note that the Fourier transform of the transition derivative affects only the noise level and not the spectral shape.

The terms that vary with k in (10.30) are plotted in Fig. 10.4. The spectrum varies as the head surface Fourier transform including the gap null, as does the signal. In addition, the noise power experiences the same spacing loss as the signal. Apart from head length effects, the noise voltage initially increases as k (6db/oct), much like the signal. At long wavelengths the noise power varies linearly with the depth of recording

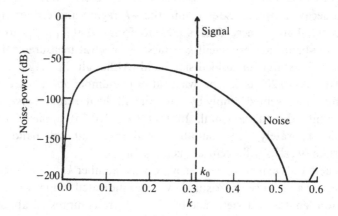

Fig. 10.4.   Complete noise power spectrum for amplitude modulation noise sources whose correlation size is small compared to the track width and depth of recording.

for this example of short vertical correlation lengths. The high frequency form of the thickness loss differs from that of the signal: the noise power experiences an extra k factor, which increases the noise with respect to the signal at short wavelengths. The noise power depends on the record current since the current determines the depth of recording. These three spectral terms occur for all medium noise mechanisms. The last term in (10.30) is the side-band term due to the longitudinal correlation length. Note that modulation noise is seen only at wavenumbers that are close to the carrier with respect to the correlation length l. This term is explicitly shown in Fig. 10.3, but is not seen in Fig. 10.4, since in the latter figure the correlation length was assumed to be extremely small.

To further compare the signal and noise power it is instructive to evaluate the slot signal-to-noise. Equation (10.21) is divided by (10.30) and is evaluated at the signal spatial frequency k_0. This power signal-to-noise ratio is written in terms of the measured frequency band width Δf as well as spatial bandwidth Δk, by utilizing $\Delta k = 2\pi\Delta f/v$:

$$\mathrm{SNR}(k_0) = \frac{4W(1 - e^{-k_0\delta})^2}{\Delta k lbs\sigma_p^2\pi^2 k_0(1 - e^{-2k_0\delta})(1 + e^{-2k_0 l})}$$

$$\mathrm{SNR}(f_0) = \frac{v^2 W(1 - e^{-2\pi f_0\delta/v})^2}{\Delta f lbs\sigma_p^2\pi^4 f_0(1 - e^{-4\pi f_0\delta/v})(1 + e^{-4\pi f_0 l/v})}$$

$$(10.31)$$

The signal-to-noise power ratio varies linearly with the track width, a result that occurs for all noise mechanisms with short cross-track correlations. Thus, halving the track width yields only a 3dB reduction in signal-to-noise ratio. At the carrier frequency the SNR does not depend on the head geometry or the recorded transition width. At long wavelengths the SNR varies linearly with the depth of recording; however, at short wavelengths, the SNR decreases in proportion to the carrier wavelength. This is a slight spectral dependence on the correlation length, but this term is unity over most frequencies of evaluation since $l \ll \lambda_0$.

It is important to note that the SNR varies linearly with the playback speed for a fixed wavenumber recorded on the tape (fixed ratio of f_0/v). Higher speeds yield increased signal-to-noise ratios. This is a consequence only of measuring the noise in a fixed frequency bandwidth. The fundamental medium signal and noise properties are solely wavelength dependent and, as can be seen from viewing (10.21) and (10.30), the signal-to-noise ratio is independent of speed when measured in a fixed wavenumber band width Δk. However, for fixed measurement band width Δf, the wavenumber bandwidth Δk decreases with increasing speed v, thus increasing the signal-to-noise ratio. All medium signal-to-noise power ratios increase linearly with playback speed. The wide band equalized SNR may be determined by integrating the inverse of (10.31). In that case the channel is assumed to be equalized 'flat' by the inverse of the transfer function. The scaling consequences are discussed in Problem 10.5. The consequences of medium noise can also be addressed directly via the autocorrelation function (10.8), which gives the total noise power directly (Problem 10.6).

It is possible to determine the noise correlation length from an analysis of the measured side-bands after recording signals at a given carrier frequency. It is perhaps simpler to analyze modulation noise mechanisms by dc magnetizing the medium or equivalently 'recording a carrier at zero frequency.' For a dc saturated medium (10.20) becomes:

$$|m(k)|^2 = 2\pi\delta(k) \tag{10.32}$$

so that (10.30) becomes:

$$P_{\text{noise}}(k > 0) = \frac{\pi^2}{4} \, \Delta k l b s \sigma_{\text{p}}^2 W (NEv\mu_0 M_{\text{r}})^2 |h_{\text{s}}(k)|^2 \mathrm{e}^{-2kd} k (1 - \mathrm{e}^{-2k\delta}) \mathrm{e}^{-kl} \tag{10.33}$$

Thus, measurement of the noise spectrum of a dc saturated tape after dividing out known factors such as the head–surface transform, spacing loss, and thickness term immediately gives the Fourier transform of the longitudinal modulation noise correlation function. If that function is a Lorentzian, as in the discussion above, the log power (dB) versus linear frequency varies linearly with a slope proprotional to the longitudinal correlation length. If the variation is not linear, a correlation length can be estimated even though a closed form of the correlation function may not always be determined. The spectral corrections assume a knowledge of the vertical correlation length relative to the depth of recording. The depth of recording, itself, can be estimated from the record current. The level of the noise voltage corrected by all known factors gives the remaining product $bs\sigma_p^2$. The rms noise variance σ_p can be estimated only if a reasonable assumption is made, for example, that the correlation lengths are the same in all directions, $l = b = s$. Spectral analysis is often complicated by the simultaneous occurrence of several noise processes.

Tape surface roughness and asperities

Tape surface roughness can occur in two forms. First, the medium can inherently possess surface roughness due to the coating process. In that case both front and back surfaces fluctuate with probably little correlation between the fluctuations at the two surfaces. The tape recording process yields optimum short wavelength recording at a depth less than full medium saturation, so that the modulation noise produced is only due to the front surface roughness. Second, a more common form of surface fluctuations is due to non-magnetic asperities between the surface of the tape and the head. Asperities cause the tape to be lifted away from the head in a tenting fashion (Fig. 10.5) where the height of the asperity gives the maximum separation and the longitudinal extent depends on the tape stiffness (Benson, 1991; Benson, *et al.*, 1992). In the case of asperities the tape thickness is fixed. Thus, the noise measured depends on the record current and the wavelength of examination. For long wavelengths, dc magnetizing the tape yields a noise that initially increases with saturating current as the surface is magnetized, allowing the fluctuations to be 'seen' by the playback head. At high currents, as the tape is completely saturated, the noise decreases since all the flux is measured and this flux does not fluctuate due to non-magnetic asperities. At short wavelengths the noise increases to a level given by the bulk medium modulation noise and asperity noise. In modern tapes asperity or

Fig. 10.5. Illustration of non-magnetic tape asperity that lifts the tape away from the head. Magnetic coating roughness is also illustrated (dashed). Both effects yield fluctuations in head–medium spacing.

surface roughness noise dominates noise due to volume non-uniformities (Coutellier & Bertram, 1987).

In this section expressions for the noise power spectra due to tape surface roughness are evaluated with focus on the effect of surface asperities. The head–medium spacing will be considered to be a two-dimensional random variable. In that case the analysis applies to both the large-scale effects of asperities as well as short-scale magnetic roughness, illustrated in Fig. 10.5. Thus, (10.10) can be written as:

$$V(x) = NEv\mu_0 \int_{-\infty}^{\infty} \mathrm{d}x' \int_{-W/2}^{W/2} \mathrm{d}z' \int_{d_1(x',z')}^{d_2(x',z')} \mathrm{d}y' h_x(x'+x,y) \frac{\mathrm{d}M_x(x',y')}{\mathrm{d}x'}$$

(10.34)

Here, the magnetization is assumed to be solely longitudinal and not subject to density fluctuations $\langle p(r') \rangle = 1, \sigma_p = 0$. In addition, fluctuations that occur solely during playback are considered: the effect of record spacing fluctuations that might cause the transition parameter to fluctuate (e.g. via (8.14)) are not included. Thus, the magnetization, as well as the playback head field, is solely a two-dimensional function of x', y':

$$M(x,y) = M_r m(x,y)$$

(10.35)

The random process enters via the medium spacing $d_1(x',z')$ as well as the depth of recording referenced to the head surface $d_2(x',z')$, as illustrated in Fig. 10.5. In the example discussed, recording to a fixed, non-fluctuating depth is assumed and only the surface near the head is taken to be a random variable. Statistics similar to those for the packing fraction, $p(r)$, as in (10.13), (10.14) are assumed:

$$\langle d_1 \rangle = d$$
$$\langle d_1^2 \rangle = d^2 + \sigma_d^2 \qquad (10.36)$$
$$\langle d_1(x',z')d_1(x'',z'') \rangle = d^2 + \sigma_d^2 R_d(x'-x'',z'-z'')$$

with $d_2(x',z')$ fixed at $d+\delta$.

Substitution of (10.35) and (10.36) into (10.34), and expansion about the fluctuation of the lower integration limit yields:

$$V(x) = V_s(x) - NE\nu\mu_0 M_r \int\limits_{-\infty}^{\infty} dx' \int\limits_{-W/2}^{W/2} dz'(d_1(x',z')-d)$$

$$(10.37)$$

$$h(x'+x,d)\,\frac{dm(x',d)}{dx'}$$

$V_s(x)$ is the average signal term, whereas the second term gives the fluctuations. Note that the integrations are only over x',z'' and the magnetization and the head field are evaluated at the medium surface at the average head–medium spacing d. The autocorrelation function can be formed as in (10.16) and yields:

$$R_s^{st}(\eta) = (NE\nu\mu_0 M_r)^2 \int\limits_{-W/2}^{W/2} dz' \int\limits_{-W/2}^{W/2} dz'' \int\limits_{-\infty}^{\infty} dx' \int\limits_{-\infty}^{\infty} dx''h(x'+x,d)$$

$$h(x''+x+\eta,d)\cdot\frac{\partial m(x',d)}{\partial x'}\frac{\partial m(x'',d)}{\partial x''}\,(\sigma_d^2 R_d(x'-x'',z'-z''))$$

$$(10.38)$$

In (10.38) only the noise term has been written: the addition term that gives the signal power is neglected. The correlation function is assumed to be Gaussian and symmetric in the recording plane with a single correlation length l_d:

$$R_d(x'-x'',z'-z'') = e^{-((x'-x'')^2+(z'-z'')^2)/l_d^2} \qquad (10.39)$$

Substitution of (10.39) into (10.38) and the carrying out of steps similar to that leading to the power spectral density (10.25) yields:

$$PSD_{\text{noise}}(k) = \frac{(NEv\mu_0 k M_r \sigma_d)^2}{2\pi} |h_s(k)|^2 e^{-2|k|d} \int_{-W/2}^{W/2} dz' \int_{-W/2}^{W/2} dz''$$

$$e^{-(z'-z'')^2/l_d^2} \cdot \int_{-\infty}^{\infty} dk' |m(k')|^2 \sqrt{\pi} l_d e^{-(k-k')^2/4l_d^2} \qquad (10.40)$$

Note that in contrast to (10.25) for bulk fluctuations, (10.40) contains no depth integration, does not depend on the depth of recording, and contains only a simple spacing loss. If it is assumed that the roughness correlation length is less than the track width, $l_d \ll W$, then using (10.20) the measured noise power can be written as:

$$P_{\text{noise}}(k > 0) = \frac{4}{\pi^2} \Delta k l_d^2 \sigma_d^2 W (NEv\mu_0 M_r)^2 |F(k_0)|^2$$

$$\cdot |h_s(k)|^2 e^{-2kd} k^2 (e^{-(k-k_0)^2/4l_d^2} + e^{-(k+k_0)^2/4l_d^2}) \qquad (10.41)$$

This form is virtually identical to that of (10.30) for the case of packing fluctuations. Side-bands occur this time with the example of a Gaussian decay in contrast to the example of a simple exponential that resulted from assuming a Lorentzian correlation function. Folded side-bands also occur; however, the noise spectrum is modulated in addition to the side-band functions only by the head field transform evaluated at the medium surface. The noise voltage decreases linearly to zero at very long wavelengths characteristic of all playback voltage expressions utilizing inductive or magnetoresistive heads. For a dc magnetized medium (10.40) takes on a particularly simple form so that l_d and σ_d can be determined from carefully calibrated experimental measurements.

Often in experimental measurements of tape modulation noise spectra the effect of surface asperities can be separated from that of volume fluctuations (Tarumi & Noro, 1982). Surface asperities cause long correlation lengths, whereas packing variations are usually short range. Thus, $l_d > l_p$ so that these two phenomena are well separated in power spectral measurements. An example is shown in Fig. 10.6 which plots the total noise due to volume packing variations (10.30) and surface asperities (10.41) including a $1/f$ background head electronics noise. It is assumed that all three noise sources are uncorrelated so that the separate noise powers are additive. It is assumed that the medium noise

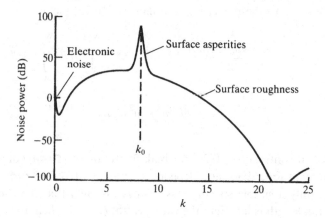

Fig. 10.6. Combined spectra due to large-scale asperities and short-scale surface roughness or density fluctuations. A $1/f$ electronics noise spectrum has been added whose level matches the medium noise at $f = vk_0/\pi$.

equals the electronics noise at a band edge of $2k_0$. Appropriate plotting parameters are: $d = 0.1\mu m$, $g = 0.25\mu m$, $\lambda_0 = 0.75\mu m$, $\delta = 0.2\mu m$, $l_d = 10\mu m$, $l_p = 1\mu m$, $b = s = l_p$, $\sigma_d = 0.1\mu m$, $\sigma_p = 1$. The side-band near the carrier is due primarily to the long surface asperity correlation, whereas the long, rather flat plateau from the carrier is due to the volume fluctuations with their short correlation length.

It is to be noted that short-scale surface roughness as illustrated in Fig. 10.6, will give a broad spectral noise power via (10.41), indistinguishable from that due to density fluctuations. In general, independent of the specific noise mechanism, short correlation lengths yield broad modulation spectra and large-scale phenomena yield narrow side-bands. If the wavelength of observation (or effective deviation from the carrier) is less than the correlation length or spatial extent of the noise source, the noise will not be seen; the medium appears uniform over that wavelength range. If the wavelength is long (corresponding to narrow band carrier deviations) with respect to the fluctuation size, then the replay voltage senses the varying magnetization.

Problems

Problem 10.1   Derive (10.17). Discuss why the x variable disappears. Carry out the details in determing (10.18).

Problem 10.2 Derive (10.20). Hint: use (10.4) and the decomposition of a square wave into harmonics.

Problem 10.3 Assume a general correlation function $R_x(x) = \mathrm{fnc}(x/l)$ to show that (10.30) becomes:

$$P_{\mathrm{noise}}(k > 0) = 2\Delta klbs\sigma_p^2 W(NEv\mu_0 M_r)^2 |F(k_0)|^2$$
$$|h_s(k)|^2 k(1 - e^{-2k\delta})e^{-2kd}(R_x((k - k_0)l) + R_x((k + k_0)l))$$
$$R_x(x) = f(x/l)$$

Use a Gaussian correlation function and plot the spectrum, including the first three signal harmonics for square wave recording at fundamental k_0.

Problem 10.4 Plot the narrow band SNR (10.31) versus record current for thick magnetic tape over a range of wavelengths. Note that (10.29) holds only at large currents and a different relation holds at low currents when the depth of recording is less than the vertical correlation length.

Problem 10.5 Show that the wide band integrated SNR for a flat channel over a pass band of $f_1 \to f_u$ is given, utilizing (10.31), by:

$$\mathrm{SNR_{WB}} = \frac{16W\lambda_u^2}{lbs\sigma_p^2 \pi^4 G(f_1, f_u, v, \delta)} = \frac{Wv^2}{lbs\sigma_p^2 \pi^2 f_u^2 G(f_1, f_u, v, \delta)}$$

where

$$G(f_1, f_u, v, \delta) = \int_{f_1/f_u}^{1} t\, dt\, \frac{(1 - e^{-2k_u \delta t})(1 + e^{-2k_u lt})}{(1 - e^{-k_u \delta t})^2}$$

Problem 10.6 Integrate (10.30) from dc to k_u to obtain an expression for the total noise. Assume $\lambda_u \sim 2g$ and approximate the gap-loss by an exponential.

Problem 10.7 Determine directly the modulation noise autocorrelation function (10.16) by following steps and assumptions similar to those leading to (10.30). Compare the total noise power with the band limited result of Problem 10.6.

Problem 10.8 Discuss an experiment to determine the vertical correlation length (b) in a tape that exhibits only packing variations.

Hint: examine the long wavelength dc magnetized medium noise spectrum plotted versus depth of magnetization (see Problem 10.4). Note that the power varies as δ for $\delta \ll b$ and as δ^2 for $\delta \gg b$.

Problem 10.9   Derive the noise power spectra for surface roughness starting with (10.34). Assume both surfaces fluctuate and consider two cases of (1) d_1, d_2 uncorrelated and (2) d_1, d_2 completely correlated and equal to each other except for a constant separation as would occur with surface asperities. Plot noise power versus current at long and short wavelengths with respect to the coating thickness.

11

Medium noise mechanisms:
Part 2 – Particulate noise

Introduction

The fundamental noise in magnetic recording is due to the granularity of the medium. If amplitude as well as phase modulation noise sources are not present, particulate noise remains. This noise can exhibit a different character in tapes than in thin films, because thin films have strong magnetic interactions and are densely packed. Nonetheless, particulate noise is basic to all recording media. The structure of this chapter will be to discuss first granularity noise neglecting particle correlations. The total noise power is simply a sum of the independent noise power from each particle or grain. Next, correlation effects are discussed that can involve spatial as well as magnetic correlations. A general formalism will be given, but only simple examples will be examined. The difficulty is that particulate noise modeling is based on Poisson statistics, which are valid only for point or infinitesimally small particles. The effects of finite particle size for moderately dilute systems, which leads to non-overlap effects, can only be included approximately. Granularity noise in thin films, where the grains are tightly packed, must be analyzed differently. A simple approach will be discussed in Chapter 12. In this chapter signal-to-noise ratios are estimated, and wherever appropriate, comparison will be made with the results for continuum fluctuations calculated in Chapter 10. The chapter begins with a calculation of the replay voltage pulse and spectrum of a single particle following previous analyses (Thurlings, 1980, 1983; Arratia & Bertram, 1984). A simplified model of particle clustering is presented at the end of this chapter. Only stationary correlations are discussed in this chapter. Non-stationary effects are the focus of Chapter 12.

Single particle replay expressions

A particle of magnetization M^p, length l and volume and v_p is considered. The cross-sectional area v_p/l is assumed to be infinitesimally small compared to all other dimensions. This particle is assumed to be magnetized uniformly, but not necessarily parallel to its long direction. The particle center is located at point x', y', z' with axis orientation given by the spherical coordinates θ, ϕ and magnetization orientation given by θ^p, ϕ^p as illustrated in Fig. 11.1. These orientations are confined to the hemisphere with axis along positive x. The particle magnetic moment is given by $\mu^p = M^p v_p$. The particle vector magnetization is written as:

$$\boldsymbol{M} = M^p \alpha(\cos \theta^p, \sin \theta^p \cos \phi^p, \sin \theta^p \sin \phi^p) \qquad (11.1)$$

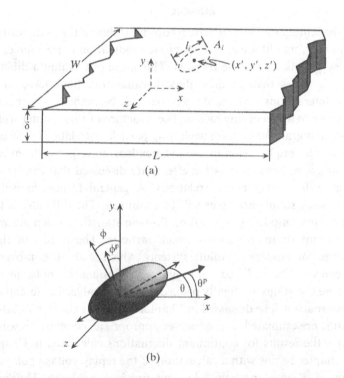

(a)

(b)

Fig. 11.1. Coordinate system for calculation of replay voltage due to a single particle. The particle is at location (x', y', z') and is assumed to be elongated with length l and (infinitesimal) cross-section A. The particle axis orientation is given in spherical coordinates by (θ, ϕ). The particle magnetization is assumed saturated at M^p and oriented at (θ^p, ϕ^p).

where $\alpha = \pm 1$ denotes the two possible magnetization orientations. From (5.30) the replay flux due to a single particle may be written as:

$$\Phi(x) = NE\mu_0 \iiint\limits_{particle} d^3 r' \boldsymbol{h}(\boldsymbol{r}' + \boldsymbol{x}) \cdot \boldsymbol{M}(\boldsymbol{r}') \tag{11.2}$$

where the integration is over the particle volume. The integration is easily performed by expressing (11.2) as a single integration along the particle length and multiplying by the infinitesimal cross-sectional area. Only 2D head fields will be utilized; edge track noise is considered in Problem 11.11. Equation (11.2) simplifies to:

$$\Phi(x) = NE\mu_0\mu^P\alpha \int_{-l/2}^{l/2} \frac{dl'}{l} \left(\cos\theta^P h_x(x + x' + l'\cos\theta, y' + l'\sin\theta\cos\phi) \right. \tag{11.3}$$
$$\left. + \sin\theta^P \cos\phi^P h_y(x + x' + l'\cos\theta, y' + l'\sin\theta\cos\phi)\right)$$

Since finite track effects are not considered here, only the components of the particle magnetization in the longitudinal and vertical direction enter. Using the general form for the head fields (2.63), (11.3) could be integrated to be written as a single integration over the head surface field (Problem 11.1). Nonetheless, the general character of the readback flux from a single particle is seen immediately and is plotted in Figs. 11.2(a), (b). For longitudinally oriented particle magnetization the replay flux varies as an averaged longitudinal head field component, which is the bell shaped curve shown dashed in Fig. 11.2(a). For a vertically oriented particle magnetization the replay flux follows an averaged vertical head field component shown dashed in Fig. 11.2(b).

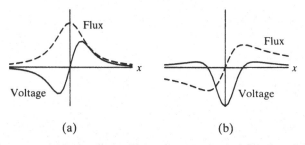

(a)   (b)

Fig. 11.2. Characteristic replay flux (dashed) and voltage pulse shapes (solid) from a single particle (a) longitudinally magnetized, and (b) vertically magnetized.

The replay voltage involves a longitudinal derivative of the flux and therefore (11.3) holds with x derivatives on the head-field functions. For a longitudinal particle the replay voltage is:

$$V(x) = - NEv\mu_0\mu^P\alpha(h_x(x + x' + l/2, y') - h_x(x + x' - l/2, y'))/l$$

$$\approx -NEv\mu_0\mu^P\alpha \frac{\partial h_x(x + x', y')}{\partial x}$$

$$(11.4)$$

and for a vertical particle:

$$V(x) = - NEv\mu_0\mu^P\alpha(h_y(x + x' + l/2, y') - h_y(x + x' - l/2, y'))/l$$

$$\approx -NEv\mu_0\mu^P\alpha \frac{\partial h_y(x + x', y')}{\partial x}$$

$$(11.5)$$

In (11.4) and (11.5) the approximate versions are for particles whose length is small compared to the spatial variation of the head fields. Note that voltage depends to first order only on the particle moment μ^P. Characteristic voltage shapes are shown in Figs. 11.2(a), (b). An arbitrary magnetization direction yields, approximately, a linear combination of the curves in Fig. 11.2(a), (b). Note that the replay flux and voltage shapes depend only on the particle magnetization orientation. The particle long axis orientation simply alters the specific averaging of the particular head-field component.

The replay flux is readily Fourier transformed to yield:

$$\Phi(k) = NE\mu_0\mu^P\alpha h_s(k)e^{-|k|y'+ikx'}(\cos \theta^P - i\,\text{sgn}(k)\sin \theta^P \cos \phi^P)$$

$$\cdot \int_{-l/2}^{l/2} \frac{dl'}{l}\, e^{ikl'\cos\theta - |k|l'\sin\theta\cos\phi}$$

$$(11.6)$$

Note that this simple form occurs since the field components in 2D are Hilbert transforms of each other as discussed in Chapter 3. The integration in (11.6) is readily performed to give:

$$\Phi(k) = NE\mu_0\mu^P\alpha h_s(k)e^{-|k|y'+ikx'}(\cos\theta^P - i\,\text{sgn}(k)\sin\theta^P\cos\phi^P)$$

$$\cdot \frac{\sin kl(\cos\theta + i\,\text{sgn}(k)\sin\theta\cos\phi)/2}{kl(\cos\theta + i\,\text{sgn}(k)\sin\theta\cos\phi)/2} \tag{11.7}$$

$$V(k) = -ikv\Phi(k)$$

In general, the flux and voltage spectra follow the replay head transform and thereby exhibit the same gap loss and head length undulations as do the signal transforms discussed in Chapter 5. The spectra exhibit a spacing loss $(\exp(-ky'))$ and a phase term $(\exp(ikx'))$ corresponding to the particle center location coordinates (x', y') in the recording plane. The cross-track coordinate does not enter because 2D head fields are presumed. As occurred for recorded transitions in Chapter 5, the phase of the replay spectra varies by $\pi/2$ as the particle magnetization rotates from longitudinal to vertical. In addition there is a particle 'sinc function' length-loss similar in form to that of the head gap-loss. In (11.7), this term is complicated because the sinc argument is complex and varies in phase depending on the particle axis orientation. This last term therefore combines the effects of spacing and phase loss due to the finite length of the particle relative to the wavelength. Two examples are given: (1) particle axis along the longitudinal direction $\theta = 0$ and (2) particle axis vertical $\theta = \pi/2$, $\phi = 0$ with magnetization along the particle axis in each case. In these cases the voltage spectrum is given and (11.7) simplifies to:

Case 1: Particle axis longitudinal, magnetization longitudinal

$$V(k) = -iNEv\mu_0\mu^P\alpha h_s(k)ke^{-|k|y'+ikx'}\frac{\sin kl/2}{kl/2} \tag{11.8}$$

Case 2: Particle axis vertical, magnetization vertical

$$V(k) = -\frac{\text{sgn}(k)NEv\mu_0\mu^P\alpha h_s(k)}{l}\,e^{-|k|(y'-l/2)+ikx'}(1 - e^{-|k|l}) \tag{11.9}$$

In Case 1, the particle is along the x axis so that the effect of a finite length is to give only a distributed replay phase. This circumstance results in a particle length-loss term that is identical in form to the gap-loss for a Karlqvist head (3.17): the replay voltage spectrum vanishes at a wavelength equal to the particle length as well as at suitable higher wavenumber multiples. In Case 2 the particle is vertical so that no phase loss occurs. However, here there is a 'thickness loss' term due to the extent of the particle in the vertical direction and the spacing loss term is referred to the end of the particle closest to the head. (11.9) should be compared with (6.38) for the general replay spectra, where the magnetization is uniform through the medium thickness. Note that (11.9) applies, apart from a phase factor, to a vertically oriented particle, but with magnetization at any angle.

General particulate noise expression

The medium is assumed to contain particles whose centers are located at random with average density n. To utilize (10.2), a distance L along the longitudinal direction is considered first (Fig. 11.1), so that the average number of particles in a volume of length L, track width W and thickness δ is $LW\delta n$. The replay voltage from this volume is simply the sum of the replay voltages from all the particles:

$$V(x) = \sum_i V_i(x) \tag{11.10}$$

where $V_i(x)$ is the replay voltage from the ith particle given for an inductive head by the derivative of (11.3) (e.g. (11.4), (11.5)). Utilizing (10.2) the power spectral density is simply:

$$PSD_s(k) \equiv \lim_{L\to\infty} \frac{\left\langle \left| \sum_i V(k) \right|^2 \right\rangle}{L} \tag{11.11}$$

which can be expanded to yield:

$$PSD_s(k) \equiv \lim_{L\to\infty} \frac{\left\langle \sum_i |V_i(k)|^2 \right\rangle + \left\langle \sum_{i\neq j} V_i(k)V_j^*(k) \right\rangle}{L} \tag{11.12}$$

The first term is the uncorrelated sum of the power from each particle, whereas the second yields the correlation terms including the signal power. The voltage expressions contain a variety of random variables such as particle orientation, particle size, magnetization. If the average number of particles in the volume $LW\delta$ is assumed independent of these variables then (11.12) may be written (Problem 11.3):

$$PSD_s(k) \equiv \lim_{L\to\infty} \frac{\displaystyle\sum_{i=1}^{LW\delta n} \langle |V_i(k)|^2 \rangle + \sum_{i\neq j}^{(LW\delta n)^2} \langle V_i(k)V_j^*(k) \rangle}{L} \tag{11.13}$$

Equation (11.13) can be evaluated directly if the particle locations are known exactly without assuming random locations. For a random array it is convenient to integrate over the volume by the standard procedure:

$$\sum_i^{LW\delta n} \rightarrow n \iiint_{LW\delta} \mathrm{d}^3 r'$$

$$\sum_{i \neq j}^{(LW\delta n)^2} \rightarrow n^2 \iiint_{LW\delta} \mathrm{d}^3 r' \iiint_{LW\delta} \mathrm{d}^3 r'' G(r' - r'')$$

(11.14)

The single-sum term reflects the assumption of Poisson statistics that the probability of finding a particle in an infinitesimal volume is proportional to that volume. The second term reflects the joint statistics that the probability of finding a particle in volume $\mathrm{d}^3 r'$ simultaneously with finding a particle in volume $\mathrm{d}^3 r''$ is the product of the volumes times a spatial coupling term $G(r' - r'')$. The physical requirements are such that two particles cannot be at the same location, the particles are uncorrelated if infinitely far from each other, and the joint probability reduces to the single particle probability if one of the particles can be anywhere. Thus, the conditions on $G(r' - r'')$ are:

$$G(0) = 0,$$
$$G(\infty) = 1,$$
$$\iiint_{LW\delta} \mathrm{d}^3 r'' G(r' - r'')$$

(11.15)

The evaluation of (11.13) is simplified by observing (11.8) and noting that $\langle |V_i(k)|^2 \rangle$ is not a function of x' or z'. Further, since correlations will be taken in this chapter to be stationary and thus a function only of relative separation, $\langle V_i(k)V_j^*(k) \rangle$ varies as $(x' - x'')$ in the longitudinal direction. Thus, (11.13), with a change of variable of $x' \rightarrow x' - x''$ in the correlation term, simplifies so that utilizing (11.14), the limit may be taken immediately to yield:

$$PSD_s(k) = Wn \int_d^{d+\delta} \mathrm{d}y' \langle |V_i(k)|^2 \rangle + n^2 \int_{-W/2}^{W/2} \mathrm{d}z' \int_{-W/2}^{W/2} \mathrm{d}z''$$

$$\int_d^{d+\delta} \mathrm{d}y' \int_d^{d+\delta} \mathrm{d}y'' \int_{-\infty}^{\infty} \mathrm{d}(x' - x'') \langle V_i(k)V_j^*(k) \rangle G(r' - r'')$$

(11.16)

The nomenclature is such that the ith particle is at x', y', z' and the jth particle at x'', y'', z''. The two terms in (11.16) are discussed separately. For simplification only perfectly oriented media will be considered. The effect of orientation is considered in Problem 11.5.

Uncorrelated noise power

The mean square voltage of the ith particle for longitudinal orientation from (11.8) is:

$$\langle |V_i(k)|^2 \rangle = (NEv\mu_0)^2 |kh_s(k)|^2 e^{-2|k|y'} \left\langle \left(\frac{\mu^p \sin kl/2}{kl/2} \right)^2 \right\rangle \qquad (11.17)$$

The orientation of the magnetization does not enter since $\alpha^2(r') = 1$. Substitution of (11.17) into the first term of (11.16) and integrating yields:

$$PSD_p^{uc}(k) = \frac{Wn(NEv\mu_0)^2}{2} \left\langle \left(\frac{\mu^p \sin kl/2}{kl/2} \right)^2 \right\rangle |k||h_s(k)|^2 e^{-2|k|d}(1 - e^{-2|k|\delta})$$

$$(11.18)$$

For discussion, the particle density is replaced by $n = p/\langle v_p \rangle$ and the moment is replaced by $\mu^p = M^p v_p$. It is assumed that the particle magnetization does not fluctuate and that the particle volume is correlated with the particle length l. For oriented media $M_r = pM^p$. With these substitutions (11.18), in terms of the measured power spectrum in bandwidth Δk for positive k, becomes:

$$P_{noise}^{uc}(k > 0) = \frac{\Delta k W \langle v_p \rangle (NEv\mu_0 M_r)^2}{2\pi p} \left\langle \left(\frac{v_p \sin kl/2}{\langle v_p \rangle kl/2} \right)^2 \right\rangle$$

$$\cdot |h_s(k)|^2 k e^{-2kd}(1 - e^{-2k\delta}) \qquad (11.19)$$

Equation (11.19) gives the contribution to the total power of the sum of the uncorrelated powers of the individual particles. This expression should be compared with (10.30) for the case of a continuum magnetization with density fluctuations with very small correlation lengths. The noise power due to uncorrelated particles varies linearly in the track width W since the particles are small with respect to the track width. The usual reproduce constants occur as in all playback expressions, but here the noise increases linearly with the particle volume: *smaller particle volumes yield less noise*. In (10.30) the equivalent is the correlation volume 'lbs'. In (11.20) the packing fraction p enters explicitly in the denominator. But since the medium remanent magnetization is proportional to the packing, this noise power actually increases linearly with the packing fraction.

The particle length-loss term:

$$\left\langle \left(\frac{v_p \sin kl/2}{\langle v_p \rangle kl/2} \right)^2 \right\rangle$$

differs from unity only when either there is a distribution in particle volumes or when the particle length is on the order of the measured wavelength. In the case of a distribution of particle volumes with very small particle lengths, this term becomes:

$$\left\langle \left(\frac{v_p \sin kl/2}{\langle v_p \rangle kl/2} \right)^2 \right\rangle = 1 + \frac{\sigma_v^2}{\langle v_p \rangle^2} \qquad (11.20)$$

Since the variance of the particle volume distribution σ_v is usually approximately equal to the average volume (Daniel, 1960; Azarian, *et al.*, 1987), a distribution in particle volumes generally doubles the uncorrelated noise power. The effect of particle length on the spectrum is discussed in Problem 11.8.

The wavelength-dependent terms of (11.19), apart from the dependence on the particle length, vary as the surface field magnitude and spacing loss squared as do *all* playback spectra. Thus, the uncorrelated noise power, similar to the signal spectra, exhibits a gap loss due to the replay head. The thickness loss term is identical to that in (10.30). In both cases a sum or integration of independent noise powers is made through the depth of the medium weighted by the square of the spacing loss for each lamina in the integration. The wavelength-dependent terms of the noise power spectra (11.19) are plotted in Fig. 11.3 versus normalized wavenumber g/λ, assuming a fixed particle length $l = g/4$ and $d = g/4$, $\delta = 4g$. (Note that for a thick tape the thickness that contributes to the total uncorrelated noise is the total medium thickness, even though the recording depth is generally much smaller ($\delta_{\text{rec}} \sim g$).) The dashed curves show the various effects of spacing, gap and particle length loss. The noise voltage initially increases linearly with wavenumber. The variation is as the square root of the frequency for wavelengths shorter than the medium coating thickness coupled with additional spectral dependencies due to head geometry and finite particle length.

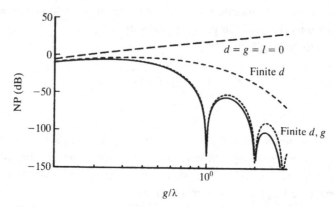

Fig. 11.3.   Noise power spectra for uncorrelated particulate noise plotted versus g/λ on a log–log scale. The dashed curves show the spectral shape changes by successively including spacing, gap and particle length loss. The relative parameters are $d = l = g/4$, $\delta = g$.

Correlation terms

The correlation terms in the power spectrum (11.16) requires $\langle V_i(k) V_j^*(k) \rangle$ for $i \neq j$. Utilizing (11.8) for a longitudinally oriented particle:

$$\langle V_i(k) V_j^*(k) \rangle = (NEv\mu_0)^2 |kh_s(k)|^2 e^{-|k|(y'+y'')+ik(x'-x'')} \langle \alpha(\mathbf{r}')\alpha(\mathbf{r}'') \rangle$$
$$\cdot \left\langle \mu_i^p \frac{\sin kl_i/2}{kl_i/2} \right\rangle \left\langle \mu_j^p \frac{\sin kl_j/2}{kl_j/2} \right\rangle \tag{11.21}$$

In (11.21) the spin correlation function $\langle \alpha(\mathbf{r}')\alpha(\mathbf{r}'') \rangle$ is assumed to be uncorrelated with the particle dimensions. The distribution of particle dimensions of one particle is assumed to be uncorrelated with that of another. Assuming only particle size fluctuations contribute to the moment fluctuations, the total power due to correlations in (11.16) can be written as:

$$PSD_p^{corr}(k) = (NEv\mu_0 M_r)^2 |h_s(k)|^2 e^{-2|k|d} \left(\left\langle \frac{v_p \sin kl/2}{\langle v_p \rangle kl/2} \right\rangle \right)^2 \int\limits_{-W/2}^{W/2} dz' \int\limits_{-W/2}^{W/2} dz''$$

$$\cdot \int\limits_0^\delta dy' \int\limits_0^\delta dy'' k^2 e^{-|k|(y'+y'')} \int\limits_{-\infty}^{\infty} d(x'-x'') e^{ik(x'-x'')}$$

$$\langle \alpha(\mathbf{r}')\alpha(\mathbf{r}'') \rangle G(\mathbf{r}'-\mathbf{r}'') \tag{11.22}$$

Notice the similarity in form to (10.18): the spectral forms depend on the correlation lengths relative to the tape thickness δ and trackwidth W. Equation (11.22) contains the signal power as well as noise terms. Thus, particle length loss terms occur in the signal as well as in the noise (Problems 11.6, 11.7, 11.8).

The spin correlation function $\langle \alpha(r')\alpha(r'') \rangle$ depends on the state of magnetization of the system, the particle interaction fields, and particle clustering via $G(r' - r'')$. In general, the spin correlation term in the presence of recorded magnetization patterns yields a non-stationary noise. The relations developed here give an average power spectrum. For dilute magnetic tapes ($p \sim 0.4$) the effect is not pronounced. For thin polycrystalline films, non-stationarity and the effect of dense packing ($p \sim 1$) alters even the uncorrelated noise power. In Chapter 12 the effects of non-stationarity as well as dense packing are discussed in detail.

Signal to particulate noise ratios

Expressions of the signal-to-noise spectra are given here, which are similar to those presented in Chapter 10 ((10.31) and Problem 10.5). The ac erased noise is utilized so that the effects of dense packing on the particulate noise may be neglected. In addition particle magnetization direction correlations will be neglected. It will be assumed that the wavelengths of interest are long compared to the particle length in the recording direction. The additional effect of particle length loss is examined in Problem 11.8. Two cases are considered: (1) particles that are small compared to the medium coating thickness and depth of recording as in tape, and (2) grains that are single domain and extend through the entire coating thickness, as in thin films.

Case 1: Thick tape

For longitudinal media the signal power is given by (10.21). Including the particle signal length-loss of Problem 11.6 and the noise given by (11.19), the slot SNR is:

$$\text{SNR}(k_0) = \frac{4Wp(1 - e^{-k_0\delta_{\text{rec}}})^2 e^{-2k_0 a}}{\pi\langle v_p \rangle k_0 \Delta k_0 (1 - e^{-2k_0\delta})(1 + \sigma_p^2 \langle v_p^2 \rangle)} \tag{11.23}$$

A square wave signal at fundamental k_0 has been recorded with assumed magnetization arctangent transition of parameter a. The SNR is given in

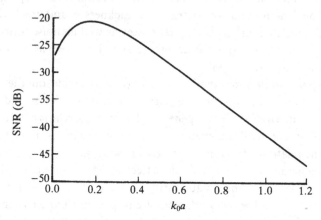

Fig. 11.4. Slot SNR for particulate tape versus normalized wavenumber $k_0 a$. The depth of recording is $\delta_{rec} = 4a$ and the medium thickness is $\delta = 4\delta_{rec}$.

terms of the slot wavenumber and therefore is independent of speed. Translation to a frequency bandwidth follows identically the discussion of (10.31). In (11.23) δ_{rec} represents the depth of recording and δ the total tape thickness. The SNR increases linearly with track width W and packing fraction p. The particulate SNR varies inversely with the mean grain size and the distribution in grain sizes. The variation with grain size and grain size distribution arises entirely from the noise power (11.19).

The wavelength dependence of the SNR is shown in Fig. 11.4. The scaled parameters are $\delta_{rec} = 4a$, $\delta = 4\delta_{rec}$. At long wavelengths the SNR is a constant and varies as the thickness (for $\delta = \delta_{rec}$). The slot SNR decreases at short wavelengths due to the signal transition breadth and the factor of k_0 in the denominator. This factor occurs because the signal is correlated with depth and the noise is not.

Case 2: Thin films

In this case (11.19) is rederived utilizing the voltage spectrum of a single domain particle whose thickness equals the film thickness ($l = \delta$). Apart from a phase, (11.7) applies assuming the dimension in the plane is small compared to wavelengths of interest. The sum over the individual particle powers, compared to (11.14), is now an integral over the film plane at a fixed spacing with an areal density $n_A = p/\langle A_g \rangle$ where p is the packing fraction of the grains in the film plane and $\langle A_g \rangle$ is the average grain area in the film plane. Evaluation of the first term of (11.13) or a modified (11.19) yields:

$$P_{\text{noise}}^{\text{uc}}(k > 0)$$
$$= \frac{\Delta k W \langle A_{\text{g}} \rangle (1 + \sigma_{\text{A}}^2/(\langle A_{\text{g}} \rangle)^2)(NEv\mu_0 M_{\text{r}})^2 |h_{\text{s}}(k)|^2 e^{-2kd}(1 - e^{-k\delta})^2}{\pi p}$$

$$(11.24)$$

Dividing (10.21) by (11.24) yields:

$$\text{SNR}(k_0) = \frac{8 W p e^{-2k_0 a}}{\pi \Delta k_0 \langle A_{\text{g}} \rangle (1 + \sigma_{\text{A}}^2/(\langle A_{\text{g}} \rangle)^2)} \qquad (11.25)$$

Equation (11.25) is remarkably simple. The film thickness does not enter since the grains are assumed to extend vertically through the film. As is typical for spectral SNR, the head–medium spacing does not enter directly (the transition parameter depends on the record spacing). The noise varies inversely as the grain area in the film plane as well as the dispersion in grain sizes. The spectrum is constant except for spectral signal decay due to a finite transition length.

The wide-band signal-to-noise ratio for a flat channel (as in Problem 10.5) is obtained by integrating the inverse of (11.25) and then re-inverting:

$$\text{SNR}_{\text{W-B}} = \frac{16 W p a e^{-2k_u a}}{\pi \langle A_{\text{g}} \rangle (1 + \sigma_{\text{A}}^2/(\langle A_{\text{g}} \rangle)^2)(1 - e^{-2k_u a})} \qquad (11.26)$$

where k_{u} is the upper band-edge wavenumber. It is clear from (11.26) that increasing the density by reducing the track width yields a smaller reduction in SNR than would be achieved by an increase in recording density of the same proportion. This statement is generally true for all medium SNR ratios that do not include other phenomena, for example adjacent track signal interference. This point is emphasized by Fig. 11.5, which plots track density (TPI = $1/W$) versus linear (kBPI = $1/B$) for constant SNR (solid) and constant areal density ($D = 1/WB$). In this figure $p = 1$, $a = 500\text{Å}$, and $\langle A_{\text{g}} \rangle = (250\text{Å})^2$. A distribution in grain areas is neglected. For this example, the upper band edge for the SNR integration is chosen to be twice the recording density ($\lambda_{\text{u}} = B$).

The SNR relations developed in this section apply for any orientation of grain magnetizations, as long as the grain size is assumed to be small compared to measured wavelengths. For thin films where the grains can be tightly coupled by magnetostatic as well as exchange interactions

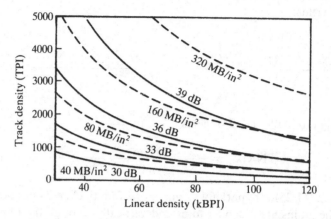

Fig. 11.5. Track density versus linear density for constant thin film wide-band granularity SNR (solid) and constant linear density (dashed). In this figure $p = 1$, $a = 500$Å, and $\langle A_g \rangle = (250$Å$)^2$.

(Bertram & Zhu, 1992), (11.25) applies approximately if the grain area is replaced by a correlation area that comprises many grains (Bertram & Arias, 1992). In general, however, signal-to-noise ratios give only guidelines to system performance or error rates in a digital channel. In particular, for thin films the non-stationarity of the noise is important; therefore the general noise voltage autocorrelation in the presence of a recorded pattern of transitions must be considered (Bertram & Che, 1993).

Erased noise spectra and correlations

The noise power spectrum of an ac-erased medium, in general consists of two contributions: the uncorrelated noise power (11.19) and a correlation term (11.22). The correlation term depends on a non-vanishing correlation of particle magnetizations, expressed by the expectation: $\langle \alpha(r')\alpha(r'') \rangle$. Even in the ac-erased state this term may not vanish. Because of the strong magnetostatic coupling in polycrystalline thin films the erased noise in polycrystalline thin films exhibits correlations larger than the grain size (Katti, *et al.*, 1988; Silva & Bertram, 1990). However, even in magnetic tape and oxide disks, the erased noise has been shown to be different from that due solely to uncorrelated particles (Bertram, 1986b; Thurlings, 1980; Nunnelly, *et al.*, 1987; Satake & Hokkyo, 1974; Anzaloni & Barbosa, 1984). An example of measured noise flux spectra

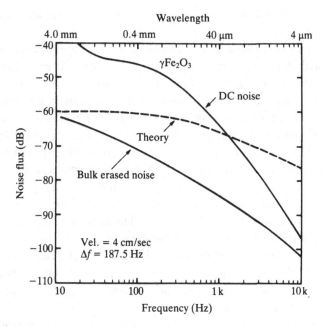

Fig. 11.6. Measured noise spectra for a γFe_2O_3 tape for erased and dc saturated states. The uncorrelated noise spectrum (11.19) is shown for comparison. Taken from Bertram (1986), © *IEEE*.

in magnetic tape is given in Fig. 11.6. The measured spectra, corrected by the replay gap loss, decrease more rapidly with increasing wavenumber than that predicted by uncorrelated particulate noise ((11.19) divided by a factor of $(kv)^2$ to obtain flux).

It is clear from Fig. 11.6 that correlations are required to explain the measured spectra. Particle interaction modeling to determine $\langle \alpha(r')\alpha(r'') \rangle$ has been made that successfully predicts a decrease in noise below that of the measured erased spectra (Arratia & Bertram, 1984; Che & Suhl, 1991). Spatial correlations or particle clustering are probably occurring in these media. In Fig. 11.6 the dc saturated noise flux is shown. At long wavelengths the noise level is high due to modulation phenomena discussed in Chapter 10. At short wavelengths, presumably smaller than the cluster sizes, the noise decreases below that of the uncorrelated particulate noise to approach the erased noise. If the noise correlation were due to magnetization orientation or spin–spin correlations $\langle \alpha(r')\alpha(r'') \rangle$, the dc noise should approach that of the uncorrelated noise. The term $\langle \alpha(r')\alpha(r'') \rangle$ is everywhere unity for a dc saturated medium. Thus, spatial as well as magnetization correlations are required to explain these spectra.

Modeling of particulate media assuming the formation of chains of particles with a common magnetization direction has been utilized to explain spectra, as in Fig. 11.6 (Nunnelly, *et al.*, 1987; Satake & Hokkyo, 1974). In this section a generalization of these models is presented in order to elucidate the relative importance of particle length and volume in the uncorrelated noise spectrum. Each particle is assumed to have length l_0, to have infinitesimal cross-section, to be longitudinally oriented and to form chains of length $l = sl_0$, where s is the number of particles in a chain (Fig. 11.7). The number of particles in a chain will be given by a distribution function. Thus, the tape will be assumed to have a collection of particles with a wide length distribution. The increased volume of long chains raises the level of the noise, whereas the length of the chains reduces the noise at wavelengths below the chain length. Thus, data can be fitted by choosing appropriate percentages of chains of various lengths. The model does not explain the phenomenon, but gives a plausible data fitting function.

The noise power spectrum is utilized in the regime where the wavelength is long with respect to the tape thickness and spacing. The head-field transform is taken to be unity. In this case (11.19) reduces to:

$$P_{\text{noise}}^{\text{uc,flux}}(k > 0) = \frac{\Delta k W (N E \mu_0 M_r)^2 \delta}{\pi p \langle v_p \rangle} \left\langle \left(\frac{v_p \sin kl/2}{kl/2} \right)^2 \right\rangle \qquad (11.27)$$

In this expression the voltage power has been divided by $(kv)^2$ to give the flux. The flux is preferable for this discussion, because it is constant at long wavelengths.

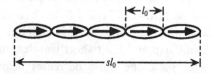

Fig. 11.7. Illustration of particle chaining. l_0 is the fixed particle length and s is the number of particles in each chain. In each chain all the particles have the same magnetization orientation. Each chain magnetization direction can vary, so that the average gives the tape magnetization state.

The distribution of chain lengths is given by $n(l)$, the probability density for chains of length l. Thus,

$$\langle v_{\mathrm{p}} \rangle = \int v n(l) \mathrm{d}l = v_0 \int \left(\frac{l}{l_0}\right) n(l) \mathrm{d}l \qquad (11.28)$$

where a continuum distribution is assumed for simplicity rather than a discrete sum. Thus, the noise power (11.27) can be written as:

$$P_{\mathrm{noise}}^{\mathrm{flux}} = \frac{P^0 \displaystyle\int_{l_{\mathrm{min}}}^{l_{\mathrm{max}}} n(l) \mathrm{d}l \left(\frac{l}{l_0}\right)^2 \left(\frac{\sin kl/2}{kl/2}\right)^2}{\displaystyle\int_{l_{\mathrm{min}}}^{l_{\mathrm{max}}} n(l) \mathrm{d}\left(\frac{l}{l_0}\right)} \qquad (11.29)$$

P^0 is the noise that would occur if the particles did not chain. The particle sizes range from l_{min} to l_{max}. For example, suppose that all the chains are of length l, so that:

$$n(l) = \delta(l - sl_0) \qquad (11.30)$$

In this case, (11.29) reduces simply to:

$$P_{\mathrm{noise}}^{\mathrm{flux}} = P^0 s \left(\frac{\sin ksl_0/2}{ksl_0/2}\right)^2 \qquad (11.31)$$

Equation (11.31) is plotted in Fig. 11.8 for two cases of long and short chain length. The long wavelength noise flux increases linearly as the chain length (or particle volume). However, the length loss term causes the spectrum to decrease rapidly at wavelengths near the particle length. Thus, longer particles have less spectral extent than do shorter particles.

For a distribution of chain lengths, the spectral shape of the noise will be the envelope of the length loss weighted by the distribution function. Assume a distribution function that decreases inversely as the square of chain length:

$$n(l) = \frac{C}{l^2} \qquad (11.32)$$

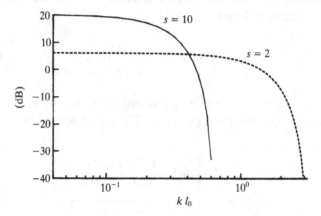

Fig. 11.8. Particle length loss spectrum for two fixed chain lengths.

With this distribution function the noise flux power may be written as:

$$P_{\text{noise}}^{\text{flux}} = \frac{P^0 \int\limits_{l_{\min}}^{l_{\max}} dl \left(\frac{\sin kl/2}{kl/2}\right)^2}{l_0 \int\limits_{l_{\min}}^{l_{\max}} dl/l} \tag{11.33}$$

The spectral behavior of (11.33) becomes evident if the variable substitution $x = kl/2$ is made. Let S be the number of particles in the longest chain and let the shortest chain length be a single particle. Then (11.33) becomes:

$$P_{\text{noise}}^{\text{flux}} = \frac{2P^0}{kl_0 \ln S} \int\limits_{kl_0/2}^{Skl_0/2} dx \left(\frac{\sin x}{x}\right)^2 \tag{11.34}$$

This form is plotted in Fig. 11.9 for $S = 200$. The spectrum varies inversely as k (-3dB/octave) over a wide range, but reaches a plateau at extremely long wavelengths.

First consider wavelengths that are short with respect to the longest chain, $Skl_0/2 \gg 1 (\sim \infty)$, and long with respect to an individual particle, $kl_0/2 \ll 1 (\sim 0)$. The integral in (11.34) becomes independent of wavenumber and equal to $\pi/2$. Thus, in this range the noise power decreases inversely as k, which, as discussed in Problem 11.10, is a

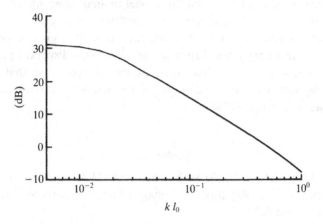

Fig. 11.9. Particle length loss power spectrum for a distribution of particle chains from l_0 to $100l_0$.

variation dependent only on the distribution function and independent of the particular spectral form for an individual particle. At long wavelengths where $Skl_0/2 \ll 1$, the noise power becomes:

$$P_{\text{noise}}^{\text{flux}} = \frac{P^0(S-1)}{\ln S} \qquad (11.35)$$

Chaining of the particles increases the noise by the factor $(S-1)/\ln S$. The minimum size of S can be estimated from the data in Fig. 11.6. The noise increases with increasing wavelength at least to the longest wavelength of measurement ($\lambda \approx 50$mils). If the particle length is approximately 0.25μm, and if one uses $Skl_0/2 = 1$ for the spectral leveling point, then the longest chain length is at least $S \sim 2000$. The difficulty with this type of heuristic fitting is that not only must the shape of the spectrum be correctly predicted, but the level must be correctly predicted as well. If the longest chain length is $S \sim 2000$, then (11.35) predicts a level increase of a factor of 263. Since Fig. 11.6 indicates that the long wavelength level is about equal to the uncorrelated particulate noise power, it must be concluded that the majority of the particles form clusters that do not contribute to the noise. Satake & Hokkyo (1974) concluded from their measurements that 85% of the particles formed into flux-closed rings or anti-parallel adjacent pairs. The data of Fig. 11.6 suggest that only 1/263 or 0.38% of the particles contribute to the noise and that the remainder form flux-closed groups in the ac-erased state.

Nunnelly, *et al.* (1987) utilized this model to analyze ac-erased noise in particulate discs. Since their measured spectrum intersected that due to uncorrelated particles at mid-wavelengths, closed flux groups were not required to adjust the level. Tapes comprised of well dispersed particles, as for example modern video tapes, exhibit erased noise that agrees reasonable well with the predictions of uncorrelated particulate noise (Coutellier & Bertram, 1987).

Problems

Problem 11.1 Utilize (2.63) with (11.3) and (11.4) to derive a general expression for the replay flux and voltage of a single particle in terms of the surface head field.

Problem 11.2 Repeat Problem 11.1 to develop an explicit replay voltage expression for a differential MR head (discussed in Chapter 7). Sketch the voltage waveforms as a function of particle orientation.

Problem 11.3 Assume that the number of particles in volume V obeys Poisson statistics so that the probability of finding N particles in V is

$$P_v(N) = \frac{(Vn)^N}{N!}\, e^{-Vn}$$

where n is the particle density. Show that for a tape of volume $LW\delta$ that

$$\left\langle \sum_i \right\rangle = LW\delta n$$

$$\left\langle \sum_{i\neq j} \right\rangle = (LW\delta n)^2$$

Problem 11.4 Show that $n = p/\langle v_p\rangle$ where p is the volume packing fraction. Hint: Consider a volume V of particles and write the effective moment as $\mu^{\mathrm{eff}} = pM^p V = nV\langle \mu^p\rangle$.

Problem 11.5 Calculate noise for spherical distribution $f(\theta)$ of particle orientations versus degree of orientation. Let the magnetizations lie along the particle axes. Choose a simple distribution function with one adjustable parameter (e.g. $f(\theta)$ constant out to a fixed angle and zero beyond, or $f(\theta) = A(\alpha)e^{\alpha\cos^2\theta}$. Plot noise power versus longitudinal squareness. The squareness in the longitudinal and transverse directions S_l and S_t, respectively, are given, with some symmetries assumed, by

$$S_l = \frac{M_r^l}{M_s} = \frac{\displaystyle\int_0^{2\pi} d\phi \int_0^{\pi/2} d\theta f(\theta, \phi) \cos\theta \sin\theta}{\displaystyle\int_0^{2\pi} d\phi \int_0^{\pi/2} d\theta f(\theta, \phi) \sin\theta}$$

$$S_t = \frac{M_r^t}{M_s} = \frac{\displaystyle 4\int_0^{2\pi} d\phi \int_0^{\pi/2} d\theta f(\theta, \phi) \sin^2\theta \cos\phi}{\displaystyle\int_0^{2\pi} d\phi \int_0^{\pi/2} d\theta f(\theta, \phi) \sin\theta}$$

Assume that the particle length is short compared to any wavelengths considered; therefore, that particle length loss need not be included.

Problem 11.6 Utilizing (11.22) evaluate the signal power, assuming that the spin correlation term $\langle \alpha(r')\alpha(r'') \rangle$ arises only for signal terms:

$$\langle \alpha(r')\alpha(r'') \rangle = m^2(r')\delta^3(r' - r'')$$

where $m(r')$ is the average normalized recorded magnetization pattern. Show that (10.21) for square wave recording at k_0 is obtained, but with the extra particle length loss factor:

$$\left(\left\langle \frac{v_p \sin(k_0 l/2)}{\langle v_p \rangle (k_0 l/2)} \right\rangle \right)^2$$

Note that if the wavelengths are long compared to the particle length, a spread in particle sizes does not affect the signal.

Problem 11.7 Use (11.3) to calculate directly the average signal pulse shape for an arctangent transition including the effect of particle length. Do both longitudinal particle with longitudinal magnetization and perpendicular particle with perpendicular magnetization.

Problem 11.8   Here the effect of particle length on signal and noise spectra is investigated for longitudinal particle orientation. Two cases are examined: (1) the particle widths and lengths are uncorrelated, and (2) the particles have a fixed ellipticity. Assume a log-normal distribution of the form:

$$P(t) = \frac{e^{-\beta^2/4}e^{(-(\ln(t/t_0)/\beta)^2)}}{\beta t_0 \sqrt{\pi}}$$

where t represents a random variable that is restricted to positive values with a distribution peak of t_0. Show that the variance for this distribution is given by:

$$\sigma^2 = t_0^2 e^{\beta^2/2}(1 - e^{-5\beta^2/4})$$

If it is assumed that the variance equals the mean square, then $\beta = 2\sqrt{(\ln 2)/5}$.

Case 1:

$$\text{Noise Power } \alpha \left\langle \left(\frac{v_p \sin kl/2}{\langle v_p \rangle kl/2}\right)^2 \right\rangle = \frac{8}{k^2}\left\langle \left(\frac{\sin kl/2}{\langle l \rangle}\right)^2 \right\rangle$$

$$\text{Signal Power } \alpha \left(\left\langle \frac{v_p \sin kl/2}{\langle v_p \rangle kl/2}\right\rangle\right)^2 = \frac{4\langle \sin kl/2\rangle^2}{k^2\langle l\rangle^2}$$

$$\text{SNR } \alpha \frac{\langle(\sin kl/2)^2\rangle}{2\langle \sin kl/2\rangle^2}$$

Case 2:

$$\text{Noise Power } \alpha \left\langle \left(\frac{v_p \sin kl/2}{\langle v_p \rangle kl/2}\right)^2 \right\rangle = \frac{4}{k^2}\left\langle \left(\frac{l^2 \sin kl/2}{\langle l^3 \rangle}\right)^2 \right\rangle$$

$$\text{Signal Power } \alpha \left(\left\langle \frac{v_p \sin kl/2}{\langle v_p \rangle kl/2}\right\rangle\right)^2 = \frac{4\langle l^2 \sin kl/2\rangle^2}{k^2\langle l^3\rangle^2}$$

$$\text{SNR } \alpha \frac{\langle(l^2 \sin kl/2)^2\rangle}{2\langle l^2 \sin kl/2\rangle^2}$$

Evaluate these expressions assuming a variance equal to the mean square; plot versus frequency and compare. Comment on how the SNR ratios are affected by perpendicular particle magnetization orientation.

Problem 11.9 The magnetization pair correlation function $\langle \alpha(r')\alpha(r'')\rangle$ for any state of magnetization of a medium depends on the magnetizing process that leads to that state. As a physically plausible form for the ac-erased state, assume that the correlation function follows the longitudinal component of the particle interaction field. Since wavelengths compared to the particle length are of interest, the dipole field is utilized:

$$H_x(r) = \frac{\mu_p(3\cos^2\theta - 1)}{r^3}$$

where the nomenclature is given in Fig 11.1. This field is the driving term for the complete many-particle interaction problem as analyzed in noise modeling. It gives positive correlation for particles aligned in the recording direction and negative correlation for transverse particles with all correlation strengths decreasing with distance between any two particles. Further neglect any spatial correlations. Let us write, therefore:

$$\langle \alpha(r')\alpha(r'')\rangle \propto \frac{3\hat{x} \cdot (r' - r'')(x' - x'') - |r' - r''|^2}{|r' - r''|^5}$$

Show that at long wavelengths the noise power is given only by the non-correlated term (11.19). Show that at finite wavelengths the effect of correlations that follow the dipole field reduces the noise below that due to uncorrelated particles.

Problem 11.10 Assume that the distribution of particle chain lengths for the discussion at the end of this chapter is given by

$$n(l) = Cl^{-n}$$

Show that for wavelengths in the range in between the longest and shortest chain, the envelope of particle length loss yields a noise power that varies as:

$$NP \propto k^{1-n}$$

Problem 11.11 Rederive (11.19) for a tape where the medium is much wider than the head track width. Assume that the head field transform off the edge of the head is given by (Lindholm, 1978):

$$h_s(z) = \frac{1}{2}e^{-kz}(1 + e^{2kz}\mathrm{erfc}(\sqrt{2kz}))$$

12
Medium noise mechanisms: Part 3 – Transition noise

Introduction

This chapter will address noise arising from fluctuations localized at the transition. Transition noise is dominant in metallic thin films where the average magnetization lies in the longitudinal or recording direction (Bertram, *et al.*, 1992). Transition noise can also occur in perpendicular film media. These media are prepared either by sputtering or by plating processes and are extremely uniform so that conventional sources of amplitude modulation noise, as discussed in Chapter 10, are not present. The fundamental feature of thin films that gives rise to transition noise is the almost completely dense packing of these polycrystalline media (Fig. 1.2). With dense packing the medium noise depends strongly on the state of magnetization.

In Figs. 12.1(a), (b) illustrations of the magnetization configurations for grains with solely longitudinal orientation are shown for a saturated medium (a) and an erased medium (b). For the saturated case the 'poles' at the ends on one grain cancel those of each adjacent grain. In this case no noise voltage will occur. In the erased case the average magnetization vanishes. To achieve this state, adjacent grains with opposite magnetization will occur, giving rise to localized 'poles' at the grain interface of twice the magnitude of that of individual grains. Even with magnetization correlation, a random distribution of these 'double' poles will occur, of both polarities, leading to a replay noise voltage. In general, noise voltage sources will occur for all configurations where the medium is not saturated. The greatest number of noise sources will occur for states where the average magnetization vanishes. Media uniformly magnetized at various states along the major loop will exhibit fluctuations that maximize at the remanent coercive state. In the case of recorded

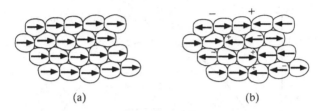

(a)   (b)

Fig. 12.1. Illustration of magnetization configurations for a film with only longitudinal magnetization orientation in the (a) saturated state and (b) erased state. Only in non-saturated configurations will replay 'poles' occur as shown in (b) by the ' + ' and '−' notation for grain pairs that have opposing magnetizations.

transitions the density of noise sources will increase for locations varied along the transition from each saturation remanent state to the transition center where the cross-track average magnetization vanishes.

A simple introductory view of magnetization fluctuations is given by considering the magnetization variance of a uniformly magnetized medium. The variance σ^2 is the squared volume average of the difference of the magnetization at each point from the mean (Davenport & Root, 1958).

$$\sigma^2 = \langle (M - \langle M \rangle)^2 \rangle = \langle M^2 \rangle - \langle M \rangle^2 \qquad (12.1)$$

In (12.1) the averages are, in general, over the volume containing the magnetic material. For grains with magnetization that can be oriented only ± along a single (e.g. longitudinal) direction, the averages in (12.1) are:

$$\begin{aligned} \langle M^2 \rangle &= pM_s^2 \\ \langle M \rangle &= pM_s \langle n \rangle \end{aligned} \qquad (12.2)$$

In (12.2) p is the medium volume grain packing fraction, M_s is the grain magnetization and $\langle n \rangle$ is the volume average of the grain magnetization orientations. $\langle n \rangle$ varies in this example from −1 to +1. Substitution of (12.2) into (12.1) yields:

$$\sigma^2 = pM_s^2(1 - p\langle n \rangle^2) \qquad (12.3)$$

For dilute systems $p \ll 1$ the variance is independent of the magnetization level and linearly proportional to the packing fraction, and yields, in a replay noise calculation, the simple uncorrelated particulate noise spectrum (11.19). As the packing increases the variance decreases

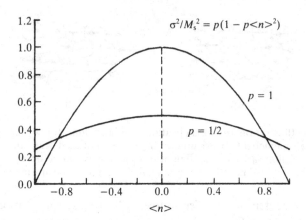

Fig. 12.2. Normalized magnetization fluctuation variance $(\sigma/M_s)^2$ versus average magnetization orientation $\langle n \rangle$ for a system with uniaxial grain orientation. Equation (12.2) is utilized for $p = 0.5$ and 1.0.

depending on the state of magnetization (independent of the net direction).

In Fig. 12.2 normalized variance $(\sigma/M_s)^2$ is plotted versus net magnetization orientation $\langle n \rangle$ utilizing (12.2) for a dense system ($p = 1$), as exemplified by thin films, and for a moderately packed medium, as typical of high density tape ($p = 0.5$). For a dense medium the variance varies in a simple quadratic manner with the net magnetization maximizing at the demagnetized state and vanishing at either saturated state. For a dilute system the variance varies only slightly with magnetization level (at most 3dB for $p = 0.5$). This variation is, therefore, generally neglected for tape media. For dense thin film media that is not perfectly oriented (12.2) is only approximate (Bertram & Che, 1993). In the transition tails the remanent magnetization is less than that at saturation and the variance does not vanish. At the transition center the magnetization directions are generally uniformly distributed in the plane so that the maximum variance is only about $(\sigma/M_s)^2 = 0.5$.

As discussed in this chapter, (12.1) also applies to recorded transitions at each location along the recording direction. In that case the averages are taken across the track width (and through the depth for thick media). Thus, for recorded transitions Fig. 12.2 applies where the variation in magnetization $-1 < \langle n \rangle < 1$ corresponds to different positions through a transition. Thus, for thin films that are densely packed the variance is maximum at the transition center and decreases substantially toward the transition boundaries where the medium becomes saturated (Fig. 10.1(b)). Thus, for a square wave recording pattern, the total noise per unit length or bit cell increases linearly with density, at least for low densities.

All replay noise power expressions for uniformly magnetized media will be proportional to (12.1). For recorded transitions the variation of the fluctuations with position along the track yield a replay noise voltage that involves, in general, a correlation of the variance with the replay head field function. These noise voltage integrals include spatial convolutions of magnetization correlation functions, as discussed in Chapter 10. For recorded patterns the down-track correlation functions are complicated, since the process is non-stationary; the correlation function depends on reference position as well as displacement.

In all noise expressions in magnetic recording, independent of how the medium is magnetized, the cross-track correlation width is a dominant factor. In general this width is less than the track width so that the noise power is proportional to the product of the track width and the cross-track correlation width (e.g. the factor 'sW' in (10.27)). An example of correlation widths is given in Fig. 12.3 using a simplified micromagnetic model (Bertram & Arias, 1992). The magnetization is in the coercive state and dark and white represent the two possible orientations. The figure shows reversed domain regions. The correlation width is approximately the width of the domains. These widths depend on the state of magnetization, the magnetic interaction parameters of the grains, and on whether the medium is uniformly magnetized or a transition has been recorded. For recorded transitions the correlation width can vary at different positions along the transition and can depend on the head–medium spacing and record current. As seen in Fig. 8.1(a) the

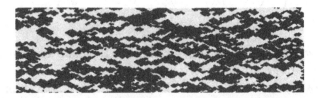

Fig. 12.3. Micromagnetic simulation of coercive state. Black and white represent two possible orientations in this model. The domain widths correspond to magnetization correlation widths. Taken from Bertram and Arias (1992).

magnetization vortex formation at the transition center of a thin film dominates the correlations.

The focus of this chapter is to explore noise mechanisms in thin film media with emphasis on the role of non-stationary correlations. The presentation will be in two parts. First will be a discussion of simplified models of transition noise that focus on position jitter and transition parameter (voltage amplitude) fluctuations. Although magnetization fluctuations occur everywhere in the transition region, it is often convenient to characterize experimental measurements in terms of simplified models. In addition noise due to inhomogeneities can be characterized by transition jitter. The effect of transition interactions on high density noise will be discussed. In the second part a general formulation for medium noise will be presented in terms of magnetization correlations. The voltage autocorrelation function will be analyzed for both uniformly magnetized media and media with recorded transitions. Signal-to-noise scaling will also be discussed.

Transition position and voltage amplitude jitter models

A transition of fixed functional form will be assumed. The transition location and parameter will be allowed to vary:

$$M(x) = M_r f\left(\frac{x - x_0 - \delta x}{a + \delta a}\right) \qquad (12.4)$$

where f represents various possible functional forms (Table 4.1) and δx, δa represent fluctuations from mean position x_0 and mean transition parameter a, respectively. For (12.1) the variance is given by (Yuan & Bertram, 1992b; Arnoldussen, 1992):

$$\sigma_M^2(x) = M_r^2\left(\sigma_x^2\left(\frac{\partial f((x - x_0)/a)}{\partial x}\right)^2 + \sigma_a^2\left(\frac{(x - x_0)\partial f((x - x_0)/a)}{a\partial x}\right)^2\right)$$

$$(12.5)$$

where σ_x^2, σ_a^2 denote the variance of position and parameter fluctuations, respectively. In Figs. 12.4(a), (b), (c) the magnetization transition shape is shown in (a); the position variance, the first term of (12.5), is given in (b); and the parameter variance, the second term in (12.5), is sketched in (c). The net magnetization variance is a weighted sum of these contributions. A physical picture of these variances can be obtained by examining the

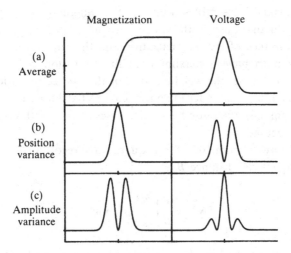

Fig. 12.4. Illustration of noise variance through a magnetization transition or corresponding voltage pulse (a) for noise sources of position jitter (b) and transition width fluctuations (c). The left and right columns are the magnetization and voltage variance versus position, respectively.

difference between two transitions perturbed by the respective fluctuation.

The variance of the voltage may be written, generally, as:

$$\sigma_V^2(x) = \sigma_x^2 \left(\frac{\partial V}{\partial x}\right)^2 + \sigma_a^2 \left(\frac{\partial V}{\partial a}\right)^2 \tag{12.6}$$

If the replay expression for an inductive head and a longitudinal thin film medium (5.34) is utilized with the scaled form for the transition (12.4), then:

$$\sigma_V^2(x) = (NWEv\mu_0 M_r\delta)^2 \left\{ \sigma_x^2 \left(\int_{-\infty}^{\infty} dx' h_x(x' + x, d + \delta/2) \right) \right.$$

$$\frac{\partial^2 f((x' - x_0)/a)}{\partial x'^2}\Big)^2 + \sigma_a^2 \left(\int_{-\infty}^{\infty} dx' h_x(x' + x, d + \delta/2) \right) \tag{12.7}$$

$$\left. \left(\frac{(x' - x_0)\partial^2 f((x' - x_0)/a)}{a\partial x'^2} + \frac{\partial f((x' - x_0)/a)}{a\partial x'}\right) \right)^2 \right\}$$

The voltage variance versus position due to position jitter and amplitude or transition parameter variation is sketched separately in Fig. 12.4. The second term in (12.4) evaluated at the transition center, $x = x_0$, gives the voltage peak amplitude fluctuations. Note that the shape of the voltage position variance resembles that of the magnetization transition parameter fluctuations; both vanish at the transition center. The positional variance of the magnetization and the amplitude variance of the voltage both have a maximum at the transition center. The net variance is, of course, the weighted sum of the two contributions. Note that since the total area under the pulse remains fixed for a long-pole ring head (Problem 5.6), the pulse height decreases as the width broadens for amplitude fluctuations.

The total power (from 10.7) for a square wave recording of transitions separated by bit cell distance B is:

$$TNP = \frac{1}{B}\int_{-B/2}^{B/2} \sigma_V^2(x)\mathrm{d}x \qquad (12.8)$$

Because the noise variance is spatially confined by the head field and transition shape width, the integral in (12.8) is a constant for all densities where the bit cell size is greater than the net variance width (see Problem 12.1 for an estimate). Thus, (12.8) yields a noise power that increases linearly with density. An increase is seen in Fig. 10.2(b), but explicit linear variation of the total power for thin-film media has been measured at low densities, as seen in Fig. 12.5.

Spectral analysis

Consider a recorded square wave at bit separation B where the transition positions are allowed to fluctuate (Barany & Bertram, 1987a). The replay voltage is given simply by:

$$V(x) = \sum_{-\infty}^{\infty}(-1)^n V_{sp}(x - nB - w_n) \qquad (12.9)$$

V_{sp} is the replay voltage of an isolated transition and w_n is the position fluctuation of the nth transition. w_n is a random variable of zero mean and variance σ_x^2 independent of specific transition. For small fluctuations (12.9) may be expanded to give:

$$V(x) = \sum_{-\infty}^{\infty}(-1)^n V_{sp}(x - nB) + \sum_{-\infty}^{\infty}(-1)^n w_n \frac{\partial V_{sp}(x - nB)}{\partial x} \qquad (12.10)$$

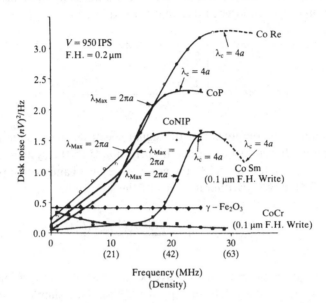

Fig. 12.5. Total noise power versus recording density for a variety of media. The noise increases linearly at low densities for the longitudinal thin film media, and then increases at a superlinear rate before peaking at high densities. The total noise was determined by integrating the noise spectrum of curves as shown in Fig. 10.2(b). The labeling 'per Hz' refers to normalization of the 20MHz integration bandwidth. Taken from Belk *et al.* (1985), © *IEEE*.

where the first term represents the signal and the second the noise. Utilization of (10.2) with the noise term of (12.10) for $s_L(x)$ yields:

$$PSD(k) = \lim_{L \to \infty} \frac{\left\langle \left| \sum_{-L/2B}^{L/2B} (-1)^n w_n \mathscr{F} \frac{\partial V_{sp}(x - nB)}{\partial x} \right|^2 \right\rangle}{L} \quad (12.11)$$

where $\mathscr{F}$ represents the Fourier transform operator (2.42). Since the Fourier transform in (12.11) is taken over a function that is spatially limited to about the width of the voltage pulse, the distinction of limiting the function to a span L is inconsequential as L becomes large. In fact the limit in (12.11) is only of importance for limitless oscillations that have finite power but infinite energy.

Since w_n is the only random variable, (12.11) simplifies, neglecting correlations between jitter in different transitions, to:

$$PSD_{\text{jitter}}(k) = \sigma_x^2 \lim_{L \to \infty} \frac{\left| \sum_{-L/2B}^{L/2B} k V_{\text{sp}}(k) \right|^2}{L} = \frac{\sigma_x^2 k^2 |V_{\text{sp}}(k)|^2}{B} \qquad (12.12)$$

The fundamental component of the signal spectrum under square wave recording is often measured, since it corresponds to the system transfer function (6.49). Utilizing (6.33), the noise power in a band Δk for positive k only (10.4) may be written in terms of the spectrum:

$$P_{\text{pos jitter}}(k > 0) = \frac{\pi \Delta k \sigma_x^2 (V_{\text{rms}}^{\text{fund}}(k))^2}{2B} \qquad (12.13)$$

This form is expressed for positive k so that an integration over $0 < k < \infty$ yields the total noise power (12.8). This simple expression states that the noise power spectral density has the same shape as the transfer function or square wave signal spectrum. Equation (12.13) is independent of the particular transition shape or perhaps complicated vector orientation. The shape of the isolated voltage pulse can be arbitrary: a Lorentzian is only one example. The noise power spectrum and the total integrated noise increase linearly with the jitter variance σ_w^2 and the recorded density $(1/B)$.

The noise power spectral shape for voltage amplitude fluctuations caused by transition parameter fluctuations is now evaluated (the second term of (12.5) and (12.7)) (Barany & Bertram, 1987b). In this case position jitter is not included, so the replay voltage pulses vary only in width and height: their position remains fixed to the bit cell center as written. The recorded transition is assumed to be scaled by a width parameter 'a' (12.4). The transition width at each nth transition is now taken to be a random variable with mean a and variance σ_a^2. Again no intertransition interactions are considered. If the voltage of an isolated pulse is considered to be a function of the transition width (12.6), then following the analysis for position jitter yields:

$$PSD_{\text{width}}(k) = \frac{\sigma_a^2}{B} \left| \frac{\partial V_{\text{sp}}(k)}{\partial a} \right|^2 \qquad (12.14)$$

or:

$$P_{\text{amp jitter}}(k > 0) = \frac{\pi \Delta k \sigma_a^2 \left(\dfrac{\partial V_{\text{rms}}^{\text{fund}}(k)}{\partial a} \right)^2}{2k^2 B} \qquad (12.15)$$

All voltage spectra are proportional to the Fourier transform of the magnetization transition (e.g. 5.36):

$$V(k) = C(k)M(ka) \qquad (12.16)$$

where $C(k)$ contains all the constants and head field transforms appropriate to the replay process. In general, if the spatial variation of the magnetization along the recording direction scales with a transition parameter a (12.4), then the transform can always be expressed as a function of the dimensionless scaled wavenumber ka. Simple manipulation of the derivative of the isolated pulse transform yields:

$$\frac{\partial V(k)}{\partial a} = kV(k) \frac{\partial \ln M(ka)}{\partial ka} \qquad (12.17)$$

Substitution of (12.17) into (12.15) yields:

$$P_{\text{amp jitter}}(k > 0) = \frac{\pi \Delta k \sigma_a^2 (V_{\text{rms}}^{\text{fund}}(k))^2}{2B} \left(\frac{\partial \ln M(ka)}{\partial ka} \right)^2 \qquad (12.18)$$

The power spectral density for transition width jitter over positive k in a band Δk is similar to that for position jitter except for an additional frequency dependent term that depends on the transition shape. Analytic examples are given in Table 12.1. Only the arctangent transition shape yields a noise power that varies as the signal. Since transitions tend to have sharper corners than an arctangent and are closer to error function in shape, width fluctuations would predict a spectrum that has less low frequency and more high frequency than the signal spectrum.

Comments on track-width dependence

The simplified model developed in the previous section yields noise power expressions that vary as the square of the replay voltage or spectrum. Thus, utilized without modification, the noise power due to these mechanisms will vary as the square of the track width (W^2). If the noise mechanisms are due to inhomogeneities that are correlated across the track (e.g., flying height variations due to substrate texturing as discussed in Bertram, *et al.*, (1986), then a squared track width dependence is

Table 12.1

Transition shape	$M(ka)/2ikMr$	$\left(\dfrac{\partial \ln M(ka)}{\partial ka}\right)^2$
Arctangent:		
$M(x) = \dfrac{2M_r}{\pi}\tan^{-1}(x/a)$	e^{-ka}	1
Error function:		
$M(x) = \dfrac{2}{\sqrt{\pi}}\displaystyle\int_0^{x/a\sqrt{\pi}} dt\, e^{-t^2}$	$e^{-\pi(ka)^2/4}$	$\pi(ka)^2$
Ramp:		
$M(x) = M_r x/a \;\; |x| < a$	$\sin(ka)/ka$	$\left(\dfrac{ka\cot ka - 1}{ka}\right)^2$

physically reasonable. However, for medium noise, as discussed in Chapters 10 and 11, the cross-track correlations are expected to be small compared to the track width. Here modification of the simple expressions is discussed. A thorough discussion of correlations in thin films is presented in the next part of this chapter.

In Fig. 12.6 a cross-track model of a transition is shown. The transition is assumed to consist of many sharp transitions of width s (Arnoldussen, 1992; Barndt, *et al.*, 1991). The average of the location of each transition is assumed to yield the cross-track average magnetization, which can be represented by the form of (12.4). The parameter s is assumed to be the cross-track correlation width; all the sub-transitions are assumed to be uncorrelated. The number of sub-transitions is simply the track width divided by the correlation width: W/s. Fig. 12.6 is an approximation to the complicated transition region as shown in Figs. 8.1(a), (b). Only position jitter is considered here.

The location of the net transition center is:

$$x_0 = \frac{1}{N}\sum_{i=1}^{N} x_i \qquad (12.19)$$

where x_i is the location of each of the sub-transitions. The relation of the net transition variance σ_x^2 to the variance of a sub-transition σ_s^2 is

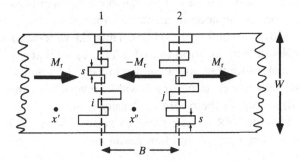

Fig. 12.6. Illustration of recorded transitions composed of subtracks of width s each with a sharp transition. The location distribution of the subtransitions gives the average transition shape. The number of subtracks is $N = W/s$. Two transitions are shown for the interbit calculation where transition 1 is written after transition 2.

obtained by squaring and averaging (12.19):

$$\sigma_x^2 = \frac{s}{W}\,\sigma_s^2 \qquad (12.20)$$

Thus, the power spectrum (12.13) is now written utilizing (12.20) as:

$$P_{\text{pos jitter}}(k > 0) = \frac{s\pi\Delta k\sigma_s^2(V_{\text{rms}}^{\text{fund}}(k))^2}{2WB} \qquad (12.21)$$

In (12.21) the voltage spectrum is that of a sharp sub-transition ($a = 0$). Utilization of (6.39) yields for thin film media:

$$P_{\text{pos jitter}}(k > 0) = \frac{4sW\Delta k\sigma_s^2(NEv\mu_0 M_r k\delta)^2 e^{-2kd}|h_{\text{sur}}(k)|^2}{\pi B} \qquad (12.22)$$

Note that the noise power varies linearly with the track width. In (12.22) $1/B$ is the recording density and k is any wavenumber of noise measurement as in Fig. 10.1(b). The broad-band integration (12.22) corresponds to the noise power in Fig. 12.5. The subscript 'sur' of the head field surface transform is included here to distinguish it from the cross-track width s. Equation (12.22) is to be compared with a similar form (11.24) as well as with (10.23).

The narrow-band SNR is, assuming a net arctangent transition:

$$\text{SNR} = \frac{2WBe^{-2ka}}{s\pi\Delta k\sigma_s^2} \qquad (12.23)$$

Equation (12.21) was utilized noting that the signal contains the factor $(\exp(-ka))$ whereas the noise of a sub-transition does not. The narrow-band SNR is identical in form to (11.25). Thus, the wide-band SNR for a flat channel has a form identical to that of (11.26), so that the discussion of density–track width trade-off as shown in Fig. 11.5 applies. In general the spectral SNR varies as the track width, inversely as the recording density $(1/B)$ as long as the recording density is not too high.

Another form of SNR is peak signal to noise. Utilizing (12.7), (12.8), and (12.20) for longitudinal thin film media yields:

$$\text{TNP} = \frac{4sW\sigma_s^2(NEv\mu_0 M_r\delta)^2}{B}\int_{-B/2}^{B/2}\left(\frac{\partial h_x(x,d+\delta/2)}{\partial x}\right)^2 dx \qquad (12.24)$$

Integration of (12.24) assuming a Karlqvist head-field function (3.16) yields (Barany & Bertram, 1987; Belk, *et al.*, 1985):

$$\text{TNP} = \frac{sW\sigma_s^2(NEv\mu_0 M_r\delta)^2}{\pi B(d+\delta/2)((g/2)^2 + (d+\delta/2)^2)} \qquad (12.25)$$

Note that both (12.24) and (12.25) contain the head–medium spacing in contrast to (12.23). The head field in (12.24) does not contain an effective spacing since the sub-transitions are assumed to be infinitely sharp. The peak voltage to total noise power SNR, assuming an arctangent transition and utilizing (5.51) and (12.25), is $(\delta \ll d)$:

$$\text{SNR} = \frac{4WBd(1+(2d/g)^2)}{\pi s\sigma_s^2}\left(\tan^{-1}\frac{g}{2(d+a)}\right)^2 \qquad (12.26)$$

Note the similarity of (12.26) to (12.23). Neither expression explicitly contains the medium flux level $M_r\delta$. However, the peak SNR depends on the head–medium spacing d in contrast to the spectral SNR. In fact, as discussed in the following section, σ_s^2 varies as d^2, so that the sub-transition positional variance decreases with decreased spacing (12.28). It is also possible that the cross-track correlation width can decrease with decreasing spacing.

Interbit interactions

The superlinear variation of the total noise at high densities (Fig. 12.5) is the result of noise discussed in (Belk, *et al.*, 1985; Barany & Bertram, 1987) with complete micromagnetic analysis in (Zhu, 1992). In the micromagnetic analysis it was noted that fluctuations of the interbit magnetostatic interactions are a dominant aspect. Here a model is presented that illuminates the role of interactions on fluctuations. For clarity the simplified model is presented first. In this analysis only interbit interactions are included. Magnetostatic interactions between the sub-transitions of a single transition are not included. Intrabit interactions have been studied for single transition noise models (Middleton & Miles, 1991; Semenov, *et al.*, 1991). The analysis involves non-linear bit-shift, as discussed in Chapter 9.

Let a simple transition uniform across the track be written in longitudinal thin film media. Assume that medium coercivity fluctuations cause position jitter. Thus, the recording position is given by:

$$H_c(x) = H_h(x_0 + w) \tag{12.27}$$

In (12.27) x_0 is the nominal writing position given by (8.10) for longitudinal media and w is the shift due to the fluctuations in the coercivity ($x = x_0 + w$). Expansion of (12.27) for small fluctuations and formation of the variance yields:

$$\sigma_x^2 = \frac{\sigma_{H_c}^2}{\left(\dfrac{\partial H_h(x_0)}{\partial x}\right)^2} = \frac{d^2 \sigma_{H_c}^2}{(Q\bar{H}_c)^2} \tag{12.28}$$

$\sigma_{H_c}^2$ is the variance of the coercivity fluctuations. The last term is the variance utilizing (8.11) for the longitudinal head-field gradient where $\bar{H}_c$ is the average coercivity. If $\sigma_{H_c}^2$ denotes the coercivity of a sub-track in Fig. 12.6, then (12.28) yields the sub-transition variance σ_s^2. Substitution into expressions in the previous section, such as (12.23) and (12.26), show that the SNR increases with decreasing head–medium spacing for a fixed coercivity fluctuation. Explicitly, (12.26) becomes:

$$\text{SNR} = \frac{4WBQ(1 + (2d/g)^2)}{d\pi s (\sigma_{H_c}^2/\bar{H}_c^2)} \left(\tan^{-1} \frac{g}{2(d+a)} \right)^2 \tag{12.29}$$

Even when a variation of s with d is not included, (12.29) agrees with experiment (Johnson, *et al.*, 1992).

Consider the writing of a transition as part of a previously written square wave pattern of transition separation B. Assuming a simple transition uniform across the track, (12.27) becomes:

$$H_c(x) = H_h(x_0 + w_1) + H_D(x_0 + B + w_2 - w_1) \tag{12.30}$$

In (12.30) '1' denotes the transition being written and '2' denotes the adjacent transition previously written. H_D is the magnetostatic field due to transition '2' evaluated at the center of transition '1'. For simplicity only the magnetostatic field from the adjacent previous transition is included, although imaging does significantly reduce the effect of the others. Expansion of (12.30) yields:

$$\delta H_c = w_1 H_h'(x_0) + (w_2 - w_1) H_D'(B) \qquad (12.31)$$

where δH_c is the variation of the coercivity and the prime on the fields denotes the derivative with respect to x. Grouping the terms (12.31) in two ways and forming the variance yields (see Problem 12.4):

$$\sigma_x^2 = \frac{\sigma_{H_c}^2}{(H_h')^2 \left(1 - \dfrac{2H_D'}{H_h'}\right)}$$

$$\langle w_1 w_2 \rangle = -\frac{\sigma_{H_c}^2 H_D'}{(H_h')^3 \left(1 - \dfrac{2H_D'}{H_h'}\right)\left(1 - \dfrac{H_D'}{H_h'}\right)}$$

$$(12.32)$$

In the derivation of (12.32) it is assumed that $\sigma_{w_1}^2 = \sigma_{w_2}^2 = \sigma_x^2$. The effect of interactions alters the variance and produces a correlation term $\langle w_2 w_2 \rangle$. Utilization of (8.11) for the head-field gradient and (9.6) for the imaged interbit interaction field yields a variance of:

$$\sigma_x^2 = \frac{d^2 \sigma_{H_c}^2}{(Q\bar{H}_c)^2 \left(1 - \dfrac{8 M_r \delta d^2}{\pi B^3 Q \bar{H}_c}\right)} \qquad (12.33)$$

Note that including interbit interactions results in a variance that is constant only at low densities. At high densities, decreasing B causes the variance to increase rapidly. Even neglecting the correlation term and utilizing (12.33) in (12.6) and (12.8) yields a noise power that exhibits the superlinear behavior exhibited in Fig. 12.5.

The simplified analysis given above is repeated assuming a transition composed of microtracks (Fig. 12.6). In that case the longitudinal interaction field between two sub-transitions (i on transition 1, and j on transition 2), is given in Problem 12.3. The field and its derivative (magnitudes) for two sub-transitions on the same track ($i = j$) is:

$$H_D \approx \frac{3sM_r\delta d^2}{\pi B^4}$$

$$H'_D \approx \frac{12sM_r\delta d^2}{\pi B^5}$$

(12.34)

The equivalent form of (12.31) for interbit interacting sub-transitions is:

$$\delta H_c = w_i \left(H'_h(x_0) - \sum_j H'_{D,ij}(B) \right) + \sum_j H'_{D,ij}(B)w_j \qquad (12.35)$$

where $H'_{D,ij}$ denotes the derivative of the interaction field between the ith sub-transition on bit 1 and the jth sub-transition on bit 2. The positional variance of each sub-transition is:

$$\sigma_s^2 = \frac{\sigma_{H_c}^2}{(H'_h)^2 \left(1 - \dfrac{2\sum\limits_j H'_{D,ij}}{H'_h} - \dfrac{\sum\limits_j (H'_{D,ij})^2}{(H'_h)^2} + \dfrac{\left(\sum\limits_j H'_{D,ij} \right)^2}{(H'_h)^2} \right)} \qquad (12.36)$$

Utilizing (12.20) and neglecting intrabit correlations yields a net transition variance of:

$$\sigma_x^2 = \frac{s\sigma_{H_c}^2}{W(H'_h)^2 \left(1 - \dfrac{2\sum\limits_j H'_{D,ij}}{H'_h} - \dfrac{\sum\limits_j (H'_{D,ij})^2}{(H'_h)^2} + \dfrac{\left(\sum\limits_j H'_{D,ij} \right)^2}{(H'_h)^2} \right)} \qquad (12.37)$$

The net position variance depends on the cross-track correlation width s, not only directly by the numerator but via the interbit interaction field derivative $H'_{D,ij}$ (Problems 12.3 and 12.4). Thus, in agreement with experiment (Baugh, et al., 1983; Yogi, et al., 1990) and micromagnetic simulation (Zhu, 1992), a reduction of the cross-track correlation width not only reduces the noise in the low density linear region, but reduces the high density superlinear behavior as well.

Generalized microscopic formulation

A generalized formulation of noise power in longitudinal thin film media in the presence of a data pattern is presented. The noise is assumed to arise from fluctuations in grain magnetization orientation produced by inherent micromagnetic phenomena (Zhu & Bertram, 1988). The presentation is similar to that given by (Bertram & Arias, 1992; Bertram & Che, 1993), but here the discrete nature of the polycrystalline films is emphasized, in a manner similar to the treatment of particulate noise in Chapter 11. The use of correlation functions is similar to that presented in Chapter 10, but here the spatial voltage autocorrelation function is calculated rather than the noise power spectra. The medium is assumed to be densely packed (as in Fig. 1.2) with identical grains of magnetization M_s, height equal to the film thickness δ, and with area A in the film plane. When the distinction is necessary, the grains are assumed to be D wide and l long. In fact, from Fig. 1.2, the grains are generally isotropic in the film plane, so that $l \sim D$. Each grain magnetization will be assumed to lie in the film plane.

The replay voltage for a single grain, assuming a medium that is thin with respect to vertical variations of the inductive replay head ($\delta < d, g$) is (11.4):

$$V_i(x) = NEv\mu_0 M_s \delta A \alpha_x(x_i, z_i)(h_x(x + x_i + l/2, d)$$

$$- h_x(x + x_i - l/2, d))/l \approx NEv\mu_0 M_s \delta A \alpha_x(x_i, z_i) \frac{\partial h_x(x + x_i, d)}{\partial x}$$

$$(12.38)$$

$\alpha_x(x_i, z_i)$ is the grain magnetization component in the longitudinal x direction. The location of the grain in the film plane (x_i, z_i) is specifically noted in the magnetization component $\alpha_x(x_i, z_i)$. A head field independent of cross-track direction is assumed for simplicity. The last term in (12.38) is a short particle approximation. For simplicity the approximate term will be used. It should be noted that the approximation in (12.38) represents a field difference in the recording direction and an average through the film thickness (in particular when reformulated with (11.5) for thick perpendicular media). For thin films ($\delta \ll d$) the center value of the field (d or $d + \delta/2$) can be used.

As in Chapter 11 the net voltage is the sum of (12.38) over all grains in the film. The spatial autocorrelation function is given, utilizing (10.8), by:

$$
\begin{aligned}
R(x, x + \chi) &= \langle V(x)V(x + \chi)\rangle \\
&= (NEv\mu_0 M_s \delta A)^2 \sum_{i,j} \frac{\partial h_x(x + x_i, d)}{\partial x} \frac{\partial h_x(x + x_j + \chi, d)}{\partial x} \\
&\quad \langle \alpha_x(x_i, z_i)\alpha_x(x_j, z_j)\rangle
\end{aligned}
\tag{12.39}
$$

Each sum includes down-track as well as cross-track grain placement. The statistical average is over the random magnetization orientations. With a recorded magnetization pattern in the medium, the voltage auto-correlation is non-stationary; it depends on x as well as displacement χ.

The correlation function of the magnetization component is written in a general form, similar to that in (10.14):

$$
\langle \alpha_x(x_i, z_i)\alpha_x(x_j, z_j)\rangle = \langle \alpha_x(x_i)\rangle \langle \alpha_x(x_j)\rangle + \sigma(x_i)\sigma(x_j)\rho(x_j, x_j, z_i - z_j)
\tag{12.40}
$$

where σ is the variance and ρ is the normalized correlation function $\rho(x_i = x_j, z_i = z_j) = 1$. $\langle \alpha_x(x)\rangle$ is the averaged magnetization component, which in the presence of a recorded pattern depends on location x along the track. Equation (12.40) expresses the non-stationarity of the correlations along the recording direction. Because the average magnetization (apart from edge-track effects) does not vary across the track the correlation function depends only on the relative cross-track position $z_i - z_j$. Along the track, however, the statistics are, in general, non-stationary, and the correlation function depends specifically on the two coordinates x_i, x_j. The variance $\sigma^2(x)$ depends only on location along the track:

$$
\sigma^2(x) = \langle \alpha_x^2(x)\rangle - (\langle \alpha_x(x)\rangle)^2
\tag{12.41}
$$

in a manner similar in form to (12.1). Note that statistical averaging over ensembles of identical recorded patterns at a given location is equivalent to averaging across a very wide track.

Since only noise properties are considered, the fluctuation term of (12.40) is substituted into (12.39):

$$R_v^n(x, x + \chi) = \langle V(x)V(x + \chi) \rangle$$

$$= (NEv\mu_0 M_s \delta A)^2 \sum_{i,j} \frac{\partial h_x(x + x_i, d)}{\partial x} \frac{\partial h_x(x + x_j + \chi, d)}{\partial x}$$

$$\sigma(x_i)\sigma(x_j)\rho(x_i, x_j, z_i - z_j)$$

(12.42)

R_v^n denotes the autocorrelation of the noise voltage. As in Chapter 10, it is convenient to separate the sums into down-track and cross-track:

$$R_v^n(x, x + \chi) = \langle V(x)V(x + \chi) \rangle$$

$$= (NEv\mu_0 M_s \delta A)^2 \sum_{i_x, j_x} \frac{\partial h_x(x + x_i, d)}{\partial x} \frac{\partial h_x(x + x_j + \chi, d)}{\partial x}$$

$$\sigma(x_i)\sigma(x_j) \sum_{i_z, j_z} \rho(x_i, x_j, z_i - z_j)$$

(12.43)

Note that in (12.43) the sums denote summing over all the grains in the planar array as illustrated in Fig. 1.2 where i_x, i_z (and j_x, j_z) denote longitudinal and transverse sums over the grains, respectively. Evaluation of (12.43) requires knowledge of the correlation function. A typical cross-track variation of the correlation function is plotted in Fig. 12.7 versus relative displacement $(z_i - z_j)/D$. The three functions are for different values of intergranular exchange (Bertram, *et al.*, 1992). In general, the correlation function decreases with separation, becomes negative and eventually vanishes, with perhaps additional small oscillations. The oscillatory behavior results from the magnetostatic grain interaction field, which from (2.16) favors anti-parallel magnetizations in a direction orthogonal to the magnetization direction (Fig. 12.3). The correlations plotted in Fig. 12.7 were computed at the center of a recorded transition, but the decay with oscillation occurs generally as seen in uniformly magnetized media (Bertram, *et al.*, 1992) and in models of the ac-erased state in particulate media (Arratia & Bertram, 1984).

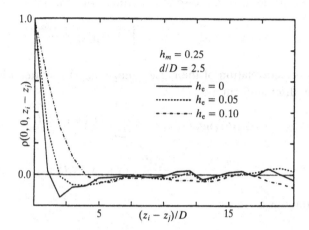

Fig. 12.7. Cross-track correlation function at the center of a recorded transition, $\langle \alpha_x \rangle = 0.0$, from numerical analysis. The three curves are for a fixed intergranular magnetostatic interaction $h_m = M_s/H_K$ and varying intergranular exchange $h_e = A^*/KD^2$. Taken from Bertram, Beardsley and Che; presented at MMM conference, Houston TX, Dec. 1993. Published in *J. Appl. Phys.*, 1993.

The first sum in (12.43) may be written as:

$$N_s(x_i, x_j) = \sum_{j_z} \rho(x_i, x_j, z_i - z_j) \qquad (12.44)$$

This sum yields a value independent of the cross-track coordinate if the correlation function decay is less than the track width. The sum is generally larger than unity since the correlation function extends many grains, depending on the interactions (Fig. 12.7). Physically N_s is the effective number of grains in a correlated unit or domain with common magnetization across the track. In general the sum depends on the coordinates x_i, x_j. The second sum over the track width yields the number of grains across the track N_w so that:

$$N_w N_s(x_i, x_j) = \sum_{i_z, j_z} \rho(x_i, x_j, z_i - z_j) \qquad (12.45)$$

N_w is related to the track width by $N_w = W/D$. A correlation width function can be defined in a discrete or continuous form as:

$$s(x_i, x_j) = DN_s(x_i, x_j) = \int_{-\infty}^{\infty} dz \rho(x_i, x_j, z) \qquad (12.46)$$

Thus, the autocorrelation of the noise voltage may be written (assuming $A = D^2$) in discrete form as:

$$R_v^n(x, x + \chi) = WAD(NEv\mu_0 M_s \delta)^2 \sum_{i_x, j_x} \frac{\partial h_x(x + x_i, d)}{\partial x}$$
$$\frac{\partial h_x(x + x_j + \chi, d)}{\partial x} \sigma(x_i)\sigma(x_j)N_s(x_i, x_j) \qquad (12.47)$$

or in continuous form as:

$$R_v^n(x, x + \chi) = W(NEv\mu_0 M_s \delta)^2 \int_{-\infty}^{\infty} dx' \int_{-\infty}^{\infty} dx''$$

$$\frac{\partial h_x(x + x', d)}{\partial x} \frac{\partial h_x(x + x'' + \chi, d)}{\partial x} \sigma(x')\sigma(x'')s(x', x'') \qquad (12.48)$$

Two cases of uniformly magnetized media and media with a recorded transition are now examined.

Uniformly magnetized media

For media uniformly magnetized along the recording direction, stationarity applies so that $s(x_i, x_j) = s(x_i - x_j)$: there is no absolute reference point for the correlations along the track. The correlation function is written in terms of the difference coordinates only: $\rho(x_i - x_j, z_i - z_j)$. The correlation function will decay along the track in monotonic form, generally without the oscillations seen in the transverse correlation (Bertram & Arias, 1992; Arratia & Bertram, 1984). The variance (12.41) depends not on location but only on the state of magnetization: $\sigma^2 = \langle \alpha_x^2 \rangle - (\langle \alpha_x \rangle)^2$. Since integrals over the correlation function are independent of absolute location, a correlation or cluster area may be defined in discrete or continuous form:

$$A_{cl} = D^2 N_{cl} = \sum_{i_x, j_z} \rho(x_i, z_j) = \int_{-\infty}^{\infty} dx \int_{-\infty}^{\infty} dz \rho(x, z) \qquad (12.49)$$

A_{cl} is the cluster area of grains that are correlated to have a common magnetization orientation. N_{cl} is the number of grains in the cluster. In general, the cluster size will be a function of the state of magnetization $\langle \alpha_x \rangle$ (Bertram & Arias, 1992). The cluster size relative to the grain size, N_{cl}, depends on the grain interactions; and, in particular, for intergranular exchange coupling, N_{cl} can increase as the grain size is decreased so that the cluster area may not decrease in proportion to the grain area (Bertram & Arias, 1992; Bertram & Che, 1993).

The autocorrelation function in discrete form (12.47) simplifies to:

$$R_v^n(x, x+\chi) = WAD\sigma^2(NEv\mu_0 M_s\delta)^2 \sum_{i_x,j_x} \frac{\partial h_x(x+x_i,d)}{\partial x}$$

$$\frac{\partial h_x(x+x_j+\chi,d)}{\partial x} N_s(x_i-x_j) \tag{12.50}$$

Changing variable sum: $x_m = x_i - x_j$ yields:

$$R_v^n(x, x+\chi) = WAD\sigma^2(NEv\mu_0 M_s\delta)^2 \sum_{i_x} \frac{\partial h_x(x+x_i,d)}{\partial x}$$

$$\sum_{m_x} \frac{\partial h_x(x+x_m+x_i+\chi,d)}{\partial x} N_s(x_m) \tag{12.51}$$

If the correlation decay along the track is rapid compared to the head-field function, then the sum over m_x may be taken solely over $N_s(x_m)$:

$$R_v^n(x, x+\chi) =$$
$$WA_{cl}\sigma^2(NEv\mu_0 M_s\delta)^2 \sum_{i_x} D\frac{\partial h_x(x+x_i, d)}{\partial x} \frac{\partial h_x(x+x_i+\chi, d)}{\partial x} \tag{12.52}$$

Since the sum with the factor of D is independent of D (as seeen easily by conversion to the integral form), the noise power scales as the cluster area. It is impossible to distinguish, with small clusters, between clusters and large grains equal in size to the cluster area.

For stationary processes (10.6) applies. The noise power spectrum can be useful for analysis for the down-track correlation length (Silva & Bertram, 1990; Tarnopolosky, *et al.*, 1991). With stationarity the continuous form of the autocorrelation (12.48) becomes:

$$R_v^n(x, x+\chi) = W\sigma^2(NEv\mu_0 M_s\delta)^2 \int_{-\infty}^{\infty} dx' \int_{-\infty}^{\infty} dx''$$

$$\frac{\partial h_x(x+x',d)}{\partial x} \frac{\partial h_x(x+x''+\chi,d)}{\partial x} s(x'-x'') \tag{12.53}$$

Utilizing (10.6), the power in a band with Δk for $k > 0$ is (Silva, *et al.*, 1990; Bertram, *et al.*, 1992):

$$P_n(k) = \frac{\Delta k}{\pi} \, W\sigma^2 (NEv\mu_0 M_s k\delta)^2 |h_{\mathrm{sur}}(k)|^2 e^{-2kd} s(k) \qquad (12.54)$$

A measurement of the power spectral density, corrected by known head and spacing parameters, gives $s(k)$. Back transformation yields the variation of the (cross-track integrated) correlation with distance along the track, from which an effective down-track correlation length can be defined. Note that $s(k)$, for stationary processes, is given by:

$$s(k) = \int_{-\infty}^{\infty} dz \rho(k,z) = \int_{-\infty}^{\infty} dz \int_{-\infty}^{\infty} dx e^{-ikx} \rho(x,z) \qquad (12.55)$$

so that at long wavelengths $s(0) = A_{\mathrm{cl}}$. Thus, at long wavelengths

$$P_n(k \to 0) = \frac{\Delta k}{\pi} \, W A_{\mathrm{cl}} \sigma^2 (NEv\mu_0 M_s k\delta)^2 \qquad (12.56)$$

Note that the product $A_{\mathrm{cl}}\sigma^2$ enters so that a determination of the cluster size requires an estimate of the variance. Since the spectral measurement gives the down-track correlation length, the unknown quantity is the product $s\sigma^2$ where s denotes the cross-track correlation width. For example, if the cluster size at the remanent coercive state where $\langle \alpha_x \rangle = 0$ is desired, then the variance $\sigma^2 = \langle \alpha_x^2 \rangle = \langle \cos^2 \theta \rangle$ must first be determined, where θ is the angle of the magnetization to the recording direction (Fig. 10.1(b)).

The variance for uniformly magnetized media, including packing effects as discussed at the beginning of this chapter, may be given by generalizing (12.3):

$$\sigma^2 = p(\langle \cos^2 \theta \rangle - p\langle n \rangle^2 \langle |\cos \theta| \rangle^2) \qquad (12.57)$$

where in this section σ^2 is normalized by M_s^2. In (12.57) it is assumed that the angular distributions are symmetric about $\pm\theta$. Due to interaction effects, the angular distributions vary from reasonably well-oriented near saturation remanence to reasonably isotropic near the demagnetized state $\langle n \rangle = 0$. For media with typical magnetostatic interactions, somewhat independent of the degree of orientation, the magnetizations are approximately uniformly distributed at the coercive state so that $\sigma^2 \sim 0.5$. The variation of variance with magnetization is reduced from that shown in Fig. 12.2 due to orientation distributions that depend on the magnetizing state. The noise voltage autocorrelation (12.53) may be written utilizing (12.57) in terms of the remanent saturation magnetization $M_r = pM_s\langle\cos\theta\rangle_r$:

$$R_v^n(x, x+\chi) = \frac{W(NEv\mu_0 M_r\delta)^2\langle\cos^2\theta\rangle}{p(\langle\cos\theta\rangle_r)^2}\left(1 - p\frac{\langle n\rangle^2\langle|\cos\theta|\rangle^2}{\langle\cos^2\theta\rangle}\right)$$

(12.58)

$$\int_{-\infty}^{\infty}dx'\int_{-\infty}^{\infty}dx''\frac{\partial h_x(x+x',d)}{\partial x}\frac{\partial h_x(x+x''+\chi,d)}{\partial x}s(x'-x'')$$

This form is useful for SNR estimates, since the factor $NEv\mu_0 M_r\delta$ is common to the signal.

Noise in the presence of a recorded transition

Equation (12.48) will be utilized for discussion of noise in the presence of a transition. The coordinates of the autocorrelation are related to the temporal pair t_1, t_2 by $x = t_1 v$, $\chi = (t_2 - t_1)v$. Evaluation requires knowledge of the magnetization fluctuations:

$$\sigma(x')\sigma(x'')s(x',x'') = \sigma(x')\sigma(x'')\int_{-\infty}^{-\infty}dz\rho(x',x'',z)$$

$$= \int_{-\infty}^{-\infty}dz(\langle\alpha_x(x',z')\alpha_x(x'',z'')\rangle - \langle\alpha_x(x')\rangle\langle\alpha_x(x'')\rangle)$$

(12.59)

Numerical micromagnetic simulation has been utilized to determine the magnetization correlations, specifically the integrand of the last term in (12.59) at $z' = z''$ (Bertram & Che, 1993). In Fig. 12.8 the correlation function is plotted versus displacement $(x'' - x')/D$ for three reference positions $x' = 0, a, 2a$. Using any of the functional forms in Table 4.1, these positions are at the transition center, approximately half-way to saturation and near saturation in the transition shoulder region. The correlation function decreases with relative displacement, from which a correlation length or cluster length can be defined. Non-stationarity occurs because the correlation function depends on reference point x', not simply on the relative displacement $x'' - x'$. As the correlation reference point is moved away from the transition center, the function

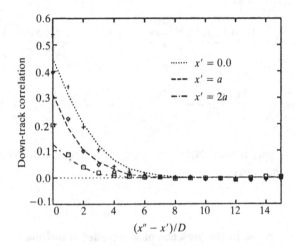

Fig. 12.8.   Down-track magnetization fluctuation correlation function in the presence of a recorded transition from numerical analysis plotted versus relative displacement at three different reference positions with respect to the transition center. Only the displacement in the positive direction is shown, but the asymmetry is only slight for $x' \neq 0$. The solid curves represent the analytic approximation of (12.64). Taken from Bertram, Beardsley and Che; presented at MMM conference, Houston TX, Dec. 1993. Published in *J. Appl. Phys.*, 1993.

decreases substantially in magnitude, almost vanishing in the saturated region. Thus the integrals in (12.48) will be confined to the region around the transition.

The non-stationary variation of the correlation function in the recording direction is, in general, complicated. Here a simplified analysis of a functional form is presented that captures, to first order, the dependence of the correlation function on relative position in the transition. The model can be visualized by referring to Fig. 12.6, which illustrates a fluctuating transition center separating regions of approximately oppositely saturated magnetization. An analytic form for (12.59) is given by directly evaluating the magnetization joint probability at locations x' and x''. An expression is required of the probability that the magnetization changes sign or crosses the transition boundary between x' and x''. In this simple model the average magnetization will be separated into the (one assumes, uncorrelated,) effect of the component of magnetization times the number of grains switched (12.3):

$\alpha_x = n|\cos\theta|$. Discounting, initially, z', z'' variations, the correlation of the longitudinal magnetization component is:

$$\langle\alpha_x(x', z')\alpha_x(x'', z')\rangle$$
$$= \langle|\cos\theta(x'')|\rangle\langle|\cos\theta(x')|\rangle$$
$$x(\text{Probability of no sign change} - \text{Probability of a sign change})$$
$$= \langle|\cos\theta(x'')|\rangle\langle|\cos\theta(x')|\rangle \qquad (12.60)$$
$$x(1 - 2\text{Probability of a sign change})$$

The probability distribution function of grains reversing is simply $(1/2)$ the derivative of $\langle n(x)\rangle$:

$$\text{PD} = \frac{1}{2}\frac{\mathrm{d}\langle n(x)\rangle}{\mathrm{d}x} \qquad (12.61)$$

The variation of $\langle n(x)\rangle$ through the transition is very closely the transition shape, except for the orientation factor. Thus, the probability that the magnetization has changed sign betwen x' and x'' is the integral of (12.61). Therefore:

$$\langle\alpha_x(x', z')\alpha_x(x'', z')\rangle = \langle|\cos\theta(x'')|\rangle\langle|\cos\theta(x')|\rangle(1 - \langle n(x'')\rangle - \langle n(x')\rangle)$$
$$(12.62)$$

In (12.62) it is assumed that the transition magnetization increases from $-M_r$ to $+M_r$ with increasing x and that $x' < x''$ and $z' = z''$. Thus, (12.62) may be written as:

$$\langle\alpha_x(x', z')\alpha_x(x'', z'')\rangle - \langle\alpha_x(x^<)\rangle\langle\alpha_x(x^>)\rangle$$
$$= \langle|\cos\theta(x^<)|\rangle\langle|\cos\theta(x^>)|\rangle(1 + \langle n(x^<)\rangle)(1 - \langle n(x^>)\rangle) \qquad (12.63)$$
$$= (\langle|\cos\theta(x^<)|\rangle + \langle\alpha(x^<)\rangle)(\langle|\cos\theta(x^>)|\rangle - \langle\alpha(x^>)\rangle)$$

where $x^<, x^>$ are respectively the larger and smaller of x', x''. Equation (12.63) is plotted as the dashed curves in Fig. 12.8. Both $\langle\alpha_x(x)\rangle$ and $\langle|\cos\theta(x)|\rangle$ were taken from the micromagnetic simulation. The agreement is quite good. It is to be noted that the major variation is due to the average transition magnetization shape $\langle\alpha_x(x)\rangle$. Note that the theory in Fig. 12.8 is below that of numerical analysis at zero displacement $x' = x''$. Part of the explanation is that the simplified model does not take into account correlations of the magnetization angles. For example, in (12.63) at $x^> = x^< = 0$, the magnetization fluctuation is given by $\langle|\cos\theta(0)|\rangle\langle|\cos\theta(0)|\rangle$ when, in fact, the limit should be $\langle\cos^2\theta(0)\rangle$. For a uniform distribution of orientations, the former is $(2/\pi)^2 \sim 0.4$, whereas the latter equals 0.5.

A simplified, completely analytic form is desired in order to illustrate the correlation function. Equation (12.63) is evaluated for $z' = z''$. To include the cross-track correlation $z' \neq z''$, and perform the integration in (12.59), it is assumed that one may simply multiply (12.63) by a constant s equal to the cross-track correlation width ($s(0,0)$ in (12.46)). In addition the angular factors in (12.63) will be placed outside the brackets and $\langle \cos^2 \theta(0) \rangle$ will be utilized. Thus, (12.59) is approximately:

$$\sigma(x')\sigma(x'')s(x',x'') \approx s\langle \cos^2 \theta(0) \rangle (1 + M(x^<)/M_r)(1 - M(x^>)/M_r)$$

$$(12.64)$$

Substitution into (12.48) yields:

$$R_v^n(x, x+\chi) = sW(NEv\mu_0 M_r \delta)^2 \frac{\langle \cos^2 \theta(0) \rangle}{(\langle \cos \theta \rangle_r)^2} \int_{-\infty}^{\infty} dx' \int_{-\infty}^{\infty} dx''$$

$$\frac{\partial h_x(x+x',d)}{\partial x} \frac{\partial h_x(x+x''+\chi,d)}{\partial x} \qquad (12.65)$$

$$(1 + M(x^<)/M_r)(1 - M(x^>)/M_r)$$

This form may be utilized for illustrative examples since once the transition is specified (e.g. a tanh shape with length parameter a), the autocorrelation may be directly computed. The fluctuations arise from the correlation width s. Here s is a parameter that sets the noise level but not the functional forms. It is important to note that (12.62) and thus (12.65) apply to a single transition.

The autocorrelation function in (12.65) may be represented by a 3D plot of voltage covariance versus the coordinates $t_1 = x/v$ and $t_2 = (x+\chi)/v$. In Fig. 12.9 the covariance is plotted for a single transition utilizing a long pole head with spacing to gap ratio $d/g = 0.25$ and a tanh transition shape (Table 4.1) with relative transition length $a/g = 0.375$. The noise power versus longitudinal position is given by the diagonal with $t_1 = t_2$ and the transition center is at $t_1 = t_2 = 0$. The peaks that occur either side of the transition center are due approximately to the squared derivative of the longitudinal head field. The negative regions on the line $t_1 = -t_2$ are due to displacements of the two head-field derivatives about $g/2$ (e.g. $x = g/4$ and $\chi = -g/2$).

The functional form of the autocorrelation in Fig. 12.9 agrees well with measurement (Tang, 1986). In the measurement the autocorrelation function was fitted to the sum of position and amplitude jitter of a fixed

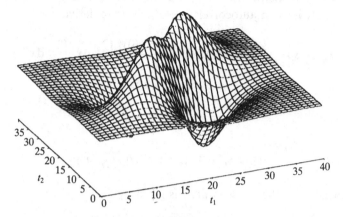

Fig. 12.9. Noise power autocorrelation function versus variables t_1, t_2 denoting respective locations along a recorded transition.

transition shape. Such fitting is possible by comparison of Fig. 12.4 and the diagonal line in Fig. 12.9. The two forms of voltage variance can be superimposed to give the double peak and center local minimum of Fig. 12.9.

Equation (12.65) can be utilized to determine a variety of noise effects as well as error rates (Bertram & Che, 1993). The total power, neglecting interbit interactions, for square wave recording, is:

$$\text{TNP} = \frac{1}{B} \int_{-B/2}^{B/2} dx R_v^n(x, \chi = 0) \tag{12.66}$$

which increases linearly with density $(1/B)$ at low densities. The amplitude jitter of the peak voltage is:

$$R_v^n(0,0) \propto sW(NEv\mu_0 M_r\delta/g)^2 \tag{12.67}$$

where the proportionality constant is a function of d/g and a/g. The ratio of the amplitude fluctuations to the square peak voltage of the isolated pulse (5.49) is:

$$\frac{R_v^n(0,0)}{V^2(0)} \propto \frac{s}{W} \tag{12.68}$$

The position jitter may be estimated by the ratio of the noise derivative variance to the second derivative of the voltage squared at the transition center (Katz & Campbell, 1979). Equation (12.65) may be transformed immediately into an autocorrelation of the noise derivative:

$$R_{\dot{v}}^{n}(x, x+\chi) = sW(NEv\mu_0 M_r\delta)^2 \frac{\langle\cos^2\theta(0)\rangle}{(\langle\cos\theta\rangle_r)^2} \int_{-\infty}^{\infty} dx' \int_{-\infty}^{\infty} dx''$$

$$\cdot \frac{\partial^2 h_x(x+x', d)}{\partial x^2} \frac{\partial^2 h_x(x+x''+\chi, d)}{\partial x^2} \tag{12.69}$$

$$\cdot (1 + M(x^<)/M_r)(1 - M(x^>)/M_r)$$

The variance of the position jitter is given by:

$$\sigma_{\Delta x}^2 \approx \frac{R_{\dot{v}}^{n}(0,0)}{\ddot{V}^2} \approx K\left(\frac{a}{g}, \frac{d}{g}\right)g^2\,\frac{s}{W} \tag{12.70}$$

Equations (12.68) and (12.70) are in the form of noise-to-signal ratios and show both the dominant importance of decreasing the cross-track correlation width s and the independence for thin films of $M_r\delta$ on these ratios. Of course a reduction in $M_r\delta/H_c$ will reduce s and a and will hence increase, in general, signal-to-noise ratios.

The focus of this and previous chapters has been to give the reader a thorough grounding in the fundamentals of signals and noise in magnetic recording as applied to inductive as well as magnetoresistive heads. In system design, in particular for digital magnetic recording channels, a major development question is: How should media and heads be best prepared to minimize the final error rate? Or in fact, how do various channel detection schemes affect system performance from a medium-manufacturing point of view? A thin-film disk system that utilizes PRML detection provides an example of how these questions can be answered. The non-stationary noise correlations developed in this chapter may be directly implemented into a PRML error estimate model (Che, *et al.*, 1993c). Thus, for example, cross-track correlation width and transition shape measured during medium manufacture can be used to directly obtain final error rates. Of course, other error-producing mechanisms occur, other than intrinsic medium noise, such as non-linear bit shift and head–media spacing fluctuations. However, the former has been discussed in Chapters 8 and 12, and a well-founded understanding of the material in this book should greatly assist the reader in answering general system questions.

Problems

Problem 12.1   Assume a Lorentzian pulse shape (Problem 5.8) and derive explicit expressions for the noise voltage variance sketched in Fig. 12.4. Assume a fluctuating PW_{50} and recall that the area under the pulse is a constant. Assume equal variances for the two noise sources and derive an explicit expression for (12.8).

Problem 12.2   Rederive (12.7) for perpendicular recording with keepered media and replay with a shielded MR head. Do two cases: (1) assume a general transition shape with scaling parameter 'a' and (2) use the result from Chapter 8 and assume that the transition shape follows the surface head field. For the second case let the fluctuations be due to variation in record spacing. For a fixed record current this assumption yields correlated fluctuations in transition position as well as shape.

Problem 12.3   Using (2.16) show that the interaction field between two sub-transitions on different main transitions (i on transition 1 and j on transition 2 in Fig. 12.6) is given, including imaging, by:

$$H_D \approx \frac{3M_r\delta d^2}{\pi B^3}\left(\sin\theta^+ - \sin\theta^- + \frac{\sin^3\theta^+ - \sin^3\theta^-}{3}\right)$$

where

$$\sin\theta^\pm = \frac{s(i-j\pm\frac{1}{2})}{\sqrt{B^2 + s^2(i-j\pm\frac{1}{2})^2}}$$

Problem 12.4   Derive (12.32) from (12.31). Hint: first square and average (12.31) to derive an expression relating the variances and the correlation $\langle w_1 w_2 \rangle$. Then, reform (12.31) by placing the w_2 term on the left-hand side and square and average. Note the coercivity fluctuation is for transition 1 and therefore is uncorrelated with w_2.

Problem 12.5   Utilize (12.10) to derive a relation for a total noise power due to position fluctuations including nearest neighbor correlations:

$$TP = \frac{\langle w_n^2 \rangle}{B} \int\limits_{-\infty}^{\infty} dx \left(\frac{\partial V_{sp}(x)}{\partial x} \right)^2 + \frac{\langle w_n w_{n+1} \rangle}{B}$$

$$\int\limits_{-\infty}^{\infty} dx \, \frac{\partial V_{sp}(x)}{\partial x} \, \frac{\partial V_{sp}(x+B)}{\partial x}$$

Show that for longitudinal thin film media with arctangent transitions and Karlqvist replay head:

$$TP(B) = \frac{(NEvW\mu_0 M_r \delta)^2 \langle w_n^2 \rangle}{\pi B (a+d)((a+d)^2 + g^2/4)}$$

$$- \frac{2(NEvW\mu_0 M_r \delta)^2 \langle w_n w_{n+1} \rangle (a+d)((a+d)^2 + g^2/4 - 3B^2/4)}{\pi B ((a+d)^2 + B^2/4) \, ((a+d)^2 + (B-g)^2/4) \, ((a+d)^2 + (B+g)^2/4)}$$

Problem 12.6   Derive the positional variance for amplitude fluctuations from (12.7) assuming a zero gap head and an arctangent transition. Plot the positional variance for a range of a/d from $a/d \ll 1$ to $a/d \gg 1$. Does the center variance ever vanish as in the case of jitter noise?

Problem 12.7   Rederive (12.64) by utilizing the simplified model for perpendicular recording developed in Chapter 8. Utilize the simple model for the fluctuations in (Bertram, *et al.*, 1992); that is, the number of reversed particles and hence the noise varies as the ratio of $M(x)/H_c$. Assume no particle correlations beyond a single particle.

References and bibliography

Aharoni, A. & Jakubovics, J. P. (1982). Factors affecting domain wall mobility in thin ferromagnetic metal films. *Philos. Mag. B*, **46**, 253–72.

Anderson, R. L., Bajorek, C. H. & Thompson, D. A. (1972). Numerical analysis of a magnetoresistive transducer for magnetic recording applications. *AIT Conf. Proc.* **10**, Part 2.

Anzaloni, A. & Barbosa, L. C. (1984). The average power density spectrum of the readback voltage from particulate media. *IEEE Trans. Magn.*, **MAG-20** (5), 693–97.

Aoi, H., Saitoh, M., Nishiyama, N., Tsuchiya, R. & Tamura, T. (1986). Noise characteristics in longitudinal thin-film media. *IEEE Trans. Magn.*, **MAG-22** (5), 895–97.

Armstrong, A. J., Bertram, H. N., Barndt, R. D. & Wolf, J. K. (1991). Non-linear effects in high-density tape recording. *IEEE Trans. Magn.*, **MAG-27** (5), 4366-76.

Arnoldussen, T. C. & Tong, H. C. (1986). Zigzag transition profiles, noise, and correlation statistics. *IEEE Trans. Magn.*, **MAG-22** (5), 889–91.

Arnoldussen, T. C. (1987). Film head read resolution and frequency response analysis. *IEEE Trans. Magn.*, **MAG-23** (5), 3152–4.

Arnoldussen, T. C. (1992). Theoretical considerations of media noise. In *Noise in Digital Magnetic Recording*, eds. T. C. Arnoldussen and L. L. Nunnelley. World Scientific Publishing Co.

Arratia, R. A. & Bertram, H. N. (1984). Monte Carlo simulation of particulate noise in magnetic recording. *IEEE Trans. Magn.*, **MAG-20** (2), 412–20.

Ash, K. P., Wachenschwanz, D., Brucker, C., Olson, J., Trcka, M. & Jagielinski, T. (1990). A magnetic head for 150 MHz high density recording. *IEEE Trans. Magn.*, **MAG-26** (6), 2960–5.

Azarian, M. H., Kuin, P. N. & Holstlann. (1987). *IEEE Trans. Magn.*, **MAG-23** (1), 192–4.

Bajorek, C. H., Krongelb, S., Romankiw, L. T. & Thompson, D. A. (1974). An integrated magnetoresistive read, inductive write high density recording head. *20th Annu. AIP Conf. Proc.*, **24**, 548–9.

Baker, B. R. (1977). Long wavelength response to elliptical reproduce heads. *IEEE Trans. Magn.*, **MAG-12** (1), 39–43.

Barany, A. (1989). Noise and non-linear recording phenomena in thin metallic films. Ph.D. Thesis, University of California at San Diego.

337

Barany, A. & Bertram, H. N. (1987a). Transition noise model for longitudinal thin film media. *IEEE Trans. Magn.*, **MAG-23** (2), 1776–87.

Barany, A. & Bertram, H. N. (1987b). Transition position and amplitude fluctuation noise model for longitudinal thin film media. *IEEE Trans. Magn.*, **MAG-23** (5), 2374–6.

Barndt, R. D. & Wolf, J. K. (1992). Modeling and signal processing for the nonlinear thin film recording channel. *IEEE Trans. Magn.*, **MAG-28** (5), 2710–12

Barndt, R. D., Armstrong, A. J., and Bertram, H. N. (1991). A simple model of partial erasure in thin film disk magnetic recording systems. *IEEE Trans. Magn.*, **MAG-27** (6), 4978–80.

Barndt, R. D., Armstrong, A. J. & Wolf, J. K. (1993). Media selection for high density recording channels. *IEEE Trans. Magn.*, **MAG-29** (1), 183–8.

Baugh, R. A., Murdock, E. S. & Natarajan, B. R. (1983). Measurement of noise in magnetic media. *IEEE Trans. Magn.*, **MAG-19** (5), 1722–4.

Beardsley, I. A. (1982a). Effect of particle orientation on high density recording. *IEEE Trans. Magn.*, **MAG-18** (6), 1191–6.

Beardsley, I. A. (1982b). Self consistent recording model for perpendicularly oriented media. *J. Appl. Phys.*, **53**, 2582–4.

Beardsley, I. A. (1986). Modeling the record process. *IEEE Trans. Magn.*, **MAG-22** (5), 454–9.

Beardsley, I. A., Armstrong, A. J. & Bertram, H. N. (1992). Improved calibration technique for high density tape recording system. *IEEE Trans. Magn.*, **MAG-28** (6), 3417–22.

Beardsley, I. A. & Bertram, H. N. (1987). Imaging effects in contact recording on particulate media. *IEEE Trans. Magn.*, **MAG-23** (6), 3922–6.

Beardsley, I. A. & Sperious, V. S. (1990). Determination of thin film media model parameters using DPC imaging and torque measurements. *IEEE Trans. Magn.*, **MAG-26** (6), 2718–20.

Belk, N. R., George, P. K. & Mowry, C. S. (1985). Noise in high performance thin film longitudinal magnetic recording media. *IEEE Trans. Magn.*, **MAG-21** (5), 1350–5.

Benson, R. C. (1991). Plate tenting with a one-sided constraint. *ASME J. Appl. Mech.* **58** (2), 484–92.

Benson, R. C., Smith, D. P., Madsen, D. D. & Fung, S. (1992). Magnetic tape tenting: Modeling and experiment. In *Advances in Information Storage Systems*, Vol. 4, ed. B. Bhushan, pp. 63–72. ASME Press.

Berkowitz, A. (Guest Editor) (1990). Magnetic recording materials. *MRS Bulletin* **15** (3) 23–72.

Bertero, G. A., Bertram, H. N. & Barnett, D. M. (1993). Fields and transforms for thin film heads. *IEEE Trans. Magn.*, **MAG-29** (1), 67–76.

Bertram, H. N. (1971). Monte Carlo calculation of magnetic anhysteresis. *J. de Phys.* **32**, 684.

Bertram, H. N. (1974). Long wavelength AC bias recording theory. *IEEE Trans. Magn.*, **MAG-10** (4), 1039–48.

Bertram, H. N. (1975a). On the convergence of iterative solutions of the integral magnetic field equation. *IEEE Trans. Magn.*, **MAG-11** (3), 928–33.

Bertram, H. N. (1975b). Wavelength response in AC biased recording. *IEEE Trans. Magn.*, **MAG-11** (5), 1176–78.

Bertram, H. N. (1978). Anisotropic reversible permeability effects in the magnetic reproduce process. *IEEE Trans. Magn.*, **MAG-14** (3), 111–18.

Bertram, H. N. (1984a). Geometric effects in the magnetic recording process. *IEEE Trans. Magn.*, **MAG-20** (3), 468–78.

Bertram, H. N. (1984b). The effect of the angular dependence of the particle nucleation field on the magnetic recording process. *IEEE Trans. Magn.*, **MAG-20** (6), 2094–2104.

Bertram, H. N. (1985). Interpretation of spectral response in perpendicular recording. *IEEE Trans. Magn.*, **MAG-21** (5), 1395–7.

Bertram, H. N. (1986a). Fundamentals of the magnetic recording process. *Proc. IEEEE*, **74** (11), 1494–1512.

Bertram, H. N. (1986b). Particle interaction phenomena. *IEEE Trans. Magn.*, **MAG-22** (5), 460–5.

Bertram, H. N. (1986c). The magnetic limits of magnetic recording. *Proc. of SMART Conference*, San Jose, CA, May.

Bertram, H. N. (1993). The Physics of Magnetic Recording. In *Applied Magnetism*, eds. R. Gerber, D. Wright and G. Asti eds. Kulwer Academic Publishers.

Bertram, H. N. & Arias, R. (1992). Magnetization correlations and noise in thin film recording media. *J. Appl. Phys.*, **71** (7), 3439–54.

Bertram, H. N., Armstrong, A. J., Wolf, J. K. & Beardsley, I. A. (1992). Theory of non-linearities and pulse asymmetry in high density tape recording. (Invited). *IEEE Trans. Magn.*, **MAG-28** (5), 2701–6.

Bertram, H. N. & Beardsley, I. A. (1988). The recording process in longitudinal particulate media. *IEEE Trans. Magn.*, **MAG-24** (6), 3234–48.

Bertram, H. N., Beardsley, I. A. & Che, X. (1993). Magnetization correlations in thin film media in the presence of a recorded transition. *J. Appl. Phys.*, **73** (10), 5545–7.

Bertram, H. N. and Bhatia, A. K. (1974). The effect of interactions on the saturation remanence of particulate assemblies. *IEEE Trans. Magn.*, **MAG-9** (2), 127–33.

Bertram, H. N. & Che, X. (1993). General analysis of noise in recorded transitions in thin film recording media. *IEEE Trans. Magn.*, **MAG-29** (1), 201–8.

Bertram, H. N. & Fielder, L. D. (1983). Amplitude and bit shift spectra comparisons in thin metallic media. *IEEE Trans. Magn.*, **MAG-29** (5), 1606–7.

Bertram, H. N., Hallamasek, K. & Madrid, M. (1986). DC modulation noise in thin metallic media and its application for head efficiency measurement. *IEEE Trans. Magn.*, **MAG-22** (4), 247–52.

Bertram, H. N. & Lindholm, D. L. (1982). Dependence of reproducing gap null on medium permeability and spacing. *IEEE Trans. Magn.*, **MAG-18** (3), 893–7.

Bertram, H. N. & Mallinson, J. C. (1976). A theory of eddy current limited heads. *IEEE Trans. Magn.*, **MAG-12** (6), 713–15.

Bertram, H. N. & Neidermeyer, R. (1978). The effect of demagnetization fields on recording spectra. *IEEE Trans. Magn.*, **MAG-14** (5), 743–5.

Bertram, H. N. & Niedermeyer, R. (1982). The effect of spacing on demagnetization in magnetic recording. *IEEE Trans. Magn.*, **MAG-18** (6), 1206–8.

Bertram, H. N. & Steele, C. W. (1976). Pole-tip saturation in magnetic recording heads. *IEEE Trans. Magn.*, **MAG-12** (6), 702–6.

340   References and bibliography

Bertram, H. N. & Zhu, J-G. (1992). Fundamental magnetization processes in thin film recording media. In *Solid State Physics Review*, Vol. 46, eds. H. Ehrenreich and D. Turnbull. Academic Press, New York, NY.

Beusekamp, M. F. & Fluitman, J. H. (1986). Simulation of the perpendicular recording process including image charge effects. *IEEE Trans. Magn.*, **MAG-22** (6), 364–6.

Bhattacharyya, M. K., Gill, H. S. & Simmons, R. F. Jr. (1989). Determination of overwrite specification in thin film head/disk systems. *IEEE Trans. Magn.*, **MAG-25** (6), 4479–89.

Bhattacharyya, M. K., Tarnopolsky, G. J. & Tran, L. T. (1991). 3D analysis of MR readback on perpendicular medium. *IEEE Trans. Magn.*, **MAG-27** (6), 4707–9.

Bloomberg, D. S., Hughes, G. F. & Hoffman, R. J. (1979). Analytic determination of overwrite capability in magnetic recording systems. *IEEE Trans. Magn.*, **MAG-15** (6), 1450–52.

Brown, W. F. (1962). *Magnetostatic Principles in Ferromagnetism*, ed. E. P. Wohlfarth. North-Holland Publishing Co.

Cannon, M. R. & Seger, P. J. (1989). Data Storage on Tape. In *Magnetic Recording Handbook*, eds. C. D. Mee and E. D. Daniel, Part 2, Chapter 4, p. 812–83, McGraw-Hill.

Che X., Barbosa, L., & Bertram, H. N. (1993c). PRML performance estimation considering medium noise down-track correlations. *IEEE Trans. Magn.*, MAG-29 (6), 4062–4

Che, X. & Bertram, H. N. (1993a). Dynamics of non-linearities in high density recording. *IEEE Trans. Magn.*, **MAG-29** (1), 317–23.

Che, X. & Bertram, H. N. (1993b). Study of magnetic recording transitions at very low flying heights. *J. Appl. Phys.*, **73** (10), 6004–6.

Che, X. & Suhl, H. (1991). Analysis of correlation of magnetization and noise after AC erasing. *J. Appl. Phys.* **69**, 2440–2.

Chen, T. (1981). The micromagnetic properties of high-coercivity metallic thin films and their effects on the limit of packing density in digital recording. *IEEE Trans. Magn.*, **MAG-17** (2), 1181–3.

Chi, C. S. (1980). Spacing loss and non-linear distortion in digital magnetic recording. *IEEE Trans. Magn.*, **MAG-16** (5), 976–8.

Chi, C. S. & Speliotis, D. E. (1974). Dynamic self-consistent interactive simulation of high bit density digital magnetic recording. *IEEE Trans. Magn.*, **MAG-10** (3), 765–8.

Chiba, K., Sato, K., Ebine, Y. & Sasaki, T. (1989). Metal evaporated (ME) tape for high band 8mm video system. *IEEE Trans. Cons. El.* **35** (3), 421-8.

Chikazumi, S. (1964). *Physics of Magnetism*. Robert E. Krieger Publishing Co., Florida.

Coleman, C., Lindholm, D., Peterson, D. & Wood, R. (1984). High data rate magnetic recording in a single channel. In *Int Conf Video and Data Recording, IERE. Proc.* **59**, Southampton, England, p. 151–7.

Comstock, R. L. & Moore, M. L. (1974). Ferrite film recording surfaces for disk recording. *IBM J. of Res. & Dev.* **18** (6), 556–62.

Comstock, R. L. & Williams, M. L. (1973). Frequency response in digital magnetic recording. *IEEE Trans. Magn.*, **MAG-9** (3), 342–5.

Coutellier, J-M & Bertram, H. N. (1987). Depth profiling of modulation noise. *IEEE Trans. Magn.*, **MAG-23** (1), 195–7.

Cullity, B. D. (1972). *Introduction to Magnetic Materials*, Addison-Wesley, MA.

Daniel, E. D. (1960). A basic study of tape noise. *Ampex Res. Rep. AEL-1.*

Daniel, E. D. (1964). A preliminary analysis of surface induced tape noise. *Trans. Comm. Electr.*, **83**, 250–3.

Daniel, E. D. (1972). Tape noise in audio recording. *J. Audio Eng. Soc.*, **20**, 92–9.

Davenport, W. B. Jr. & Root, W. L. (1958). *An Introduction to the Theory of Random Signals and Noise*. McGraw-Hill Publishers, New York.

Duinker, S. & Geurst, J. A. (1964). Long-wavelength response of magnetic reproducing heads with rounded outer edges. *Philips Res. Repts.* **12**, 1–28.

Fan, G. J. (1961). A study of the playback process of a magnetic ring head. *IBM J. of Res. & Dev.* **5**, 321–325.

Fayling, R., Szczech, T. J. & Wollack, E. F. (1984). A model for overwrite modulation in longitudinal recording. *IEEE Trans. Magn.*, **MAG-20** (5), 718–20.

Freiser, M. J. (1979). On the zigzag form of charged domain walls. *IBM J. of Res. & Dev.* **23** (3), 330–8.

Fritzsch, K. (1968). Long-wavelength response of magnetic reproducing heads. *IEEE Trans. Audio and Electroacoustics*, **AU-16** (4), 486–94.

Fujiwara, T. (1979). Non-linear distortion in long wavelength ac bias recording. *IEEE Trans. Magn.*, **MAG-15** (1), 894–8.

Futamoto, M., Kugiya, F., Suzuki, M., Takano, H., Matsuda, Y., Inaba, N., Miyamura, Y., Akagi, K., Nakao, T., Sswaguchi, H., Fukuoka, H., Munemoto, T. & Takagaki, T. (1991). Investigation of 2 Gb/in² magnetic recording at a track density of 17 kTPI. *IEEE Trans. Magn.*, **MAG-27** (6), 5280–5.

Gill, H. S., Hesterman, V. W., Tarnopolsky, G. J., Tran, L. T., Frank, P. D. & Hamilton, H. (1989). A magnetoresistive gradiometer for detection of perpendicularly recorded magnetic transitions. *J. Appl. Phys.* **65** (1), 402–4.

Haas, C. W. & Callen, H. B. (1963). Ferromagnetic relaxation and resonance line widths. In *Magnetism*, G. T. Rado and H. Suhl, eds. Academic Press.

Haynes, M. K. (1976). Write-separation loss and phase functions in ac bias recording. *IEEE Trans. Magn.*, **MAG-12** (6), 761–3.

Haynes, M. K. (1977). Experimental determination of the loss and phase transfer functions of a magnetic recording channel. *IEEE Trans. Magn.*, **MAG-13** (5), 1284–6.

Haynes, M. K. (1984). Density-response and modulation-noise testing of digital magnetic recording tapes. *IEEE Trans. Magn.*, **MAG-20** (5), 897–9.

Hokkyo, J., Hayakawa, K., Saito, I., Satake, S., Shirane, K., Honda, N., Shimamura, T. & Saito, T. (1982). Reproducing characteristics of perpendicular magnetic recording. *IEEE Trans. Magn.*, **MAG-18** (6), 1203–5.

Howell, T. D., McGowan, D. P., Diola, T. A., Tang, Y-S., Hense, K. R. & Gee, R. L. (1990). Error rate performance of experimental gigabit per square inch recording components. *IEEE Trans. Magn.*, **MAG-26** (5), 2298–2302.

Hoyt, R. F., Heim, D. E., Best, J. S., Horng, C. T. & Horne, D. E. (1984). Direct measurement of recording head fields using a high-resolution inductive loop. *J. Appl. Phys.* **55** (6), 2241–4.

Hudson, V. N., Loze, M. K. & Middleton, B. K. (1986). Measurement of error rates in a digital recording system. *IERE Conference Proceedings*, **67**, 177–83.

Hughes, G. F. (1983a). Magnetization reversals in cobalt–phosphorus films. *J. Appl. Phys.*, **54** (9), 5306–13.

Hughes, G. F. (1983b). Thin film recording head efficiency and noise. *J. Appl. Phys.*, **54** (7), 4168–73.

Hunt, Robert, P. (1971). A magnetoresistive readout transducer. *IEEE Trans. Magn.*, **MAG-7** (1), 150–4.

Ichiyama, Y. (1975). Reproducing characteristics of thin film heads. *IEEE Trans. Magn.*, **MAG-11** (5), 1203–5.

Ichiyama, Y. (1979). Theoretical analysis of bit error rate considering intertrack crosstalk in digital magnetic recording equipment. *IEEE Trans. Magn.*, **MAG-15** (1), 899–906.

Iizuka, M., Kanai, Y., Abe, T., Sengoku, M. & Mukasa, K. (1988). Analysis and design of metal-in-gap head for rigid disk files. *IEEE Trans. Magn.*, **MAG-24** (6), 2623–5.

Indeck, R. S., Judy, J. H. & Iwasaki, S. (1988). A magnetoresistive gradiometer. *IEEE Trans. Magn.*, **MAG-24** (6), 2617–19.

Ishida, S. & Seki, K. (1991). Thin film disk technology. *Fujïtsu Sci. Techn. J.* **24** (4), 337–52.

Iwasaki, S-I. & Ouchi, K, (1978). Co–Cr recording films with perpendicular magnetic anisotropy. *IEEE Trans. Magn.*, **MAG-14** (5), 849–51.

Iwasaki, S-I. & Nakamura, Y. (1977). An analysis for the magnetization mode for high density magnetic recording. *IEEE Trans. Magn.*, **MAG-13** (5), 1272–7.

Iwasaki, S-I. & Takemura, K. (1975). An analysis for the circular mode of magnetization in short wavelength recording. *IEEE Trans. Magn.*, **MAG-11** (5), 1173–5.

Iwasaki, S-I. & Suzuki, T. (1968). Dynamical interpretation of magnetic recording process. *IEEE Trans. Magn.*, **MAG-4** (5), 269–76.

Jackson, J. D. (1975). *Classical Electrodynamics*. 2nd edition. John Wiley & Sons, Inc. Publishers, New York.

Jacob, I. S. & Bean, C. P. (1963). Fine particles, thin films and exchange anisotropy (effects of finite dimensions on the basic properties of ferromagnets). In *Magnetism*, G. T. Rado and H. Suhl, eds, Vol. 3, Chapter 6, p. 271–350. Academic Press.

Jagielinski, T. (1990). Materials for future high performance magnetic recording heads. *MRS Bulletin*. **15** (3), 36–44.

Jeffers, F. (1986). High-density magnetic recording heads. *Proc. IEEE*, **74** (11), 1540–56.

Jeffers, F. & Wachenschwanz, D. (1987). A measurement of signal-to-noise ratio versus track width from 128 mm to 4 mm. *IEEE Trans. Magn.*, **MAG-23** (5), 2088–90.

Johnson, K. E., Ivett, P. R., Timmons, D. R., Mirzamaani, M., Lambert, S. E. & Yogi, T. (1990). The effect of Cr underlayer thickness on magnetic and structural properties of CoPtCr thin films. *J. Appl. Phys.*, **67**, 4686–8.

Johnson, K. E., Wu, E. Y., Palmer, D. C. & Zhu, J-G. (1992). Media noise improvement through head-disk spacing reduction. *IEEE Trans. Magn.*, **MAG-28** (5), 2713–15.

Jones, R. E., Jr. (1978). Analysis of the efficiency and inductance of multiturn thin film magnetic recording heads. *IEEE Trans. Magn.*, **MAG-14** (4), 509–11.

Jones, R. E. & Mee, C. D. (1990). Recording heads. In *Magnetic Recording Handbook*, eds. C. D. Mee and E. D. Daniel, Part 1, Chapter 4, p. 258–356, McGraw-Hill.

Jorgensen, F. (1988). *The Complete Handbook of Magnetic Recording*. TAB Books, Inc., Blue Ridge Summit, PA.

Kahn, M. R., Lee, S. Y., Duan, S. L., Pressesky, J. L., Heiman, N., Speliotis, D. E. & Scheinfein, M. R. (1991). Correlations of modulation noise with magnetic microstructure and intergranular interactions for CoCrTa and CoNi thin film media. *J. Appl. Phys.*, **69** (8), 4745–7.

Kahn, M. R., Lee, S. Y., Pressesky, J. L., Williams, D., Duan, S. L., Fisher, D. R., Heiman, N., Scheinfein, M. R., Unguris, J., Pierce, D. T., Celotta, R. J. & Speliotis, D. E. (1990). Correlations of magnetic microstructure and anisotropy with noise spectra for CoNi and CoCrTa thin film media. *IEEE Trans. Magn.*, **MAG-26** (5), 2715–17.

Kajiwara, K., Hayakawa, M., Kunito, Y., Ikeda, Y., Hayashi, K., Aso, K. & Ishida, T. (1988). Auger spectroscopy analysis of metal/ferrite interface layer in metal-in-gap magnetic head. *IEEE Trans. Magn.*, **MAG-24** (6), 2620–2.

Karlqvist, O. (1954). Calculation of the magnetic field in the ferromagnetic layer of a magnetic drum. *Trans. Roy. Inst. Technol., Stockholm*, **86**, 3–27.

Kasiraj, P. & Holmes, R. D. (1990). Effect of magnetic domain configuration on readback amplitude variation in inductive film heads. *IBM Res. Rept.* **RJ-7805**.

Kasiraj, P., Shelby, R. M., Best, J. S. & Horne, D. E. (1986). Magnetic domain imaging with a scanning Kerr effect microscope. *IEEE Trans. Magn.*, **MAG-22** (5), 837–9.

Katti, R. R, Veeravalli, V. V., Kryder, M. H. & Vijaya Kumar, B. V. K. (1988). Model for demagnetization-induced noise in thin film magnetic recording media. *IEEE Trans. Magn.*, **MAG-24** (4), 2150–8.

Katz, E. R. (1978). Finite element analysis of the vertical multi-turn thin-film head. *IEEE Trans. Magn.*, **MAG-14** (5), 506–8.

Katz, E. R. (1980). Numerical analysis of ferrite recording heads with complex permeability. *IEEE Trans. Magn.*, **MAG-16** (6), 1404–9.

Katz, E. R. & Campbell, T. G. (1979). Effect of bitshift distribution on error rate in magnetic recording. *IEEE Trans. Magn.*, **MAG-15** (3), 1050–3.

Kelley, G. V. (1988). Write-field analysis of metal-in-gap heads. *IEEE Trans. Magn.*, **MAG-24** (6), 2392–4.

Kelley, G. V. & Ketcham, R. A. (1978). An analysis of the effect of shield length on the performance of magnetoresistive heads. *IEEE Trans. Magn.*, **MAG-14** (5), 515–17.

Kelley, G. V. & Valstyn, E. P. (1980). Numerical analysis of writing and reading with multiturn film heads. *IEEE Trans. Magn.*, **MAG-16** (5), 788–90.

Kittel, C. (1949). Physical theory of ferromagnetic domains. *Rev. Mod. Phys.*, **21**, 541–83.

Kneller, E. (1968). Magnetic-interaction effects in fine-particle assemblies and in thin films. *J. Appl. Phys.* **39** (2), 945–55.

Kneller, E. & Köester, E. (1977). Relation between anhysteretic and static magnetic tape properties. *IEEE Trans. Magn.*, **MAG-13** (5), 1388–90.

Knowles, J. E. (1978). Measurements on single magnetic particles. *IEEE Trans. Magn.*, **MAG-14** (5), 858–60.

Köester, E. & Arnoldussen, T. C. (1990). Recording media. In *Magnetic Recording Handbook*, eds. C. D. Mee and E. D. Daniel, Part 1, Chapter 3, p. 101–257, McGraw-Hill.

Köester, E. & Pfefferkorn. (1980). The effect of remanence and coercivity on short wavelength recording. *IEEE Trans. Magn.*, **MAG-16** (1), 56–8.

Koren, N. L. (1981). A simplified model of non-linear bit shift in digital magnetic recording. Presented at INTERMAG Conference, Grenoble, France.

Lathi, B. P. (1968). *An Introduction to Random Signals and Communication Theory*. International Textbook Company, Scranton, PA.

Lean, M. H. & Bloomberg, D. S. (1983). BEM analysis of spectral response from finite pole-length heads on perpendicular media. *IEEE Trans. Magn.*, **MAG-17** (6), 2360–3.

344   *References and bibliography*

Lean, M. H. & Wexler, A. (1982). Accurate field computation with the boundary element method. *IEEE Trans. Magn.*, **MAG-16** (2), 331–5.

Lemke, J. U. (1979). Ultra-high density recording with new heads and tapes. *IEEE Trans. Magn.*, **MAG-15** (16), 1561–3.

Lemke, J. U. (1982). An isotropic particulate medium with additive Hilbert and Fourier field components. *J. Appl. Phys.*, **53** (3), 2561–6.

Lin, H. G., Barndt, R., Bertram, H. N. & Wolf, J. K. (1992). Experimental studies of non-linearities in high density disk recording. *IEEE Trans. Magn.*, **MAG-28** (5), 3279–81.

Lin, H. G., Zhao, Y. & Bertram, H. N. (1992). Overwrite in thin film disk recording system. *IEEE Trans. Magn.*, **MAG-29** (6), 4215–23.

Lindholm, D. A. (1973). A phenomenological model for self-demagnetization in recording media. *IEEE Trans. Magn.*, **MAG-9** (3), 339–42.

Lindholm, D. A. (1975a) Dependence of reproducing gap null on head geometry. *IEEE Trans. Magn.*, **MAG-11** (6), 1692–6.

Lindholm, D. A. (1975b). Magnetic fields of finite track width heads. *IEEE Trans. Magn.*, **MAG-11** (6), 1692–6.

Lindholm, D. A. (1976a). Frequency response function of a semi-infinite parallel plate head. *IEEE Trans. Magn.*, **MAG-12** (1), 45–6.

Lindholm, D. A. (1976b). Reproduce characteristics of rectangular magnetic heads. *IEEE Trans. Magn.*, **MAG-12** (6), 710–21.

Lindholm, D. A. (1977). Image fields for two-dimensional recording heads. *IEEE Trans. Magn.*, **MAG-13** (5), 1463–1465. (Note in equation #5, the factor of 4 in front of the equation should be replaced by the number 2.)

Lindholm, D. A. (1978a). Power spectra of channel codes for digital magnetic recording. *IEEE Trans. Magn.*, **MAG-14** (5), 321–3.

Lindholm, D. A. (1978b). Spacing losses in finite track width reproducing systems. *IEEE Trans. Magn.*, **MAG-14** (2), 55–9.

Lindholm, D. A. (1980a). Effect of track width and side shields on the long wavelength response of rectangular magnetic heads. *IEEE Trans. Magn.*, **MAG-16** (2), 430–5.

Lindholm, D. A. (1980b). Notes on boundary integral equations for three-dimensional magnetostatics. *IEEE Trans. Magn.*, **MAG-16** (6), 1409–13.

Lindholm, D. A. (1980c). Secondary gap effect in narrow and wide track reproduce heads. *IEEE Trans. Magn.*, **MAG-16** (5), 893–5.

Lindholm, D. A. (1981). Application of higher order boundary integral equations of two-dimensional magnetic head problems. *IEEE Trans. Magn.*, **MAG-17** (6), 2445–52.

Lindholm, D. A. (1984). Three-dimensional magnetostatic fields from point-matched integral equations with linearly varying scalar sources. *IEEE Trans. Magn.*, **MAG-20** (5), 2025–32.

Lopez, O. (1983). Reproducing vertically recorded information – double layer media. *IEEE Trans. Magn.*, **MAG-19** (5), 1614–16.

Lopez, O. (1984). Analytic calculation of write induced separation losses. *IEEE Trans. Magn.*, **MAG-20** (5), 715–17.

Luitjens, S. B. (1990). Magnetic recording trends: Media developments and future (video) recording systems. *IEEE Trans. Magn.*, **MAG-62** (1), 6–11.

McGuire, T. R. & Potter, R. I. (1975). Anisotropic magnetoresistance in ferromagnetic 3d alloys. *IEEE Trans. Magn.*, **MAG-11** (4), 1018–38.

McKnight, J. G. (1977). The permeability of laminations for magnetic recording and reproducing heads. *Audio Eng. Soc.* Reprint No. 1265 (G-5).

McKnight, J. G. (1978). How the magnetic characteristics of a magnetic tape head are affected by gap length and a conducting spacer. *J. Audio Eng. Soc.*, **26** (12) 930–4.

McKnight, J. G. (1979). Magnetic design theory for tape recorder heads. *J. Audio Eng. Soc.* **27** (3), 106–20.

Madrid, M. & Wood, R. (1986). Transition noise in thin film media. *IEEE Trans. Magn.*, **MAG-22** (5), 892–984.

Maller, V. A. J. & Middleton, B. K. (1973). A simplified model of the writing process in saturation magnetic recording. *IERE Conf. Proc.*, **26**, 137.

Mallinson, J. C. (1969). Maximum signal-to-noise ratio of a tape recorder. *IEEE Trans. Magn.*, **MAG-5** (5), 182–6.

Mallinson, J. C. (1973). One-sided fluxes: A magnetic curiosity. *IEEE Trans. Magn.*, **MAG-9** (4), 678–82.

Mallinson, J. C. (1990). Achievements in rotary head magnetic recording. *IEEE Proc.* **76** (6), 1004–16.

Mallinson, J. C. & Bertram, H. N. (1984a). A theoretical and experimental comparison of the longitudinal and vertical modes of magnetic recording. *IEEE Trans. Magn.*, **MAG-20** (3), 461–7.

Mallinson, J. C. & Bertram, H. N. (1984b). On the characteristics of pole-keeper head fields. *IEEE Trans. Magn.*, **MAG-20** (5), 721–3.

Mallinson, J. C. & Miller, J. W. (1976). Optimal codes for digital magnetic recording. *IERE Conf. Proc.*, **35**, 161–9.

Mallinson, J. C. & Minuhin, V. B. (1984). The reciprocity integral: Convolution or correlation? *IEEE Trans. Magn.*, **MAG-20** (2), 456–8.

Mansuripur, M. (1989). Demagnetizing field computation for thin films: Extension to the hexagonal lattice. *J. Appl. Phys.*, **66** (8), 3731–3.

Mansuripur, M. & Giles, R. (1988). Demagnetizing field computation for dynamic simulation of the magnetization reversal process. *IEEE Trans. Magn.*, **MAG-24** (6), 2326–8.

Marcus, B. H., Siegel, P. H. & Wolf, J. K. (1992). Finite-state modulation codes for data storage. *IEEE Journal on Selected Areas in Communiation* **10**, (1), 5–37.

Mayergoyz, I. D. (1990). Mathematical Models of Hysteresis. Springer Verlag Publishers.

Mayergoyz, I. D. & Bloomberg, D. S. (1986). Analytical solution for side-fringing fields in perpendicular recording. *IEEE Trans. Magn.*, **MAG-22** (3), 163–7.

Megory-Cohen, I. & Howell, T. D. (1988). Exact field calculations for asymmetrical finite-pole-tip ring heads. *IEEE Trans. Magn.*, **MAG-24** (3), 2074–80.

Melas, M., Arnett, P. & Moon, J. (1983). Noise in a thin metallic medium: The connection with non-linear behavior. *IEEE Trans. Magn.*, **MAG-24** (6), 2712–14.

Middleton, B. K. (1966). The dependence of recording characteristics of thin metal tapes on their magnetic properties and on the replay head. *IEEE Trans. Magn.*, **MAG-2** (1), 225–9.

Middleton, B. K. & Miles, J. J. (1991). Recording magnetization distribution in thin film media. *IEEE Trans. Magn.*, **MAG-27** (6), 4954–9.

Middleton, B. K. & Wisely, P. L. (1976). The development and application of a simple model of digital magnetic recording to thick oxide media. *IERE Conf. Proc.* **35**, 33–42.

Middleton, B. K. & Wisely, P. L. (1978). Pulse superposition and high-density recording. *IEEE Trans. Magn.*, **MAG-14** (5), 1043–50.

Middleton, B. K. & Wright, C. D. (1982). Perpendicular recording. *4th Int. IERE Conf. Proc.*, **54**, 181–92.

Middleton, B. K. & Wright, D. C. (1983). An analytical model of the write process in perpendicular magnetic recording. *IEEE Trans. Magn.*, **MAG-19** (3), 1486–88.

Minnaja, N. & Nobile, M. (1972). Stability conditions for saw-tooth walls between head-on domains. *AIP Conf. Proc.* **10**, 1001–5.

Minuhin, V. B. (1984). Comparison of sensitivity functions for ideal probe and ring-type heads. *IEEE Trans. Magn.*, **MAG-20** (3), 488–94.

Minuhin, V. B. (1987). Theoretrical comparison of readback, harmonic responses for longitudinal recording and perpendicular recording with probe head over a medium with permeable underlayer. *IEEE Trans. Magn.*, **MAG-22** (5), 388–90.

Monson, J. E. (1972). Field analysis for non-linear magnetic heads. *IEEE Trans. Magn.*, **MAG-8** (3), 533–6.

Monson, J. E. (1990). Recording measurements. In *Magnetic Recording Handbook*, eds. C. D. Mee and E. D. Daniel, Part 1, Chapter 4, pp. 396–449, McGraw-Hill.

Moon, J. J. & Carley, L. R. (1990). Detection performance in the presence of transition noise. *IEEE Trans. Magn.*, **MAG-26** (5), 2172–4.

Muller, M. W. & Murdock, E. S. (1987). Williams–Comstock type model for sawtooth transitions in thin film media. *IEEE Trans. Magn.*, **MAG-23** (5), 2367–70.

Murdock, E. S., Simmons, R. F. & Davidson, R. (1992). Roadmap of 10 Gbt/in^2 Media: Challenges. *IEEE Trans. Magn.*, **MAG-28** (5), 3078–83.

Nakamura, Y., Ouchi, K., Yamaoto, S. & Watanabe, I. (1990). Recording characteristics of perpendicular magnetic hard disk measured by non-flying single-pole head. *IEEE Trans. Magn.*, **MAG-26** (5), 2436–8.

Nakamura, Y. & Tagawa, I. (1989). An analysis of perpendicular magnetic recording using a newly-developed 2D-FEM combined with a medium magnetization model. *IEEE Trans. Magn.*, **MAG-25** (5), 4159–61.

Nakanishi, T., Koshimoto, Y. & Ohara, S. (1982). Recording characteristics of 3.2 GByte multi-device disk storage. *Rev. Electr. Commun. Lab.* (Japan), **30** (1), 14–23.

Nunnelly, L. L., Heim, D. E. & Arnoldussen, T. C. (1987). Flux noise in particulate media: Measurement and interpretation. *IEEE Trans. Magn.*, **MAG-28** (2), 1767–75.

Ohtake, N., Isshiki, M., Endoh, K. & Kotoh, T. (1986). Magnetic recording characteristics of R-DAT. *IEEE Trans. Consum. Electron.*, **CE32** (94), 707–12.

Ortenburger, I. B. & Potter, R. I. (1979). A self-consistent calculation of the transition zone in thick particulate recording media. *J. Appl. Phys.*, **50** (3), 2393–5.

Palmer, D., Coker, J., Meyer, M. & Ziperovich, P. (1988). Overwrite in thin media measured by the method of pseudo-random sequences. *IEEE Trans. Magn.*, **MAG-24** (6), 3096–8.

Palmer, D., Ziperovich, P., Wood, R. & Howell, T. (1987). Identification of non-linear write effects using pseudo-random sequences. *IEEE Trans. Magn.*, **MAG-23** (5), 2377–9.

Parker, F. T. & Berkowitz, A. E. (1990). Anisotropy field distributions in magnetic recording particles. *J. Appl. Phys.* **67** (9), 5158–60.

Patel, A. M. (1990). Signal and error-control coding. In *Magnetic Recording Handbook*, eds. C. D. Mee and E. D. Daniel, Part 2, Chapter 8, pp. 1115–1209, McGraw-Hill.

Paton, A. (1971). Analysis of the efficiency of thin film magnetic recording heads. *J. Appl. Phys.* **42** (13), 5868–70.

Patton, C. E., McGill, T. C. & Wilts, C. H. (1966). Eddy-current-limited domain-wall motion in thin ferromagnetic films. *J. Appl. Phys.*, **37**, 3594–8.

Perlov, C. M. & Bertram, H. N. (1987). A simple iterative model of perpendicular recording. *IEEE Trans. Magn.*, **MAG-23** (5), 2859–61.

Peterson, E. & Wrathall, L. R. (1977). Eddy currents in composite laminations. *J. Audio Eng. Soc.* **26** (12), 1026–32.

Poncet, C. (1981). Principles of three-dimensional recording model for short wavelength mangetic recording. *IEEE Trans. Magn.*, **MAG-17** (3), 1262–7.

Potter, R. I. (1970). Analysis of saturation magnetic recording based on arctangent magnetization transitions. *J. Appl. Phys.*, **41**, 1647–51.

Potter, R. I. (1974). Digital magnetic recording theory. *IEEE Trans. Magn.*, **MAG-10** (3), 502–8.

Potter, R. I. (1975). Analytic expression of the fringe field of finite pole-tip length recording heads. *IEEE Trans. Magn.*, **MAG-11** (1), 80–1.

Potter, R. I. & Beardsley, I. A. (1980). Self-consistent computer calculations for perpendicular magnetic recording. *IEEE Trans. Magn.*, **MAG-16** (5), 967–72.

Potter, R. I. & Schmulian, R. J. (1971). Self-consistently computed magnetization patterns in thin magnetic recording media. *IEEE Trans. Magn.*, **MAG-7** (4), 873–80.

Pressesky, J. L., Lee, S. Y. Heiman, N., Williams, D., Coughlin, T. & Speliotis, E. D. (1990). Effect of chromium underlayer thickness on recording characteristics of CoCrTa longitudinal recording media. *IEEE Trans. Magn.*, **MAG-26** (5), 1596–8.

Ramo, S., Whinnery, J. R. & Van Duzer, T. (1984). *Fields and Waves in Communication Electronics*, 2nd edition, John Wiley & Sons Pubs., New York.

Robinson, G. M., Englund, C. D., Szczech, T. J. & Cambronne, R. D. (1985). Relationship of surface roughness of video tape to its magnetic performance. *IEEE Trans. Magn.*, **MAG-21** (5), 1386–8.

Rodé, D. & Bertram, H. N. (1989). Characterization of head saturation. *IEEE Trans. Magn.*, **MAG-25** (1), 703–9.

Ruigrok, J. J. M. (1990). *Short-Wavelength Magnetic Recording: New Methods and Analyses*. Philips Research Laboratories, Eindhoven, The Netherlands.

Salling, C., Schultz, S., McFadyen, I. & Ozaki, M. (1991). Measuring the coercivity of individual sub-micron ferromagnetic particles by Lorentz microscopy. *IEEE Trans. Magn.*, **MAG-27** (6), 5184–6.

Satake, S. & Hokkyo, J. (1974). A theoretical analysis of the erased noise of magnetic tape. In *IECE Techn. Group Meeting of Magn. Rec.*, (Japan) **MR74-23**, 39–49.

Schabes, M. E. (1991). Micromagnetic theory of non-uniform magnetization processes in magnetic recording particles. *J. Magn. Magn. Mat.* **95**, 249–88.

Schneider, R. C. (1985). Write equalization in high-linear-density magnetic recording. *IBM J. of Res. & Dev.*, **29** (6).

Schwarz, T. A. & Decker, S. K. (1979). Comparison of calculated and actual density responses of a magnetoresistive head. *IEEE Trans. Magn.*, **MAG-15** (6), 1622–4.

Semenov, V., Factorivich, A. & Gikas, M. (1991). The effect of coercive squareness S* on transition noise in thin metal media. *Proc. Jap. Mag. Society of Japan* **15** (S2), 251–6.

Shelledy, F. B. & Nix, J. L. (1992). Magnetoresistive heads for magnetic tape and disk recording. *IEEE Trans. Magn.*, **MAG-28** (5), 2283–8.

Shelor, J. R. (1986). A reluctance model for thin film magnetic recording heads. Masters Thesis, Dept. of Electrical & Computer Engineering, University of California-Santa Barbara.

Shtrikman, S. & Smith, D. R. (1992). Analytic results for the unshielded magnetoresistive head. *IEEE Trans. Magn.*, **MAG-28** (5), 2295–7.

Siegel, P. H. (1982). Applications of a peak detection model. *IEEE Trans. Magn.*, **MAG-18** (6), 1250–2.

Siegel, P. H. & Wolf, J. K. (1991). Modulation and coding for information storage. *IEEE Com.* **29** (12), 68–86.

Silva, T. J. & Bertram, H. N. (1990). Magnetization fluctuations in uniformly magnetized thin film recording media. *IEEE Trans. Magn.*, **MAG-26** (6), 3129–39.

Simmons, R. & Davidson, R. (1992). Media design for user density of up to 3 bits per pulse width. *IEEE Trans. Magn.*, **MAG-29** (1), 169–76.

Smaller, P. (1965). Reproduce system noise in wide-band magnetic recording systems. *IEEE Trans. Magn.*, **MAG-1** (4), 357–63.

Smit, J. & Wijn, H. P. J. (1959). *Ferrites.* John Wiley & Sons, New York.

Smith, N. (1987a). Reciprocity principles for magnetic recording theory. *IEEE Trans. Magn.*, **MAG-23** (4), 1995–2002.

Smith, N. (1987b). Micromagnetic analysis of a coupled thin-film self-biased magnetoresistive sensor. *IEEE Trans. Magn.*, **MAG-23** (1), 259–72.

Smith, N. (1988). A specific model for domain-wall nucleation in thin film permalloy microelements. *J. Appl. Phys.* **63** (8), 2932–7.

Smith, N. (1993). Reciprocity principles for magnetoresistive heads. *IEEE Trans. Magn.*, **MAG–29** (5), 2279–85.

Smith, N., Freeman, J., Koeppe, P. & Carr, T. (1992a). Dual magnetoresistive head for very high density recording. *IEEE Trans. Magn.*, **MAG-28** (5), 2992–4.

Smith, N., Jeffers, F. & Freeman, J. (1991). A high-sensitivity magnetoresistive magnetometer. *J. Appl. Phys.* **69** (8), 5082–7.

Smith, N., Smith, D. R. & Shtrikman, S. (1992b). Analysis of a dual magnetoresistive head. *IEEE Trans. Magn.*, **MAG-28** (5), 2295–7.

Smith, N. & Wachenschwanz, D. (1987). Magnetoresistive heads and the reciprocity principle. *IEEE Trans. Magn.*, **MAG-23** (5), 2494–6.

Smith, R. L. (1991). Use of unbiased MR sensors in a rigid disk file. *IEEE Trans. Magn.*, **MAG-27** (6), 4561–6.

Speliotis, D. E. & Chi, C. S. (1978). Computer-based modeling of the digital magnetic recording channel. *IEEE Trans. Magn.*, **MAG-14** (5), 643–8.

Speriosu, V. S., Herman, Jr., D. A., Sanders, I. L. & Yoti, T. (1990). Magnetic thin films in recording technology. *IBM J. of Res. & Dev.* **34** (6), 884–902.

Stoner, E. & Wohlfarth, E. P. (1948). A mechanism of magnetic hysteresis in heterogeneous alloys. *Phil. Trans. Roy. Soc.* **A-240**, 599–642. *Philos. Trans. R. Soc.* London, Ser. A 249, 74.

Stubbs, D. P., Whisler, J. W., Moe, C. D. & Skorjanec, J. (1985). Ring head recording on perpendicular media: Output spectra for CoCr and CoCr/NiFe media. *J. Appl. Phys.*, **57** (8), 3970–2.

Su, J. L. & Williams, M. L. (1974). Noise in disk data-recording media. *IBM J. of Res. & Dev.* **18**, 570–5.

Szczech, T. J. (1979). Analytic expressions for field components of nonsymmetrical finite pole tip length magnetic head based on measurements on large-scale model. *IEEE Trans. Magn.*, **MAG-15** (5), 1319–22.

Szczech, T. J. & Iverson, P. R. (1986). An approach for deriving field equations for magnetic heads of different geometrical configuration. *IEEE Trans. Magn.*, **MAG-22** (5), 355–60.

Szczech, T. J. & Iverson, P. R. (1987). Improvements of the coefficients in field equations for thin film recording heads. *IEEE Trans. Magn.*, **MAG-23** (5), 3866–7.

Tagami, K., Aoyama, M., Nishimoto, K. & Goto, F. (1985). Ferrite thin film disks using electroless-plated Ni-P substrates. *IEEE Trans. Magn.*, **MAG-21** (2), 1164–8.

Tagawa, I., Yamamoto, S. & Nakamura, Y. (1992). The use of reciprocity in the analysis of the magnetic reproducing process. *IEEE Trans. Magn.*, **MAG-28** (5), 2719–21.

Takano, H., Fukuoka, H., Suzuki, M., Shiiki, K. & Kitada, M. (1991). Submicron-trackwidth inductive/MR composite head. *IEEE Trans. Magn.*, **MAG-27** (6), 4678–83.

Takayama, S., Sueoka, K., Setoh, H., Schäfer, R., Argyle, B. E. & Trouilloud, P. L. (1992). A study of MIG head readout waveform asymmetry, using magnetic force and Kerr microscopy. *IEEE Trans. Magn.*, **MAG-28** (5), 2647–9.

Tang, Y-S. (1986). Noise autocorrelation in high density recording on metal film disks. *IEEE Trans. Magn.*, **MAG-22** (5), 883–5.

Tang, Y-S. & Tsang, C. (1989). Theoretical study of the overwrite spectra due to hard-transition effects. *IEEE Trans. Magn.*, **MAG-25** (1), 698–702.

Tang, Y. & Tsang, C. (1991). A technique for measuring non-linear bit shift. *IEEE Trans. Magn.*, **MAG-27** (6), Part II, 5316–18.

Tarnopolsky, G. J., Bertram, H. N. & Tran, L. T. (1991). Magnetization fluctuations and characteristic lengths for sputtered CoP/Cr thin film media. *J. Appl. Phys.*, **69** (9), 4730–2.

Tarnopolsky, G. J., Tran, L. T., Barany, A. M., Bertram, H. N. & Bloomquist, D. R. (1989). DC modulation noise and demagnetizing fields in thin metallic media. *IEEE Trans. Magn.*, **MAG-25** (4), 3160–5.

Tarumi, K. & Noro, Y. (1982). A theoretical analysis of modulation noise and dc erased noise in magnetic recording. *Apply. Phys. A.*, **28** (4), 235–40.

Thompson, D. A. (1975). Magnetoresistive transducers in high-density magnetic recording. *AIP Conf. Proc.*, **24**, 528–33.

Thompson, D. A., Romankiw, L. T. & Mayadas, A. F. (1975). Thin film magnetoresistors in memory storage, and related applications. *IEEE Trans. Magn.*, **MAG-11** (4), 1039–50.

Thornley, R. F. M. & Bertram, H. N. (1978). The effect of pole-tip saturation on the performance of a recording head. *IEEE Trans. Magn.*, **MAG-14** (5), 430–2.

Thurlings, L. (1980). Statistical analysis of signal and noise in magnetic recording. *IEEE Trans. Magn.*, **MAG-16** (3), 507–13.

Thurlings, L. (1983). On the noise power spectral density of particulate recording media. *IEEE Trans. Magn.*, **MAG-19** (2), 84–9.

Tjaden, D. L. A. & Tercic, E. J. (1975). Theoretical and experimental investigations of digital magnetic recording on thin media. *Philips Res. Rep.*, vol. 30, p. 120.

Tsang, C., Chen, M-M., Yogi, T. & Ju, K. (1990). Gigabit density recording using dual-element MR/inductive heads on thin-film disks. *IEEE Trans. Magn.*, **MAG-26** (5), 1689–93.

Tsang, C. & Tang, Y. (1991). Time-domain study of proximity-effect induced transition shifts. *IEEE Trans. Magn.*, **MAG-27** (6), 795–802.

Vajda, F. & Della Torre, E. (1992). Efficient numerical implementation of moving-total hysteresis models. *IEEE Trans. Magn.*, **MAG–28** (5), 2611–3.

van Herk, A. (1977). Analytical expressions for side fringing response and crosstalk with finite head and track widths. *IEEE Trans. Magn.*, **MAG-13** (6), 1764–6.

van Herk, A. (1980). Three-dimensional computation of the field of magnetic recording heads. *IEEE Trans. Magn.*, **MAG-16** (5), 890–2.

van Herk, A. & Wesseling, P. (1974). An analytical model of the write and read process in digital recording with a 'linear' recording medium. *IEEE Trans. Magn.*, **MAG-10** (3), 761–4.

Victora, R. N. & Peng, J. P. (1989). Micromagnetic predictions for signal and noise in barium ferrite recording media. *IEEE Trans. Magn.*, **MAG-25** (3), 2751–60.

Wachenschwanz, D. & Bertram, H. N. (1991). Modeling the effect of head saturation in high density tape recording. *IEEE Trans. Magn.*, **MAG-27** (6), 4981–3.

Wachenschwanz, D. & Carr, T. (1991). Modulation noise measurements of high density recording channels using MR heads. *IEEE Trans. Magn.*, **MAG-27** (6), 5310–12.

Wachenschwanz, D. & Jeffers, (1985). Overwrite as a function of record gap length. *IEEE Trans. Magn.*, **MAG-21**, 1380–82.

Wallace, R. L., Jr. (1951). The reproduction of magnetically recorded signals. *Bell Syst. Techn. J.*, **30**, 1145–73.

Wang, H. S. C. (1966). Gap loss function and determination of certain critical parameters in magnetic data recording instruments and storage systems. *Rev. Sci. Instrum.*, Vol. 37, p. 1124.

Wessel-Berg, T. & Bertram, H. N. (1978). A generalized formula for induced magnetic flux in a playback head. (Ltr.) *IEEE Trans. Magn.*, **MAG-14** (3), 129–31.

Westmijze, W. K. (1953). Studies on magnetic recording. *Philips Res. Rep.* Part II, vol. 8, no. 3, pp. 161–183, Part III, vol. 8, no. 4, 245–69.

White, R. L. (1984). *Introduction to Magnetic Recording*, IEEE Press, New York, NY.

White, R. L. (1992). Giant magnetoresistance: A primer. *IEEE Trans. Magn.*, **MAG-28** (5), 2482–7.

Wielinga, J., Fluitman, H. J. H. & Lodder, J. C. (1983). Perpendicular stand-still recording in Co-Cr films. *IEEE Trans. Magn.*, **MAG-19** (2), 94–103.

Wierenga, P. E., Winsum, J. A. V. & Lenden, J. H. M. vd. (1985). Roughness and recording properties of particulate tapes: A quantitative study. *IEEE Trans. Magn.*, **MAG-21** (5), 1383–5.

Williams, E. M. (1992). Recording systems considerations of noise and interference. In *Noise in Digital Magnetic Recording*, eds. T. C. Arnoldussen and L. L. Nunnelley. World Scientific Publishing Co.

Williams, M. L. & Comstock, R. L. (1971). An analytical model of the write process in digital magnetic recording. *17th Annu. AIP Conf. Proc.*, **5**, 738–42.

Wolf, J. K. (1990). Digital magnetic recording systems. *Proceedings of the 1990 IEEE International Conference on Computer Design: VLSI in Computers and Processors*, pp. 210–13, Cambridge, MA.

Wood, R. (1990). Magnetic megabits. *IEEE Spectrum* **27** (5), 32–9.

Wood, R., Lindholm, D. A. & Haag, R. M. (1985). On the bandwith of magnetic recording/reproduce heads. *IEEE Trans. Magn.*, **MAG-21** (5), 1566–8.

Wood, R. & Peterson, D. A. (1986). Viterbi detection of class IV partial response on a magnetic recording channel. *IEEE Trans. Comm*, **COM-34**, 454–61.

Wright, C. D. & Middleton, B. K. (1985). Analytical modelling of perpendicular recording. *IEEE Trans. Magn.*, **MAG-21** (5), 1398–1400.

Yamada, K., Maruyama, T., Tatsumi, T., Susuki, T., Shimabayashi, K., Motomura, Y., Aoyama, M. & Urai, H. (1990). Shielded magnetoresistive head for high density recording. *IEEE Trans. Magn.*, **MAG-26** (6), 3010–15.

Yaskawa, S. & Heath, J. (1989). Data storage on flexible disks. In *Magnetic Recording Handbook*, eds. C. D. Mee and E. D. Daniel, Part 2, Chapter 3, p. 772–811, McGraw-Hill.

Yeh, N. (1982). Asymmetric crosstalk of magnetoresistive head. *IEEE Trans. Magn.*, **MAG-18** (6), 1155–7.

Yeh, N. (1985). Supressed write spacing losses in magnetic recording. Presented at ICC'85.

Yeh, N. & Niedermeyer, R. (1990). Transition asymmetry in high density digital recording. *IEEE Trans. Magn.*, **MAG-26** (5), 2175–7.

Yogi, T., Nguyen, T. A., Lambert, S. E., Gorman, G. L. & Castillo, G. (1990a). Gigabit density recording using dual-element MR inductive heads on thin film disks. *IEEE Trans. Magn.*, **MAG-26** (5), 1689–93.

Yogi, T., Tsang, C., Nguyen, T. A., Ju, K., Gorman, G. L. & Castillo, G. (1990b). Longitudinal media for 1Gbit/in² areal density. *IEEE Trans. Magn.*, **MAG-26** (5), 2271–6.

Yoshida, H. & Shinohara, K. (1989). High-density magnetic recording properties of evaporated Co-Ni-O thin film. *J. Mag. Soc. Japan* **13** (S1), 139–43.

Yuan, S. W. (1992). Micromagnetics of domains and walls in soft ferromagnetic materials. Ph.D. thesis. University of California-San Diego.

Yuan, S. W. & Bertram, H. N. (1992a). Fast adaptive algorithms for micromagnetics. *IEEE Trans. Magn.*, **MAG-28** (5), 2031–6.

Yuan, S. W. & Bertram, H. N. (1992b). Statistical data analysis of magnetic recording noise mechanisms. *IEEE Trans. Magn.*, **MAG-28** (1), 84–92.

Yuan, S. W. & Bertram, H. N. (1993a). Magnetoresistive heads for ultra high density recording. *IEEE Trans. Magn.*, **MAG–29** (6), 3811–6.

Yuan, S. W. & Bertram, H. N. (1993b). Micromagnetics of small unshielded MR elements. *J. Appl. Phys.*, **73** (10), 6235–7.

Yuan, S. W. & Bertram, H. N. (1993c). Spatial resolution of shielded MR heads. Submitted to *IEEE Trans. Magn.*

Zhao, Y. (1992). Private communication.

Zhu, J-G. (1989). Interactive phemomena in magnetic thin films. Ph.D. Thesis, University of California at San Diego.

Zhu, J-G. (1992). Noise of interacting transitions in thin film recording media. *IEEE Trans. Magn.*, **MAG-27** (6), 5040–2.

Zhu, J-G. (1993). Transition noise properties in longitudinal thin film media. *IEEE Trans. Magn.*, **MAG-29** (1), 195–200.

Zhu, J-G. & Bertram, H. N. (1986). Computer modeling for the write process in perpendicular recording. *IEEE Trans. Magn.*, **MAG-22** (5), 379–81.

Zhu, J-G. & Bertram, H. N. (1987a). Comparison of thin film and ring heads for perpendicular recording. *IEEE Trans. Magn.*, **MAG-23** (1), 177–9.

Zhu, J-G. & Bertram, H. N. (1987b). Computer simulation of non-linear bit shift in perpendicular recording. *IEEE Trans. Magn.*, **MAG-23** (5), 2862–4.

Zhu, J-G. & Bertram, H. N. (1988a). Micromagnetic studies of thin metallic films (invited). *J. Appl. Phys.*, **63** (8), 3248–53.

Zhu, J-G. & Bertram, H. N. (1988b). Recording and transition noise simulations in thin film media. *IEEE Trans. Magn.*, **MAG-24** (6), 2706–8.

Zhu, J-G. & Bertram, H. N. (1989a). Magnetization reversal in CoCr perpendicular thin films. *J. Appl. Phys.*, **66** (3), 1291–1307.

Zhu, J-G. & Bertram, H. N. (1989b). Transition noise study in oriented longitudinal thin films. Presented at March 28–31, 1989 International Magnetics Conference, Washington, D.C.

Zhu, J-G. & Bertram, H. N. (1990). Study of noise sources in thin film media. *IEEE Trans. Magn.*, **MAG-26** (5), 2140–2.

Zhu, J-G. & Bertram, H. N. (1991a). Magnetization structures in thin-film recording media. *IEEE Trans. Magn.*, **MAG-27** (5), 3553–62.

Zhu, J-G. & Bertram, H. N. (1991b). Reversal mechanisms and domain structures in thin film recording media. *J. Appl. Phys.*, **69** (8), 6084–89.

Zhu, J-G. & Bertram, H. N. (1991c). Self-organized behavior in thin film recording media. *J. Appl. Phys.*, **69** (8), 4709–11.

Index

356

Index

Printed in the United States
By Bookmasters